Tributes
Volume 36

Logic, Philosophy of Mathematics and their History
Essays in Honor of W. W. Tait

Volume 26
Learning and Inferring. Festschrift for Alejandro C. Frery on the Occasion of his 55th Birrthday
Bruno Lopes and Talita Perciano, eds.

Volume 27
Why is this a Proof? Festschrift for Luiz Carlos Pereira
Edward Hermann Haeusler, Wagner de Campos Sanz and Bruno Lopes, eds.

Volume 28
Conceptual Clarifications. Tributes to Patrick Suppes (1922-2014)
Jean-Yves Béziau, Décio Krause and Jonas R. Becker Arenhart, eds.

Volume 29
Computational Models of Rationality. Essays Dedicated to Gabriele Kern-Isberner on the Occasion of her 60th Birthday
Christoph Beierle, Gerhard Brewka and Matthias Thimm, eds.

Volume 30
Liber Amicorum Alberti. A Tribute to Albert Visser
Jan van Eijck, Rosalie Iemhoff and Joost J. Joosten, eds.

Volume 31
"Shut up," he explained. Essays in Honour of Peter K. Schotch
Gillman Payette, ed.

Volume 32
From Semantics to Dialectometry. Festschrift in Honour of John Nerbonne.
Martijn Wieling, Martin Kroon, Gertjan van Noord, and Gosse Bouma eds.

Volume 33
Logic and Computation. Essays in Honour of Amílcar Sernadas
Carlos Caleiro, Fransciso Dionísio, Paula Gouveia, Paulo Mateus and João Rasga, eds.

Volume 34
Models: Concepts, Theory, Logic, Reasoning, and Semantics. Essays Dedicated to Klaus-Dieter Schewe on the Occasion of his 60th Birthday
Atif Mashkoor, Qing Wang and Bernhrd Thalheim, eds.

Volume 35
Language, Evolution and Mind. Essays in Honour of Anne Reboul
Pierre Saint-Germier, ed.

Volume 36
Logic, Philosophy of Mathematics and their History.
Essays in Honor of W. W. Tait
Erich H. Reck, ed.

Tributes Series Editor
Dov Gabbay   dov.gabbay@kcl.ac.uk

Logic, Philosophy of Mathematics and their History
Essays in Honor of W. W. Tait

edited by
Erich H. Reck

© Individual authors and College Publications 2018. All rights reserved.

ISBN 978-1-84890-287-9

College Publications
Scientific Director: Dov Gabbay
Managing Director: Jane Spurr

http://www.collegepublications.co.uk

Cover design by Laraine Welch

Printed by Lightning Source, Milton Keynes, UK

All rights reserved. No part of this publication may be reproduced, stored in a retrieval system or transmitted in any form, or by any means, electronic, mechanical, photocopying, recording or otherwise without prior permission, in writing, from the publisher.

William W. Tait

Contents

1	*Introduction: W. W. Tait's Career and Contributions* ERICH H. RECK	1

Part I: Proof Theory and its History

2	*Bill Tait's Exceptional Contributions to an Exceptional Decade in Mathematical Logic* SOLOMON FEFERMAN	13
3	*Conversations with Bill about Functionals and Terms* WILLIAM A. HOWARD	33

Part II: Logic and Philosophy of Mathematics

4	*A Proposition is the (Homotopy) Type of its Proofs* STEVE AWODEY	53
5	*Reflections on Reflection in a Multiverse* GEOFFREY HELLMAN	77
6	*Concepts versus Objects* CHARLES PARSONS	91
7	*On Reconstructing Dedekind Abstraction Logically* ERICH H. RECK	113

Part III: History of Logic and Philosophy of Mathematics

8 *Carnap's Philosophy of Logic and Mathematics*   **141**
   MICHAEL FRIEDMAN

9 *Wittgenstein against Logicism*   **171**
   WARREN GOLDFARB

10 *Eudoxus' Theory of Proportion and his Method of Exhaustion*   **185**
   STEPHEN MENN

11 *Frege's Introduction of Value-Ranges: A Reading of Grundgesetze §§ 29–31*   **231**
   THOMAS RICKETTS

12 *Dedekind's Structuralism: Creating Concepts and Deriving Theorems*   **251**
   WILFRIED SIEG AND REBECCA MORRIS

13 *Bibliography of Works by W. W. Tait*   **303**

14 *Information about Contributors*   **309**

15 *Name Index*   **311**

Introduction: W. W. Tait's Career and Contributions

Erich H. Reck

ABSTRACT: In a career that spans 60 years so far, W. W. Tait has made many contributions to logic, the philosophy of mathematics, and their history. The present collection of essays—contributed by former students, colleagues, and friends—is meant as a *Festschrift*, i.e., a celebration of his life and work. In this editorial introduction, a brief sketch of Tait's contributions will be provided. The sketch starts with some biographical remarks. Then Tait's contributions to four areas are outlined: proof theory; other parts of mathematical logic; the philosophy of mathematics and its history; and philosophy more broadly. Rather than explaining the technical details, the goal will be to highlight their general significance and underlying unity. Doing so will also put the essays that follow into context.

Biographical Remarks

William Walker ("Bill") Tait was born on January 22, 1929. His higher education began at Lehigh University (Bethlehem, PA), where he received a B. A. in 1952. His undergraduate teachers included Lewis White Beck and Adolf Grünbaum in philosophy, Theodore Hailperin in logic, and Samuel Goldberg in mathematics. He went on to Yale University, so as to receive a Ph.D. in philosophy in 1958 (dissertation title: "A Theory of Partial Recursive Operators"). His dissertation advisor was Frederick Fitch, but he also studied with Alan Ross Anderson and others while in graduate school. In addition, he spent the year 1954–55 at the University of Amsterdam, Netherlands, as a Fulbright Fellow. His plan in that connection was to work with L. E. J. Brouwer; but as Brouwer had just retired, Tait ended up taking logic classes with Arend Heyting, Leon Henkin, and others. Another formative event during this period was his attendance of a summer school at the Institute in Symbolic Logic, Cornell University, in 1957. A number of major logicians interacted with students at that summer school, including Alfred Tarski, Georg Kreisel, and Paul Halmos, and the event had a strong impact on other young logicians present as well, e.g., on Charles Parsons, Solomon Feferman, and Martin Davis.[1]

From Yale, Tait moved on to Stanford University, now as Assistant Professor of Philosophy (1958–65). At Stanford he joined a very strong group of logicians recently brought there by Patrick Suppes and J.C.C. McKinsey, in the Philosophy Department, and by Halsey Royden, in the Mathematics Department. This group included: Solomon Feferman, Harvey Friedman, John Myhill, Jaako Hintikka, and Dana Scott as regular faculty; and as short- or long-term visitors, Michael Dummett, Robin Gandy, William Howard, Georg Kreisel, and Dag Prawitz. Tait also received a grant to visit the Institute for Advanced Study at Princeton University in 1961–62, where he discussed logic with Kurt Gödel. Back at Stanford, he participated in a very influential seminar on proof theory in the summer of 1963, taught by Kreisel and attended by various students and faculty. These events provided the background for Tait's influential contributions to proof theory (sketched further below).

The next step in Tait's career was to go to the University of Illinois at Chicago (1965–71). "Chicago Circle", as it was known then, was a brand new campus in the University of Illinois system. Its Philosophy Department, led by Ruth Barcan Marcus, together with its Mathematics and Computer Science Departments, with members like William Howard, once again provided a conducive environment for work in logic.[2] After a one-year interlude (1971–72) at the University of Aarhus, Denmark, Tait arrived at the University of Chicago in 1972, as Full Professor. There he became part of a strong group in logic and the philosophy of science, including Leonard Linsky, David Malament, Ian Mueller, Howard Stein, and Bill Wimsatt in philosophy, also, e.g., Saunders Mac Lane and Ted Slaman in mathematics. At Chicago, where he served as Chair of the philosophy department for a number of years (1981–87), Tait's focus shifted, first from proof theory to other parts of logic, then to the philosophy of mathematics and its history. He retired in 1996, but he has remained active as Emeritus Professor, including participating in conferences worldwide. In 2005, a selection of essays from his Chicago period was published as a book, entitled *The Provenance of Pure Reason: Essays in the Philosophy of Mathematics and its History*. In publications since then, he has expanded on themes from the book, while also returning to his early work in proof theory.

Over the years, Tait has been granted a number of major fellowships and other awards for his work. These include: a Fulbright Scholarship, for studying at the Municipal University of Amsterdam, Netherlands (1954–55); a Research Grant, to work at the Institute for Advanced Study, Princeton (1961–62); a

---

[1] For Tait's own account of his education and these events, see Tait (2008); compare also the essays by Solomon Feferman and William Howard to the present collection.

[2] These were turbulent times politically. Thus, in 1966 Tait signed a tax resistance vow to protest the Vietnam War, as noted prominently on the Wikipedia page about him.

Introduction

Fellowship from the Center for Advanced Study, University of Illinois (1966–67); a Guggenheim Fellowship (1968–69); and a National Science Foundation Fellowship (1990–91). One of his most noted papers, "Finitism" (1981), was selected for inclusion in the *Philosopher's Annual* for 1982. In 2002, he was elected Fellow of the American Academy of Arts and Sciences. Finally, he was invited to give several prestigious lecture series, most recently: the *Skolem Lecture*, University of Oslo, Norway, 2010[3]; and the *Tarski Lectures*, University of California at Berkeley, 2016.[4]

Contributions to Proof Theory

Tait's scholarly career can be divided into several phases, resulting in contributions to various different parts of logic, the philosophy of mathematics, and philosophy more generally. In the first phase his focus was on proof theory.[5] The corresponding work follows Hilbert and his proof-theoretic program, as shaped further by Kurt Gödel, Gerhard Gentzen, Kurt Schütte, Clifford Spector, Gaisi Takeuti, Georg Kreisel, and others. The original goal in this tradition was to defend classical (infinitary, set-theoretic, non-constructive) mathematics by giving finitist consistency proofs for various parts of it, starting with arithmetic; and a main tool was Hilbert's epsilon-substitution method. But Gödel's Incompleteness Theorems from 1931 had undermined that original goal, at least according to some understandings of "finitism". The reaction was an extended Hilbert program, using a broader range of foundational theories, and with the added goal of calibrating their strengths. But proof theorists were pursuing other goals too, such as: rendering more precise the constructive content of intuitionistic mathematics; and "unfolding" the constructive content of various parts of classical mathematics.

When Tait entered the scene, three developments in proof theory pointed the way forward, especially within the research groups in which he worked. First, there was Gentzen's non-finitist consistency proof for arithmetic from 1936, including the techniques of cut elimination, normalization of terms, and ordinal analysis that grew out of it. Second, Gödel had presented his "Dialectica interpretation" in 1958, as a way of interpreting intuitionistic arithmetic in a finite-type extension of primitive recursive arithmetic. This brought with it a reformulation of Hilbert's substitution method in terms of primitive recursive functionals of higher type, as applied initially to classical first-order arithmetic. Third, in 1962 Clifford Spector was able to extend this approach to classical second-order arithmetic, thus to analysis. However, Spector proceeded on a basis that was not constructive. The main theme of Kreisel's proof-theory seminar

---

[3]Title: "Primitive Recursive Arithmetic: Its History and its Role in Foundations of Mathematics".

[4]Lecture I: "On Skepticism about the Ideal"; Lecture II: "Cut-Elimination for Subsystems of Second-Order Number Theory: The Predicative Case"; and Lecture III: "Cut-Elimination for Π_1^1-CA with the Omega-Rule".

[5]This phase is represented by all the articles published 1959–1971 in the bibliography.

at Stanford in 1963 was to build on this work by Gentzen, Gödel, Spector, and a few others. The more specific goal was a constructive consistency proof for second-order arithmetic, or at least for a series of its subsystems, together with the "unfolding" of their constructive contents.[6]

The way in which Tait approached these issues was in terms of functionals not only of finite but also of transfinite type. And he used a theory of terms of infinite length modeled on Kurt Schütte's analysis, from the late 1950s, of Gentzen's consistency proof for arithmetic. This led him to a number of original, substantive results. Other logicians quickly started to use ingredients from Tait's approach to address further problems, including applications of combinatory logic and the lambda calculus in this context. Without going into technical details (see the essays by Solomon Feferman and William Howard in this collection for more), two general remarks should be added: For Tait, all his research was seen as mathematical work motivated by a philosophical program, namely that of understanding classical mathematics and its foundations better. And as the limits of the approach became clearer, both from mathematical and philosophical points of view, his focus shifted away from it, but only after having established himself as one of the main proof theorists at the time.[7]

Finitism, Intuitionism, Set Theory, and Type Theory

One main reason for Tait's change of direction, i.e., his abandonment of proof theory for several decades, were technical results from the 1960s and early 70s. These results made clear the limits of the Gödel-Spector approach with respect to strong subsystems of second-order arithmetic. Careful reflections on the finitist, or more generally constructivist, basis used therein played a crucial role for Tait as well. These reflections led him to further technical work, and also to more informal, philosophical investigations, including in-depth studies of the historical background of central ideas. It took some time for these investigations to bear fruit. As an extended series of articles from the 1980s on show, the core topic for the resulting second phase of Tait's career became the supposedly radical difference between classical and constructive mathematics. The goal was now to clarify the character of both kinds of mathematics, including clearing up important misconceptions about each.

The starting point in this context was again Hilbert's work, together with Brouwer's intuitionist challenge to it, although now in a somewhat different way. Based on codifying various parts of classical mathematics axiomatically, Hilbert's suggestion had been to secure those parts by providing finitist consistency proofs. As highlighted by Tait, "classical mathematics" should here be understood as the rigorous, axiomatically grounded investigation of certain

---

[6]The approach used in this context has been called "reductive proof theory" (Feferman). In his contribution to the present volume, William Howard describes it thus: "A suitable portion of mathematical practice is formalized by means of a theory L. Then L is analyzed by means of constructive theory L^*, typically by giving a translation of L into L^*."

[7]See Tait (2005b) and Tait (2008) for his own account of that shift of focus.

idealized domains (the natural numbers, the real numbers, the universe of sets, etc.). The restriction to finitist means for consistency proofs, rooted historically in the work of Leopold Kronecker, had been based on the hope that this would provide a minimal, neutral basis from which to mediate between the classical and intuitionist sides. But as the technical results by Gentzen, Gödel, Spector, and others had shown, it matters significantly how "finitism" is understood, i.e., how narrowly or broadly. In his landmark paper, "Finitism" (1981), Tait addressed this issue directly. He argued that there are strong systematic reasons for identifying finitism with the formal system of primitive recursive arithmetic. Both very influential and not uncontroversial, there is an ongoing debate about this suggestion in the philosophy of mathematics.[8]

While finitist reasoning, in this or some related sense, might be seen as a minimal and neutral part of constructive reasoning, various questions about the latter remained; and this was especially so for its intuitionistic version. Famously, Brouwer had claimed that intuitionistic mathematics is radically different from classical mathematics, both with respect to its technical underpinnings and its philosophical justification. As a main example, he developed his theory of choice sequences, which was meant not just as an alternative approach to analysis, but also as a challenge to the Law of the Excluded Middle as assumed in classical logic. Tait responded to this challenge in two related ways: (i) He argued that, if one formalizes the theory of choice sequences carefully, constructive mathematics reveals itself, not as radically incommensurable with, but as a restricted part of classical mathematics. (ii) He questioned the philosophical foundations of intuitionism, not just in Brouwer's original form, but also in Husserl-inspired updates, and in Michael Dummett's more recent version based on a Wittgensteinian theory of meaning.[9]

Constructive mathematics can be traced back from Brouwer through Kronecker to Kant, and perhaps all the way to Euclid. But there is another, more forward looking approach to it as well. That approach consists of formalizing constructive reasoning in constructive type theory. It is a combination of constructivism, combinatory logic, and the lambda calculus, tied together in Curry-Howard type theory (based on the conception of "propositions as types"). This approach started to attract significant attention in the 1970s, especially as developed further by Per Martin-Löf, and interest in it has grown over time. Very recently, it has also been built into homotopy type theory.[10] Over the years, Tait has contributed to its exploration in two main ways: by working out a variable-free formalization of Curry-Howard type theory; and by showing how it can be used to characterize not just constructive but also classical mathematics. In particular, he showed that it is the addition of certain "ideal types" that leads from the former to the latter; and this means that

---

[8] Cf. Tait (2002b), (2012a) for his own reflections on the issue later on.
[9] Cf. Tait (1983), more recently also Tait (2005b), (2008) on this theme.
[10] Cf. the contribution by Steve Awodey to this volume.

constructive mathematics can be recognized to be part of classical mathematics also in this context.[11]

Constructive type theory is one way among several in which one can study the foundations of constructive mathematics. With respect to classical mathematics, what has played a parallel but more unified role is set theory, as rooted in Cantor's study of the transfinite and as later based on the ZFC axioms. Axiomatic set theory can, more specifically, be seen as the foundational study of the classical notions of number, function, and set, as Ernst Zermelo emphasized early on. In this context, Gödel made again several seminal contributions, especially with his results about the relative consistency of the Axiom of Choice and the Continuum Hypothesis. Together with Russell's, Burali-Forti's, and similar antinomies, these results established that there is an essential incompleteness or inexhaustibility to set theory, and thus, to classical mathematics. What is the proper reaction to this situation? This is another central theme to which Tait has devoted sustained attention since the 1980s. Self-consciously following Gödel, he contributed to the search for new set-theoretic principles, especially large cardinal axioms.[12] A particular form this has taken is the investigation of "reflection principles", building on work by Paul Bernays, W.N. Reinhardt, and others.[13]

As with his earlier work in proof theory, the contributions by Tait mentioned in the present section—concerning finitism, intuitionism, constructive type theory, and set theory—are often technical in their details. However, they were again meant as contributions to the philosophy of mathematics (not merely as parts of mathematical logic, seen either as a field of study in itself or as a non-foundational part of mathematics). In other words, the goal for Tait has always been to combine mathematics and philosophy, or better, to make progress with philosophical questions by connecting them with specific, more definite mathematical questions. This project has deep historical roots. As already mentioned, Gödel, Zermelo, Hilbert, Brouwer, and Cantor played central roles for Tait in this context. However, the list should also include, in one way or another, Kronecker, Dedekind, Frege, and other nineteenth-century thinkers. In fact, the line of influence can be traced back further, through Kant and Leibniz to Plato and Ancient Greek mathematics. This leads to the next area of Tait's contributions.

The Philosophy of Mathematics and its History

Formal results in mathematical logic can be extremely significant philosophically, as Gödel's theorems illustrate. However, they need to be seen in a larger context, including historically, since it is only then that their full significance—including the limits of this significance—becomes clearer. It is also by taking

---

[11] Cf. Tait (1995), (1998b), (2006c), and (2015b), partly also Tait (2005b), (2008).
[12] Cf. Tait (1990b), (1996), (1998a), (2000), and (forthcoming b).
[13] Geoffrey Hellman's contribution to this volume picks up on this theme.

into account historical trajectories that one can reorient oneself when a dead end seems to have been reached, as happened to Tait with his work in proof theory during the 1970s. In that case, he was led to an in-depth engagement with the philosophical ideas and arguments of crucial historical figures, from Hilbert all the way back to Plato. Tait also started to engage critically with several more contemporary philosophers of mathematics, including Michael Dummett, W. V. O. Quine, and Ludwig Wittgenstein. This resulted in a series of articles on the history of the philosophy of mathematics. It also led to many fruitful, long-lasting interactions with students, colleagues, and friends.[14]

Tait's work on the philosophy of mathematics and its history tends to have both a systematic/foundational side and a critical/therapeutic side. The former concerns the clarification and formal reconstruction of basic notions and principles in mathematics, including studying the history of attempts to formulate them, e.g., the notions of finitism and of infinity. The latter concerns the disarming of misguided or otherwise unsatisfactory philosophical views; and for Tait, that means especially views favoring the rejection or substantive revision of classical mathematics, as in intuitionism, radical finitism, and nominalism. There have been two core targets for his criticisms: the assumption that intuitionistic, more broadly constructive, or finitist mathematics has a privileged epistemological status in some strong sense (an assumption shared by him in his early work on proof theory); the assumption that classical mathematics can only be understood either in a problematic Platonist way, where mathematical truth is adjudicated by appeal to a "model in the sky", or in a radical formalist way, where it is denied that mathematics has any content.

Several arguments for the claim that constructive mathematics, and especially intuitionistic or finitist mathematics, has a special epistemological status have played a role in the history of the philosophy of mathematics. Two main ones start from Brouwerian (psychologistic) or Husserlian (phenomenological) appeals to basic "meaning-giving mental acts", on the one hand, and from Dummett's Wittgenstein-inspired (use-based) theory of meaning, on the other hand. After a detailed study of Wittgenstein's *Philosophical Investigations*, encouraged by his colleague Leonard Linsky, Tait came to reject both arguments, i.e., the supposed grounding role of mental acts for mathematics, in the sense of either Brouwer or Husserl, and the view that, along Dummett's and Wittgenstein's lines, the perspective on language one can find in the *Investigations* necessarily leads to a repudiation of classical mathematics. Together with his critique of more technical arguments by Brouwer, as mentioned above, this led Tait to the conclusion that classical and constructive mathematics are on a more equal footing than often assumed. Indeed, constructive mathematics

---

[14]This is reflected in the majority of contributions to this volume, namely those by Michael Friedman, Warren Goldfarb, Stephen Menn, Charles Parsons, Erich Reck, Thomas Ricketts, and Wilfried Sieg. All of them have interacted with Tait in sustained ways since the 1980s-90s, partly already earlier. Three are former students of his at the University of Chicago from that period, namely: Steve Awodey, Stephen Menn, and Erich Reck. For another fruit of the same kind of interactions, cf. Tait (1997).

reveals itself again as part of classical mathematics, not as an incompatible alternative. This should not, however, be seen as undermining interest in constructive mathematics altogether; it only takes on a different, philosophically less radical nature.

But what about classical mathematics itself? Why, in particular, is it misguided to think that we must chose between a problematic form of Platonism and a simple kind of formalism? The core of Tait's diagnosis in this context is that mathematical existence tends to be misunderstood. A main form this misunderstanding takes is when philosophers try to work with a univocal or "monochrome" conception of existence, where mathematical objects are supposed to exist, or not exist, in the same sense as ordinary physical objects. For Tait, examples of the latter are Quine's holism, various forms of nominalism, e.g. Hartry Field's, but also Frege's assumption of a fixed, universal universe of discourse in logic, as maintained again by Crispin Wright and other recent neo-logicists. At the same time, Frege's work contains an antidote as well, namely his "context principle", at least as developed further by Wittgenstein. Crucial here is to consider the extent to which, and the different ways in which, both physical and mathematical objects are "constituted" in our use of language. We also have to recognize that there are different contexts and correspondingly different criteria for existence. The result is a "multichrome" conception of existence. As applied to classical mathematics, this means considering the ways in which the axioms for its various parts, together with the relevant background logics, determine the sense of mathematical propositions. For constructive mathematics as well, the sense of its propositions should be seen as determined by corresponding rules of proof.

Tait's overall perspective on classical mathematics has several further aspects. To mention four of them very briefly: (i) According to him, Plato was the first to note explicitly that mathematics involves, from the very beginning, a kind of idealization; and this is what allows us to assume that mathematical concepts, unlike ordinary concepts, have "sharp edges", thereby making the strict and general application of logical laws possible. (ii) As emphasized by Cantor and Dedekind in the nineteenth century, modern mathematics involves a further, more radical distancing from the physical world, in the sense that its basic principles may not have necessary, or no direct, applications to that world any more. (iii) With the adoption of Hilbertian formal axiomatics, we should also distinguish more "abstract" parts of mathematics, such as group theory or topology, in which many different kinds of things can satisfy the relevant axioms, from more "concrete" parts, where basic mathematical objects are characterized categorically or quasi-categorically, including the natural numbers, the real numbers, and the universe of sets, thereby allowing us to construct models for the more abstract parts. (iv) Digesting the effects of Gödel's Incompleteness theorems fully means that the sense of mathematical propositions is never determined entirely, although we can try to push this determination further by searching for additional axioms.

Philosophy More Broadly

As already noted, Tait's approach to the philosophy of mathematics has both a positive and a negative side. These two sides are closely connected. For one thing, they are both grounded in detailed historical studies, and especially, investigations of how the central mathematical notions of number, function, and set have developed over time. For another, Tait's aim has been a unified perspective on mathematics and logic developed in dialog with various thinkers, both past and present. Pursuing this aim has sometimes led him beyond logic and the philosophy of mathematics understood in a narrow sense. The references to Plato and Wittgenstein above already hint at that fact; and there are several ways in which Tait has explored these broader connections further.

In the 1980s, Tait started to develop an interpretation that makes Plato not only an early precursor of Hilbert, but also a pioneering thinker with respect to *a priori*, purely conceptual thinking more generally, and thus, with respect to the power of "pure reason". One side of this interpretation is again negative, in the sense of arguing that Plato's insights have often been obscured, e.g. by Aristotle and Kant. But there is a positive side as well, in terms of attributing to Plato certain specific insights, as noted above.[15] In connection with Wittgenstein too, Tait has presented negative arguments, directed against what he takes to be misrepresentations or misappropriations of his ideas. This concerns Dummett's theory of meaning, but also Saul Kripke's book about Wittgenstein's rule-following considerations, as well as John McDowell's more recent views about the mind.[16] Of course, Wittgenstein presented himself as doing mostly "therapeutic" philosophy. Yet again, Tait has long insisted that there is something positive to learn from Wittgenstein, especially about the way in which, by building on Frege's context principle, he developed a general perspective on human language, understanding, and knowledge.[17] Beyond Plato and Wittgenstein, one can find a critical dialog with Kant in Tait's writings, often connected with finitism and constructivism, but also concerning "a priori" reasoning more generally.[18]

So far, such broader philosophical considerations have only found expression in a few articles. But occasionally Tait has indicated that he is working on a corresponding book, with the title: *Meaning, Understanding, and Knowing: A View from Outside*. In addition, he has plans to finish another book, with the working title *Lectures on the Infinite*, in which philosophical, mathematical, and historical themes concerning the infinite will be treated in an extended way. And perhaps there is more after that, e.g., a collection of Tait's articles on Gödel. These books would complete his deep and wide-ranging contribution to logic, the philosophy of mathematics, and their history.

---

[15] Cf. Tait (1986c), (2002a) for more on this theme.
[16] Cf. Tait (1986b), (2002b) for the latter two lines of thought.
[17] Cf. Tait (2012b), also (1986b), (forthcoming b).
[18] Cf. Tait (1992), (2016), and (forthcoming a) for that theme.

The Essays in this Volume

This brings us to the contributions in this volume. Part I contains two essays, by Solomon Feferman and William Howard, in which Tait's early work on proof theory is discussed in considerable detail, both with respect to technical aspects and historical context. Part II consists of four essays that are related to his contributions to other parts of logic and the philosophy of mathematics: constructive type theory, in Steve Awodey's essay; set theory, in Geoffrey Hellman's essay; views about mathematical objects and concepts, in Charles Parsons' essay; and Dedekind's notion of abstraction, in my essay. Finally, Part III contains five essays that pick up on Tait's interest in figures and developments from the history of logic and the philosophy of mathematics: an essay on Carnap, by Michael Friedman; an essay on the early Wittgenstein, by Warren Goldfarb; an essay on Eudoxos and Ancient Greek mathematics, by Stephen Menn; an essay on Frege, by Thomas Ricketts; and an essay on Dedekind, by Wilfried Sieg and Rebecca Morris. Each of them is preceded by an abstract that outlines its content further.

Acknowledgments

I am grateful to the contributors for their essays, but also for their patience during the long period when this volume was in process. I would like to thank College Publications, particularly Jane Spurr, for general support. Three people have helped me with the editorial work: Andrew Law, Hans-Christoph Kotzsch, and Steve Awodey. I am especially indebted to Dirk Schlimm, without whose generosity, LaTeX expertise, and extensive help I would not have been able to complete the book, and to Sally Ness, for moral support and much else. Finally and very sadly, Solomon Fefferman did not live to see the publication of this volume, which he supported from the start and for which he submitted his essay shortly before he died in 2016.

Part I:
Proof Theory and its History

Bill Tait's Exceptional Contributions to an Exceptional Decade in Mathematical Logic

SOLOMON FEFERMAN[*]

ABSTRACT: This article surveys the exceptional contributions that Bill Tait made in the decade of the 1960s to mathematical logic, especially in proof theory and the constructive foundations of mathematics.

1 Tait's achievements in the 1960s

Bill Tait made a series of exceptional contributions to mathematical logic in the 1960s, a decade of exceptional development in the field as a whole. His work from that period ranges over proof theory, recursive functions and functionals, combinatory and lambda calculi, and the constructive foundations of mathematics. Tait's publications in those areas are found under his name in the references below as the ten items numbered from (1961) to (1971).[1] These papers are characteristically bold and original, and in a number of cases *sui*

---

[*]Editor's note: Solomon Feferman died before this volume—which he had supported from the beginning—was finished. He passed away on July 26, 2016. A few weeks earlier, he submitted a revised and presumably final version of his contribution. This paper thus constitutes one of his very last publications. For a tribute to his life and work, see: https://philosophy.stanford.edu/news/tribute-solomon-feferman-1928-2016.

[1]Tait's first published paper (1959), gives a neat counterexample to a conjecture of Scott and Suppes concerning sentences preserved under passage to substructure in finite models. In addition to the papers of which he is the sole author in the references below, one should mention the articles Gandy, Kreisel and Tait (1961) concerning which sets occur in all models of second-order theories, and Kreisel and Tait (1961) providing a complete equational calculus for the finite definability of functions. Throughout this article, references by date without a name are to publications solely by Tait.

generis; the writing is uniformly crisp. Two of the most important ones, "Infinitely long terms of transfinite type" (1965b) and "Intensional interpretation of functionals of finite type" (1967) are treated at length by Bill Howard in his piece for this volume.

My aim here is to give a tour of the papers (1961–1971) with sketches of their content, significance, and relations to other work in the literature. While logic continued to make remarkable progress well past the 1960s into the present Tait abandoned the field following the last of these articles and turned his full attention to philosophy, especially the philosophy of mathematics. In the conclusion below I shall say something about his reasons for that and about the subsequent return of his interests in logic.

I first met Bill in 1957 at the fabled Institute in Symbolic Logic held for six weeks at Cornell University in the summer of that year, but our extended contact did not come about until he joined the faculty of the Stanford University Philosophy Department a year later. We developed a close friendship then that began with our points of common interest and that led to many enlightening conversations; these were often over bag lunches at the men's pool.[2] Our friendship widened through get-togethers of our families at picnics and dinners. Contact was diminished for a time following Bill's departure for Chicago in 1965 but later reestablished itself on a new basis; we would meet each other at conferences or during his periodic visits back to California. I especially appreciated his contribution (2002) to the 1998 conference in honor of my 70th birthday and we have had more and more interaction since then, in recent years enhanced by the ready ease of email. I am truly grateful to Bill for our long and warm friendship and his empathy in hard times.

As with logic in general, the 1960s were an exceptional period for logic at Stanford. Activity in that area was initially greatly spurred by the efforts of Patrick Suppes, who had come to the Philosophy Department in 1951; he was joined at that time by J.C.C. McKinsey with whom he collaborated on the axiomatic foundations of physics until the latter's tragic death in 1954. McKinsey had earlier done important joint work on logic with Alfred Tarski, and it was through him that Suppes became close to Tarski.[3] He was no doubt inspired by the example and influence of Tarski—who had built up a leading school in logic at U.C. Berkeley—to establish logic as an active subject at Stanford. Through his great powers of persuasion and with the help of Halsey Royden in Mathematics, Suppes worked to bring about faculty appointments in that field at the junior to senior levels, many of them jointly in the Philosophy and Mathematics departments. As a finishing PhD student of Tarski's, I came to Stanford in 1956, at first as an Instructor. A couple of years later, Georg Kreisel began spending part of each year as a Visiting Professor; his appointment was made permanent in 1964. Other additions to the faculty were those of John Myhill in the early 60s, then Dana Scott, and later Harvey Friedman, while

---

[2] The Stanford men's pool no longer exists; in those days it provided a venue for sunning and swimming in the nude.

[3] Cf. Feferman and Feferman (2004), especially pp. 232–234 and 251–254.

Bill Tait and Jaako Hintikka were brought into the Philosophy Departments. The period of the 60s saw, too, the beginning of a steady stream of visitors and the production of first-class PhD students.[4]

All of this was of course very stimulating to Tait and me, and in particular we were both to fall under the great influence of Kreisel, though along different lines for each of us. Of particular importance for Bill was the seminar on the foundations of analysis conducted by Kreisel in the summer of 1963; among the participants were Bill Howard, Verena Dyson, Rohit Parikh and, besides myself, my student Joseph Harrison.[5] As explained by Howard in his piece for this volume, the driving aim of the seminar was to see how much of Clifford Spector's (1962) extension of Gödel's (1958) quantifier free higher type functional interpretation of arithmetic and an interpretation of the 2nd order system of classical analysis in a higher type theory of bar recursive functionals could be accounted for on constructive grounds. Tait had a start on this work through his primarily autodidactic studies of proof theory and recursion theory via Kleene's *Introduction to Metamathematics* (1952) and the two volumes of Hilbert and Bernays' *Grundlagen der Mathematik* (1934, 1939).[6]

****

Since I have touched here on the history of logic at Stanford, one must mark three great losses in recent years: Patrick Suppes in November 2014 at the age of 92, Georg Kreisel in March 2015 at the age of 91, and Grigori Mints in May 2014 at the untimely age of 75. Obituaries or *Wikipedia* articles for all three can be found on the internet; see too Feferman and Lifschitz (2015) for a full piece in memory of our friend and stalwart colleague, Grisha Mints.

****

Let us turn now to Tait's distinctive contributions to our subject in the 1960s; note that these will not be presented chronologically and some topics are treated at greater length than others.

---

[4]Among the visitors in the 1960s were Verena Dyson, Andrzej Ehrenfeucht, Joseph Shoenfield, Azriel Levy, Robin Gandy, Michael Morley, Michael Dummett, Dag Prawitz and Jean van Heijenoort, and among the PhD students were Jon Barwise, Richard Platek, and Angus MacIntyre.

[5]Extensive notes for the seminar by Kreisel, Howard, Tait, Parikh and Harrison, are to be found in Kreisel (ed.) (1963), a bound volume in the Mathematics and Statistics Library at Stanford.

[6]Bill has described his rather mixed studies in logic in Tait (2008), first as an undergraduate at Lehigh University with Theodore Hailperin and then as a graduate student at Yale with Frederick Fitch and Alan Ross Anderson. Aside from his own readings none of that really prepared him for the kind of work he was to immerse himself in when he came to Stanford. I was amused to learn that he came to Lehigh on an athletic scholarship and would have continued on the football field but for an injury in his second year; this led to an academic scholarship in its place and a turn to mathematics and philosophy under the influence of the philosopher of science Adolph Grünbaum. Athletically, Bill remains an avid cyclist.

2 Constructive functions and functionals

2.1 Nested recursion

The main result of Tait's classic article (1961) that kicks off the decade seems isolated at first sight within his *oeuvre*; but in fact it connected directly with the work on second order functionals to be described next, which in turn was applied to proof theory. Moreover, what (1961) established turns out to be directly relevant to the question of the limits of finitism to which we shall return below. The main result is that for ordinals α given by a standard ordering on the natural numbers, definition of a function by nested transfinite recursion on α can be reduced to ordinary transfinite recursion on ω^α; moreover, recursions of the latter type can be reduced to those of the former type. Ackermann's function is the standard example of a function defined by nested recursion; a simplified form is provided by

$$f(0,n) = n+1, f(m+1,0) = f(m,1),$$

and

$$f(m+1, n+1) = f(m, f(m+1, n)).$$

This may be recast as nested recursion on the standard ordering of type ω^2.

2.2 Second-order functionals defined by transfinite recursion and the substitution method

The technical tools for Tait's work on the substitution method were set up in the article (1965). It concerns quantifier free second order systems R_α whose functional definitions are given by primitive recursion and transfinite recursion up to α (where the ordinal α is represented by a primitive recursive ordering). One of the main results is that if F is defined in R_α by course-of-values recursion up to β then F can be defined by ordinary recursion up to α^β in the corresponding system; this is a generalization of (1961). It is also shown that certain classes of functional equations in R_α can be solved in R_γ for some $\gamma < \varepsilon(\alpha)$ (= the first epsilon number greater than α).

The substitution method was the principal methodological tool that had been proposed by Hilbert for the advancement of his finitist consistency program. It was to be applied to derivations in which quantifiers have been formally eliminated in favor of so-called epsilon terms $\varepsilon_x A(x)$; informally, such a term represents an x such that $A(x)$ holds if there is any x at all for which that is the case. Under that interpretation and with suitable axioms concerning these terms one can replace $(\exists x)A(x)$ by $A(\varepsilon_x A(x))$ and $(\forall x)A(x)$ by $A(\varepsilon_x \neg A(x))$, and replace derivations in axiomatic systems formulated in the predicate calculus by derivations in the epsilon-calculus.[7] The idea of the sub-

---

[7] Following Hilbert, Tait uses a dual version of the epsilon symbol. See Avigad and Zach (2013) for a useful introduction to the substitution method and its historical background.

stitution method when applied to derivations in arithmetic is to make trial replacements of epsilon terms by numerals in such a way that—if they fail to satisfy a given derivation as a whole—are systematically modified until they do. Then the main aim is to show that this method always terminates. It turned out that proofs of termination were notoriously tricky and subject to error. In particular, the supposed proof in Ackermann (1924) of the consistency of Peano Arithmetic (PA) by means of the method turned out to work only for the subsystem based on quantifier free induction. Gentzen (1936) instead established the consistency of PA using his reformulation of logic via the sequent calculus and cut-elimination; this was carried out in an extension of Primitive Recursive Arithmetic (PRA) by a scheme of transfinite induction up to Cantor's ordinal ε_0. Later, Ackermann (1940) established termination of the substitution method for the epsilon-calculus version of PA by adaptation of Gentzen's argument.

Tait (1965a) undertook a more general approach to the substitution method and the question of its termination. In his formulation, the problem is to associate with a first-order system S formulated in the epsilon-calculus another system S' without any bound variables, and provide an effective transformation of derivations in S into derivations in S'. Depending on the nature of the latter system, this may no longer be a finitistic reduction. One of the main new contributions of Tait's article is to show that if S' is a suitable second order (functional) system, then the termination of the substitution method can be expressed by the solvability of certain second order functional equations. In particular, he could apply the results of (1965) to deal with the following two cases: (i) if S is number theory without induction, then S' can be taken to be the free variable part S^* of S with an added rule of definition of functionals by cases, and (ii) if S is number theory with the principle of transfinite induction up to α then S' can be taken to be S^* augmented by definition by transfinite recursion up to each ordinal $\gamma < \varepsilon(\alpha)$.

Despite the fact that Tait's functional reformulation of the substitution method seemed to provide a real conceptual and technical advance, further work in the subject by others did not (as far as I'm aware) take advantage of it.[8] In any case, Tait himself did not pursue it further.[9] Incidentally, the references to (1965a) include, as forthcoming, two articles that never actually appeared, both on the no-counterexample interpretation (n.c.i.), the first for arithmetic and the second for ramified analysis. The n.c.i. had been introduced in Kreisel (1951, 1952) as a means for extracting computational content from non-constructive proofs. The connection with the substitution method is that

---

[8] Cf. Avigad and Zach for a fairly up-to-date bibliography of the substitution method. More recently, Feferman and Lifschitz (2015) provides references to the work of my former colleague Grigori Mints, who was among the leaders in extending that method to subsystems of analysis.

[9] Tait writes in (2008) p. 253 that "[m]y own program of attempting to constructively interpret second-order number theory using the epsilon-substitution method bit the dust in 1962-63. (In my defense, I wasn't the only one naïve enough to think that such a result was obtainable ...)."

Kreisel carried out the n.c.i. for the system of arithmetic by direct use of the work in Ackermann (1940).[10] However, the details of that were quite complicated, and perhaps Tait saw his work in (1965a) as providing a means to both simplify it and extend it to other systems. In any case, it turned out that the n.c.i. for arithmetic could be extracted in a quite simple way from Gödel's (1958) functional interpretation (cf. Troelstra (1990), p. 225)), and that may be the reason that the promised publications never materialized. And it is Tait's work on that interpretation and certain of its extension to which we turn next.

2.3 Normal term models of Gödel's and Spector's functionals

Both the articles (1965b) and (1967) discussed at length in Howard's contribution to this volume consider Gödel's system T of primitive recursive functionals as a system of terms to be treated via reduction procedures.[11] Since the functions of natural numbers in T may be generated by primitive recursive functionals of arbitrary finite type, an impredicativity is involved, and Tait's main aim was to show how to reduce arbitrary terms for such to terms that are predicative in a suitable sense. In (1965b) this is accomplished by showing for a much wider system of infinite terms of possibly transfinite type how to show that every such term can be reduced to one in normal form with constructive ordinal bounds on lengths. Tait notes that this was suggested by the methods of Schütte (1960) to establish cut-elimination theorems for infinitely long derivations to provide constructive consistency proofs with ordinal bounds for arithmetic and ramified analysis, among other systems. In particular, in the case of arithmetic, in the last section of (1965b) it is shown as a special case of his normalization argument that every term for a function of natural numbers reduces to one given by ordinary transfinite recursion on an ordinal $\alpha < \varepsilon_0$.[12]

In (1967) Tait returned to the question of direct normalization of the terms of T as given in finite form. He introduced a notion of *convertibility* that informally expresses hereditary reducibility to normal form, and proved by a simple induction that every term is convertible. The argument achieves its end in practically one stroke, but is impredicative and provides no ordinal bounds. Then in his final contribution of the decade of the 60s, Tait (1971) extended his convertibility notion to prove normalizability of terms for Spector's bar recursive functionals of finite type. As a remarkable example of an idea whose time has come, in the same volume as that article, Girard (1971), Martin-Löf (1971), and Prawitz (1971) independently each adapted Tait (1967) to establish

---

[10]Cf. Feferman (1996) for an overview of Kreisel's "unwinding" program via the no-counterexample interpretation.

[11]Cf. Avigad and Feferman (1998) for an introduction to Gödel's functional interpretation and its extensions, in particular, secs. 4.2–4.4 on term models.

[12]Tait's innovative introduction of infinite terms to measure reduction procedures in functional calculi turned out to be very useful in my own work in the 1970s on functional interpretations over non-constructive functionals; cf. Avigad and Feferman (1998) sec. 8 for a survey and references.

normalization and cut-elimination results for various systems of analysis and the theory of types.

3 Finitism and constructive reasoning

The article "Constructive reasoning" (1968a) is a stand alone article in its mode of presentation: the context is that of a version of the untyped combinatory calculus, the presentation is largely informal and *prima facie* logic-free, and the notions of constructive function and constructive proof are taken for granted. Many of the topics dealt with in the papers discussed above are reconsidered within this framework, though quite briskly: intensional interpretation of Gödel's T, primitive recursion in finite types vs. transfinite recursions up to ε_0, the substitution method, and the question which bar recursive functionals are constructive. But there are also several new topics that are dealt with in this article: finitism, Brouwer's theory of ordinals *qua* well-founded trees, the Bar Theorem, and functions of free choice sequences. The essentially novel topic is Tait's thesis on finitism, and I shall restrict attention to that in the following.

Basic to the general framework of (1968a) is the notion of reduction in the combinatory calculus and the consequent notion of a normal term (nt); two terms s and t are said to be definitionally equal if $\imath st$ reduces to 0, where $\imath$ is the identity combinatory operator. Tait says (p. 185) that he uses the word *function* "in the intensional sense of referring to the nt which defines the operation, so that identity of two functions means definitional equality, not extensional equality." If u is a nt then variables b^u are introduced to range over those nt t such that ut reduces to 0, in symbols (here) $ut \Rightarrow 0$; that is re-expressed as Ut. Referring to Hilbert (1926), *finitism* is said to be concerned with free variable propositions $A(b^u)$ of the form $s(b^u) = 0$.[13] A finitist proof of $A(b^u)$ is then taken to be a description of how to transform arbitrary reductions of the form $ut \Rightarrow 0$ into a reduction of $s(t)$ to 0. That is said to immediately justify the principle of substitution, from $A(b^u)$ and $U(t)$ infer $A(t)$. Using the basic combinator π given by the immediate conversions $\pi st \Rightarrow_1 s$, the natural numbers are identified with the normal terms $0, 1 = \pi 0, 2 = \pi 1, \ldots$. Next, by the combinatory form of the recursion theorem (or fixed point theorem) a nt ν is introduced such that $\nu t \Rightarrow 0$ for t a nt just in case t is a natural number, and is undefined otherwise; then Ns can be taken to be $\nu s \Rightarrow 0$. It is shown how the principle of mathematical induction is obtained on this basis and that the finitist functions are closed under primitive recursion in the usual sense, hence that all the theorems of primitive recursive arithmetic (when suitably re-expressed in this system) are finitistically valid. It is argued that the proof

---

[13]In (1968) p. 188, it is said that "[i]t is difficult perhaps to determine what Hilbert really had in mind ...". Other than that, the historical issue as to what Hilbert and his school would have accepted as finitistic is not taken up by Tait until many years later (cf. (2005), Ch. 2 and the Appendix to Chs. 1 and 2).

of totality of the Ackermann function makes essential use of the concept of a numerical function, and that is why the function is not finitistic. Tait adds (p. 188), "if iteration of the substitution principle is the only method available for defining finitist functions—and I do not see what else there can be—then it is easy to prove that each finitist function is primitive recursive."[14] Finally, it is said that the analysis of finitism in Kreisel (1960, 1965) is based on an essentially different conception, namely that the essential feature of a finitist proof is its visualizability; iterated autonomously into the transfinite, this was shown by Kreisel to lead to a system of strength Peano Arithmetic.[15]

4 Normal derivability and constructive consistency proofs

4.1 The Tait calculus

The simplifications of the sequent calculi made by Tait in his paper, "Normal derivability in classical logic" (1968)—often called "the Tait calculus"—seem at first sight to be merely cosmetic beyond the work of Gentzen (1935) and Schütte (1960); but their great advantages become apparent as one works one's way through the proof of the main Elimination and Induction theorems and their applications. His formalism is to begin with a version of countably infinite propositional logic in which negation is driven down to the atoms. That is, each atom p has associated with it a "complement" $-p$ for which the double negation law is assumed, i.e. one takes $-(-p) = p$.[16] Propositional formulas (pf's) are generated by closure under disjunctions $\bigvee A_i$ ($i \in I$) and conjunctions $\bigwedge A_i$ ($i \in I$), where I is a finite or countably infinite set. Following de Morgan, one defines $-A$ inductively for all pf's by: $-(\bigvee A_i) = \bigwedge(-A_i)$ and $-(\bigwedge A_i) = \bigvee(-A_i)$. The axioms and rules of inference are for derivations of finite sets Γ of pf's interpreted disjunctively, i.e. if $\Gamma = \{A_0, \ldots, A_{n-1}\}$ then Γ is to be valid just in case $\bigvee A_i$ ($i < n$) is valid. I write Γ, Δ for Tait's $\Gamma + \Delta$ (i.e. $\Gamma \cup \Delta$), and Γ, A for $\Gamma \cup \{A\}$.

---

[14] Given the unusual formalism and conventions employed in (1968), the reader who wishes to consider Tait's thesis on finitism in more usual terms is best advised to see the later and more famous article (1981), reprinted as Ch. 1 in (2005), as well as the references in fn. 13.

[15] Using my general notion of the unfolding of a schematic system, Feferman and Strahm (2010) considered two schematic systems of finitist arithmetic, FA and FA + BR, for which the respective unfoldings are shown to be exactly of the strength of PRA and PA respectively. There are some points of contact with the approach to finitism in Tait (1968), including that the unfolding concept is based on an underlying partial untyped combinatory calculus (though no special attention is given to normal terms) and that the substitution rule is fundamental. However, unlike Tait's, the language of our systems admit a limited logical formalism. It has been argued by Zach (1998, 2001) and Ravaglia (2003) that the Hilbert school admitted to finitism functions given by k-fold nested recursions; by Tait (1967) this would take us up to ω^{ω^ω} in unnested transfinite recursions, thus to a system intermediate in strength between PRA and PA. It would be of interest to see if that corresponds to the unfolding of some natural system between FA and FA + BR.

[16] The reader of (1968) will note that I have made some inessential changes in the notation.

By an axiom system S is meant a collection of finite sets of atoms with the property that whenever Γ, p and $\Delta, -p$ are in S then so also is some subset of Γ, Δ. (For example, the axioms might consist of "true" atoms and all instances of excluded middle, $p, -p$.) There are just three *normal* rules: (i) Axioms (possibly weakened by side formulas), (ii) Disjunction Introduction, and (iii) Conjunction Introduction. In addition, the Cut Rule now takes the form: from Γ, A and $\Gamma, -A$, infer Γ.

Next, derivations $D \vdash \Delta$ are defined inductively; a *normal derivation* D is one without the Cut Rule. Tait notes that a non-constructive argument can be used to prove completeness of the normal rules, but his work in this paper is carried out in a completely constructive (though informal) way. The (Cut) Elimination Theorem that takes the place of Gentzen's Hauptsatz is formulated in a way that also keeps track of ordinal bounds on the length of a derivation and its cut-rank. These are referenced to a primitive recursive well-ordering $\prec$ on the natural numbers, on which are defined exponentiation to the base 2, addition, natural sum, and the Veblen critical functions χ^z for which $\chi^0(x) = 2^x$, and for $z > 0$, $\chi^z(x)$ is the xth simultaneous solution of $\chi^u(y) = y$ for all $u \prec z$. (Tait uses lower case Roman letters x, y, z, u, v to range over ordinals as given by the ordering $\prec$.) The relations $A \prec u$ and $A \preceq u$ (in words, A is of *rank* $< u$, resp. $\leq u$) are defined inductively; working constructively, we cannot in general find the least rank of A if we can give a bound for it at all. For a derivation D in which the cut-rule is applied, D is said to be of *cut-degree* $\leq u$ if $A \prec u$ for every cut formula A of a cut occurring in D. So, D is of cut-degree 0 just in case it is normal. Similarly, one defines the relations $D \prec u$ and $D \preceq u$ (in words, D is of *length* $< u$, resp. $\leq u$) inductively. Finally, we take $D \vdash \Delta[u, v]$ to mean that D is a derivation of length $\leq u$ and cut-rank $\leq v$. In the case that $v = 0$, i.e., that D is normal (cut-free), we write $D \vdash \Delta[u]$. The fundamental result in this system is now as follows.

Elimination Theorem. If $D \vdash \Delta[u, v + \omega^z]$, then $\vdash \Delta[\chi^z(u), v]$.

By induction, every derivable Δ has a normal derivation. That is then used to prove the Induction Theorem that puts a bound on the length of a provable well-ordering in terms of the length of a normal derivation of the statement of induction on that ordering relative to an arbitrary predicate. Both theorems are proved in an expeditious manner.

4.2 Applications of countably infinite propositional logic

The first applications of these theorems are to the predicate calculus and arithmetic. For the former, universally quantified formulas are translated into conjunctions of all substitution instances by terms, and in the latter by conjunctions of all substitution instances by numerals (and similarly for existentially quantified formulas and disjunctions). As expected, in the case of arithmetic one obtains the usual bound ε_0 by a combination of the Elimination and Induction Theorems. Tait then moves on to subsystems of analysis, where in each case the translations of second order universal quantifications $\forall X A(X)$ is given by conjunctions of substitution instances $A(F)$ where F ranges over a prescribed collection $\mathfrak{F}$ of analytic formulas (af's). In the case of elementary analysis and ramified analysis this collection can be chosen in advance. But in the case of the system based on the Σ_1^1 Axiom of Choice (Σ_1^1-AC), that cannot be done in advance in a constructive way, and the matter is trickier. The infinitary propositional system with the corresponding rule, from which the axiom follows directly, is denoted for a given $\mathfrak{F}$ by $\Sigma(F)$. It is shown that if $\vdash \Delta[u,v]$ in $\Sigma(\mathfrak{F})$, where Δ consists of Σ_1^1 formulas, then $\vdash \Delta[u']$ in $\Sigma(\mathfrak{F}')$ for some u' and $\mathfrak{F}'$ depending on u and v. By successive modification of the classes of af's $\mathfrak{F}$ that are used, the paper concludes with the result that the proof-theoretic ordinal of the system Σ_1^1-AC is bounded by $\chi^{\varepsilon_0}(0)$, and that is shown to be best possible by Theorem 6.19 of Feferman (1964). The ordinal bound for Σ_1^1-AC had first been obtained by a non-constructive argument in Friedman (1967), who showed that it is conservative over the transfinite iteration through all ordinals up to ε_0 of the Π_1^0 Comprehension Axiom.[17] Moreover, as was soon realized, the same holds for the system Σ_1^1-DC of the Σ_1^1 Axiom of Dependent Choices.

4.3 Applications of uncountable propositional logic to subsystems of analysis

In the paper (1970), Tait went on to expand these methods in order to obtain a constructive consistency proof of the system Σ_2^1-DC via a system of uncountable propositional logic in a sense to be explained. No proof-theoretic ordinal bound is obtained for that system, but it is rightly pointed out that that is a separate issue. This is a rather condensed and difficult paper, but one that has been stimulating in various ways. However, as will be explained below, it has subsequently been superseded both as to the constructive reduction and determination of its proof-theoretic ordinal, starting with work by Friedman and Feferman in the same volume Kino et al. (1970) in which Tait's article appeared.

Central to Tait's work in (1970) is a theory Ind of generalized inductive definitions based on the following two schemata:

Ind$_1$ $\quad \forall x[A(P_A(\underline{Q}), \underline{Q}, x) \to P_A(\underline{Q})x]$

---

[17]In Feferman (1971), I showed how to use an extension of Gödel's functional interpretation with the non-constructive minimum operator (μ) in combination with Tait's normalization procedure for infinitely long terms to obtain Friedman's conservation result and ordinal bound; cf. also Feferman (1977) and Avigad and Feferman (1998), sec. 8.2–8.3. For a direct use of Herbrand-Gentzen methods with μ see Feferman and Sieg (1981).

Ind$_2$ $\forall x[A(B,\underline{Q},x) \to B(x)] \to \forall x[P_A(\underline{Q})x \to B(x)]$,

where $A(P,\underline{Q},x)$ is an arithmetical formula in which P has only positive occurrences, and $\underline{Q} = Q_1,\ldots,Q_n$ for some $n \geq 0$ and B is an arbitrary formula. Here, P,Q range over sets of natural numbers, P_A is a function from sets to sets and $\overline{P}_A(\underline{Q})x$ expresses that x belongs to the set $P_A(\underline{Q})$; the informal interpretation is that for any given $\underline{Q}$, $P_A(\underline{Q})$ is the least P satisfying $A(P,\underline{Q},x)$.[18] However, use is made only of a special subtheory Ind$'$ of Ind in which $A(P,\underline{Q},x)$ is of the form $A_0(\underline{Q},x) \vee \forall y P(t(x,y))$, and where in Ind$_2$, B is restricted to formulas in Π_1^1 normal form. It is shown that Ind$'$ is equivalent in strength to Π_1^1-CA and moreover, that the system Σ_2^1-DC is equivalent in strength to $^*\Sigma_1^1$-DC + Ind$'$, where the * indicates that the basic formulas of the system may contain the P_A operators. Then, the uncountable propositional logic in which the proof theory of this last system is carried out makes use of a generalization of Brouwer's tree ordinals by means of the constructive inductive definition of the well-founded tree classes T_z given as follows. T_0 is the set of natural numbers, and for $z > 0$, T_z is inductively defined by: $\langle z,0 \rangle \in T_z$, and if $x \prec z$ and f is a constructive operation that maps T_x into T_z then $\langle z,x,f \rangle \in T_z$.[19] Here $\prec$ is, as above, the standard primitive recursive ordering of natural numbers of order type ε_0. The system PL^{ε_0} of uncountable propositional logic allows the formation of conjunctions and disjunctions over each T_z for z an ordinal notation in this ordering. The main result is that every derivation of an arithmetical formula in Σ_2^1-DC can be transformed into a derivation of it in PL^{ε_0}, via the proof-theoretical analysis of $^*\Sigma_1^1$-DC + Ind$'$.

As mentioned above, the work of Tait (1970) was eventually superseded and strengthened beginning with the work of Friedman (1970) and Feferman (1970), all presented independently of each other at the 1968 Buffalo conference on intuitionism and proof theory. Extending the methods from his thesis, Friedman showed in his paper that Σ_{n+1}^1-DC is of the same strength as Δ_{n+1}^1-CA and is conservative over a system for the iteration of Π_n^1-CA up to (but not including) ε_0. In my paper for the conference volume, I gave an interpretation of theories of iterated Π_1^1-CA in theories of iterated inductive definitions ID$_\alpha$.[20] Each of them is a first order system in classical logic containing arithmetic and based on the following closure and minimality schemata like Ind$_1$ and Ind$_2$ but without the free predicate parameters $\underline{Q}$. Its language L_α contains the languages L_β for all $\beta < \alpha$ together with new predicate constants P_A for each formula $A(P,x)$ in which P has positive occurrences only while the other predicate constants of A are all in L_β for some $\beta < \alpha$; in the second of these schemata, B is an

---

[18] A basic theory ID$_1$ of generalized inductive definitions given by the schemata Ind$_1$ and Ind$_2$ without set parameters Q had been introduced by Kreisel in the 1963 seminar on the foundations of analysis.

[19] If constructive operations f are taken to be indices of partial recursive functions, the classes T_z are just a form of the Church-Kleene notations for ordinals in the first and higher number classes (cf. Kleene (1938)).

[20] Both Friedman and Tait recognized that one should have such reductions.

arbitrary formula of L_α.

$\text{Ind}_{1,\alpha} \quad \forall x[A(P_A, x) \to P_A(x)]$

$\text{Ind}_{2,\alpha} \quad \forall x[A(B, x) \to B(x)] \to \forall x[P_A(x) \to B(x)]$.

By $\text{ID}_{<\lambda}$ for λ a limit ordinal is meant the union of the ID_α for $\alpha < \lambda$. What was given in my 1970 article was an interpretation, for various ordinals α, of α times iterated Π_1^1-CA in ID_α, and of Π_1^1-CA iterated up to a limit ordinal λ in $\text{ID}_{<\lambda}$ for various λ, including $\lambda = \varepsilon_0$. In particular, in combination with Friedman's results, one thus has a reduction of Σ_2^1-DC to $\text{ID}_{<\lambda}$ for $\lambda = \varepsilon_0$. However, that is not yet a constructively satisfactory reduction, since Friedman's argument for his result was non-constructive and since the iterated ID systems are based on classical logic. As to the first of these issues, it was later shown in Feferman and Sieg (1981) how to obtain Friedman's result constructively using Herbrand-Gentzen methods. In the meantime, to carry out the second part of the constructivization goal *and* add an ordinal analysis of the systems involved, I called for an ordinally informative, conceptually clear proof-theoretic reduction of classical systems of transfinitely iterated inductive definitions to corresponding intuitionistic systems. That was first achieved by Pohlers (1975) initially for finitely iterated IDs and then in Pohlers (1977) for arbitrary iterations; the same results were obtained by revised methods in Pohlers (1981) and by alternative methods in Buchholz (1981, 1981a). The history is traced in my introduction to Buchholz et al. (1981), and in its later extensions and improvements are surveyed in Feferman (2010). It would take far too much space to repeat the details here, but it suffices to say that one now has a completely satisfactory constructive treatment with ordinal bounds not only of Σ_2^1-DC but also of various of its extensions, the strongest—by the work of Jäger (1983) and Jäger and Pohlers (1983)—being its extension by the Bar Induction axiom scheme.

Following this body of work, attention was turned by proof theorists to still stronger subsystems of analysis. Stepping back for a moment, as an avenue to that, Takeuti conjectured in the early 1950s that Gentzen's Hauptsatz (Cut-elimination Theorem) for first-order logic could be extended to impredicative second-order logic. Then Schütte (1960a) showed that this is equivalent to a semantical statement of the form that every partial valuation of a suitable kind can be extended to a total valuation. And that was finally proved by Tait (1966) by a non-constructive argument, but it did not help advance the constructive attack on subsystems of analysis.[21]

---

[21]Schütte (1960a) had also reformulated the Hauptsatz for simple type theory in similar semantical terms. Following Tait's lead, that was established by Takahashi (1967) and independently by Prawitz (1968).

5 Conclusion

Despite the train of exceptional achievements that I have traced in the preceding, Tait abandoned proof theory and constructive foundations following the last of these articles because—in his own words (2008), p. 254—on the one hand, it seemed to him to have failed to have "any redeeming philosophical virtue" and, on the other hand, he aimed to define himself "as primarily a philosopher." As to the latter, Tait's philosophical work since then is largely represented in the collection, *The Provenance of Pure Reason* (2005), which received critically substantive yet sympathetic reviews by Avigad (2006) and Parsons (2009). Most of the material in that book is concerned with the philosophy of mathematics and is clearly informed throughout by his earlier work in logic at least indirectly, while some of the chapters involve direct use of his expertise in that field. Among the latter one can count the continuing interest in the issue of what finitism comes to (Chs. 1, 2, and their Appendix), the work on large cardinals in set theory (Ch. 6), and the last three chapters of the book on Frege, Cantor, Dedekind, and Gödel (Chs. 10–12). Moreover, Tait's searching contributions to Gödeliana have continued with his articles (2005a), (2006), (2006a) and (2010). Nor has Tait's interest in proof theory *per se* ceased (cf., e.g., 2015), though his remark (2008) p. 254 suggests that he is pursuing that as "a purely mathematical theory."

So it is Tait's view as to the failure of proof theory and constructive foundations to have "any redeeming philosophical virtue" that deserves separate comment. Here we should quote Tait himself at greater length:

> My work in proof theory in the 1960's and early 1970's was in aid of a philosophic program; but whatever intrinsic value that work has, the philosophic program failed. Moreover, the program presupposed a radical difference between constructive mathematics and classical mathematics, the former based on the idea of construction, the latter based on an idealized domain which we access by axiomatically describing it. As a result of subsequent philosophical reflection [...], I no longer believe that: I don't see constructive mathematics as based on a different conception of mathematics but as, basically, a subdomain of classical mathematics. (2008, p. 257)

Tait first discussed this change of view in the paper "Against intuitionism" (1983) (which he calls "badly written") and again in the introduction to (2005). In the form argued by him that is clearly controversial (see, for example, the Avigad and Parsons reviews in that respect as well as Schlimm (2005)). In any case, the constructive consistency program was a reasonable one *for the time* on philosophical grounds and can still be appreciated as such; viewed in that light, what Tait accomplished in the 60s has not only continuing technical value but also philosophically intrinsic value. Independently of that, I would speculate that the real reason for Tait's abandoning that program at the beginning of the 70s is that the prospects were bleak for its going much farther into analysis (*qua* second order arithmetic).[22] Certainly the high hopes for the program were undermined by the realization of its apparent limits that had made their first appearance in the study of Spector's interpretation. We can now say a lot more of a definite nature. On the positive side, one has such work as that of Ratjen (1995) giving an ordinal analysis for the system Π_2^1-CA that long seemed unapproachable; its constructive status though, is unsettled. On the other hand, on the negative side the hopes by some that the consistency program could be advanced much farther via a strengthened form of Martin-Löf's constructive type theory must face the limits convincingly established by Ratjen (2009).

But whatever the full reasons for Tait's disaffection, I agree with him that even given "the attraction of constructive mathematics, the game was (is) to understand classical mathematics" (2008, p. 254), and that has somehow to be played by totally different means.

References

Ackermann, W. (1924). Begründung des "tertium non datur" mittels der Hilbertschen Theorie der Widerspruchsfreiheit. *Mathematische Annalen 93*, 1–36.

Ackermann, W. (1940). Zur Widerspruchsfreiheit der reinen Zahlentheorie. *Mathematische Annalen 117*, 162–194.

Avigad, J. (2006). Review of Tait (2005b). *Bulletin of Symbolic Logic 12*, 608–611.

Avigad, J. and S. Feferman. Gödel's functional ("dialectica") interpretation. In *Buss (1998)*, pp. 337–405.

Avigad, J. and R. Zach (2013). The Epsilon Calculus. In E. N. Zalta (Ed.), *The Stanford Encyclopedia of Philosophy*: https://plato.stanford.edu/archives/win2013/entries/epsilon-calculus/.

Barwise, J. (Ed.) (1968). *The Syntax and Semantics of Infinitary Languages*, Volume 72 of *Lecture Notes in Mathematics*. Berlin: Springer.

Barwise, J. (Ed.) (1977). *The Handbook of Mathematical Logic*. Amsterdam: North-Holland.

Buchholz, W. (1981a). The $\Omega_{\mu+1}$-rule. In *Buchholz et al. (1981)*, pp. 188–233.

Buchholz, W. (1981b). Ordinal analysis of ID_ν. In *in Buchholz et al. (1981)*, pp. 234–260.

---

[22]Cf. also fn. 9.

Buchholz, W., S. Feferman, W. Pohlers, and S. W. (1981). *Iterated Inductive Definitions and Subsystems of Analysis. Recent Proof-Theoretical Studies*, Volume 897 of *Lecture Notes in Mathematics*. Berlin/Heidelberg: Springer.

Buss, S. (Ed.) (1998). *Handbook of Proof Theory*. Amsterdam: North-Holland.

Feferman, A. B. and S. Feferman (2004). *Alfred Tarski: Life and Logic*. Cambridge: Cambridge University Press.

Feferman, S. Theories of finite type related to mathematical practice. In *Barwise (1977)*, pp. 913–971.

Feferman, S. (1964). Systems of predicative analysis. *Journal of Symbolic Logic 29*, 1–30.

Feferman, S. (1970a). Formal theories for transfinite iteration of generalized inductive definitions. In *Kino et al. (1970)*, pp. 303–326.

Feferman, S. (1970b). Ordinals and functionals in proof theory. In *Proc. International Congress of Mathematicians, Nice*, Volume 1, pp. 229–233. Paris: Gauthier-Villars.

Feferman, S. (1996). Kreisel's "unwinding" programm. In P. Odifreddi (Ed.), *Kreiseliana*, pp. 247–273. Wellesley: A. K. Peters.

Feferman, S. (2010). The proof theory of classical and constructive inductive definitions. A fourty year saga, 1968–2008. In R. Schindler (Ed.), *Ways of Proof Theory*, pp. 7–30. Frankfurt: Ontos-Verlag.

Feferman, S. and V. Lifschitz (2015). In memoriam: Grigori E. Mints. *Bulletin of Symbolic Logic 21*, 31–33.

Feferman, S., C. Parsons, and S. G. Simpson (Eds.) (2010). *Kurt Gödel. Essays for his Centennial*, Volume 33 of *Lecture Notes in Logic*. Cambridge: Cambridge University Press.

Feferman, S. and W. Sieg (1981). Proof theoretic equivalences between classical and constructive theories for analysis. In *Buchholz et al. (1981)*, pp. 78–142.

Feferman, S. and T. Strahm (2010). Unfolding finitist arithmetic. *Review of Symbolic Logic 3*, 665–689.

Fenstad, J. E. (Ed.) (1971). *Proceedings of the Second Scandinavian Logic Symposium*. Amsterdam: North-Holland.

Friedman, H. (1967). Subsystems of set theory and analysis. PhD Dissertation, Massachusetts Institute of Technology.

Friedman, H. (1970). Iterated inductive definitions and Σ_2^1-AC. In *Kino et al. (1970)*, pp. 435–442.

Gandy, R. O., G. Kreisel, and W. W. Tait (1960/1961). Set Existence. *Bulletin de l'Académie Polonaise des Sciences. Série des sciences mathématiques, astronomiques et physiques 8/9*, 577–582/881–882.

Gentzen, G. (1935). Untersuchungen über das logische Schließen. *Mathematische Zeitschrift 39*, 176–210, 405–431.

Gentzen, G. (1936). Die Widerspruchsfreiheit der reinen Zahlentheorie. *Mathematische Annalen 112*, 493–565. English translation in Gentzen (1969), 132–213.

Gentzen, G. (1969). *The Collected Papers of Gerhard Gentzen*. Amsterdam: North-Holland.

Girard, J.-Y. (1971). Une extension de l'interpretation de Gödel à l'analyse, et son application à l'élimination des coupures dans la théorie des types. In *Fenstad (1971)*, pp. 63–92.

Gödel, K. (1958). Über eine bisher noch nicht genützte Erweiterung des finiten Standpunktes. *Dialectica 12*, 280–287. Reprinted with English translation in Gödel (1990), 240–251.

Gödel, K. (1990). *Collected Works, Vol. 2: Publications 1938–1974*. New York: Oxford University Press.

Hilbert, D. (1926). Über das Unendliche. *Mathematische Annalen 95*, 161–190.

Hilbert, D. and P. Bernays (1934). *Grundlagen der Mathematik, Bd. 1*. Berlin: Springer-Verlag.

Hilbert, D. and P. Bernays (1939). *Grundlagen der Mathematik, Bd. 2*. Berlin: Springer-Verlag.

Howard, W. Conversations with bill about functionals and terms.

Jäger, G. (1983). A well-ordering proof for Feferman's theory T_0. *Archive for Mathematical Logic 23*, 65–77.

Jäger, G. and W. Pohlers (1983). Eine beweistheoretische Untersuchung von (Δ^1_2-CA + BI) und verwandter Systeme. In *Bayrische Akademie der Wissenschaften, Sitzungsberichte Jahrgang 1982*, pp. 1–28. Verlag der Bayrischen Akademie der Wissenschaften.

Kino, A., J. Myhill, and R. E. Vesley (Eds.) (1970). *Intuitionism and Proof Theory. Proceedings of the Summer Conference at Buffalo, N.Y.* Amsterdam: North-Holland.

Kleene, S. C. (1938). On notation for ordinal numbers. *Journal of Symbolic Logic 3*, 150–155.

Kleene, S. C. (1952). *Introduction to Metamathematics*. Amsterdam: North-Holland.

Kreisel, G. (1951). On the interpretation of non-finitist proofs, part I. *Journal of Symbolic Logic 16*, 241–267.

Kreisel, G. (1952). On the interpretation of non-finitist proofs, part II. *Journal of Symbolic Logic 17*, 43–58. (Erratum ibid., p. iv).

Kreisel, G. (1958). Ordinal logics and the characterization of informal concepts of proof. In J. A. Todd (Ed.), *Proceedings International Congress Mathematicians 1958*, pp. 289–299. Cambridge: Cambridge University Press.

Kreisel, G. (Ed.) (1963). *Summer Seminar in the Foundations of Analysis*. Stanford University Mathematics and Statistics Library.

Kreisel, G. (1965). Mathematical logic. In T. L. Saaty (Ed.), *Lectures on Modern Mathematics*, pp. 95–195. New York: Wiley.

Kreisel, G. and W. W. Tait (1961). Finite definability of number-theoretic functions and parametric completeness of equational calculi. *Zeitschrift für mathematische Logik und Grundlagen der Mathematik 7*, 28–38.

Martin-Löf, P. (1971). Hauptsatz for the theory of species. In *Fenstad (1971)*, pp. 217–233.

Parsons, C. (2009). Review of Tait (2005b). *Philosophia Mathematica 17*, 220–272.

Pohlers, W. (1975). An upper bound for the provability of transfinite induction in systems with n-times iterated inductive definitions. In J. Diller and G. H. Müller (Eds.), *ISILC Proof Theory Symposium*, Volume 500 of *Lecture Notes in Mathematics*, pp. 271–289.

Pohlers, W. (1977). *Beweistheorie der iterierten induktiven Definitionen*. Habilitationsschrift. Ludwig-Maximilians-Universität München.

Pohlers, W. (1981). Proof-theoretical analysis of ID_ν by means of the method of local predicativity. In *Buchholz et al. (1981)*, pp. 261–357.

Prawitz, D. (1968). Hauptsatz for higher-order logic. *Journal of Symbolic Logic 33*, 452–457.

Prawitz, D. (1971). Ideas and results in proof theory. In *Buchholz et al. (1981)*, pp. 235–307.

Ratjen, M. (1995). Recent advances in ordinal analysis: Π_2^1-CA and related systems. *Bulletin of Symbolic Logic 1*, 468–485.

Ratjen, M. (2009). The constructive Hilbert programme and the limits of Martin-Löf type theory. In S. Lindström (Ed.), *Logicism, Intuitionism, and Formalism. What has become of them?*, Volume 241 of *Synthese Library*, pp. 397–433. Berlin: Springer.

Ravaglia, M. (2009). Explicating the Finitist Standpoint. PhD Dissertation, Carnegie Mellon University.

Schlimm, D. (2005). Against against intuitionism. *Synthese 147*, 171–188.

Schütte, K. (1960a). *Beweistheorie*. Berlin: Springer.

Schütte, K. (1960b). Syntactical and semantical properties of simple type-theory. *Journal of Symbolic Logic 25*, 305–326.

Spector, C. (1962). Provably recursive functions of analysis: a consistency proof of analysis by an extension of principles formulated in current intuitionistic mathematics. In J. C. E. Dekker (Ed.), *Recursive Function Theory*, Volume 5 of *Proceedings of Symposia in Pure Mathematics*, pp. 1–27. American Mathematical Society.

Tait, W. W. (1959). A counterexample to a conjecture of Scott and Suppes. *Journal of Symbolic Logic 24*, 15–16.

Tait, W. W. (1961). Nested recursion. *Mathematische Annalen 143*, 236–250.

Tait, W. W. (1965a). Functionals defined by transfinite recursion. *Journal of Symbolic Logic 30*, 155–174.

Tait, W. W. (1965b). Infinitely long terms of transfinite type. In M. Dummet and J. M. Crossley (Eds.), *Formal Systems and Recursive Functions*, pp. 176–185. Amsterdam: North-Holland.

Tait, W. W. (1965c). The substitution method. *Journal of Symbolic Logic 30*, 175–192.

Tait, W. W. (1966). A non-constructive proof of Gentzen's Hauptsatz for second-order predicate logic. *Bulletin of the American Mathematical Society 72*, 980–983.

Tait, W. W. (1967). Intensional interpretation of functionals of finite type I. *Journal of Symbolic Logic 32*, 198–212.

Tait, W. W. (1968a). Constructive reasoning. In B. van Rootselaar and F. Staal (Eds.), *Logic, Methodology and Philosophy of Science III*, pp. 185–199. Amsterdam: North-Holland.

Tait, W. W. (1968b). Normal derivability in classical logic. In *Barwise (1968)*, pp. 204–236.

Tait, W. W. (1970). Applications of the cut-elimination theorem to some subsystems of classical analysis. In *Kino et al. (1970)*, pp. 475–488.

Tait, W. W. (1971). Normal form theorem for bar recursive functionals of finite type. In *Fenstad (1971)*, pp. 353–367.

Tait, W. W. (1981). Finitism. *Journal of Philosophy* 78, 524–556. Reprinted as Ch. 1 in Tait (2005b), 21–42.

Tait, W. W. (1983). Against intuitionism. Constructive mathematics is part of classical mathematics. *Journal of Philosophical Logic* 12, 175–195.

Tait, W. W. (2002). Remarks on finitism. In W. Sieg et al. (Ed.), *Reflections on the Foundations of Mathematics. Essays in honor of Solomon Feferman*, Volume 15 of *Lecture Notes in Logic*, pp. 410–419. Reprinted as Ch. 3 in Tait (2005b), 43–53.

Tait, W. W. (2005a). Gödel's reformulation of Gentzen's first consistency proof for arithmetic: the no-counter-example interpretation. *Bulletin of Symbolic Logic* 11, 225–238. Reprinted in Feferman and Strahm (2010), 74–87.

Tait, W. W. (2005b). *The Provenance of Pure Reason*. Oxford: Oxford University Press.

Tait, W. W. (2006a). Gödel's correspondence on proof theory and constructive mathematics. *Philosophia Mathematica* 14, 76–111.

Tait, W. W. (2006b). Gödel's interpretation of intuitionism. *Philosophia Mathematica* 14, 208–228.

Tait, W. W. (2008). Philosophy of mathematics: 5 questions. In V. F. Hendricks and H. Leitgeb (Eds.), *Philosophy of Mathematics: 5 questions*, pp. 249–263. Automatic Press.

Tait, W. W. (2010). Gödel on intuitionism and on Hilbert's finitism. In *Feferman and Strahm (2010)*, pp. 88–108.

Tait, W. W. (2015). Cut-elimination for deductions in Π_1^1-CA with the ω-rule. MS.

Takahashi, M. (1967). A proof of cut-elimination theorem in simply type-theory. *Journal of the Mathematical Society Japan* 19, 399–410.

Troelstra, A. S. (1990). Introductory note to 1958 and 1972. In *Gödel (1990)*, pp. 217–241.

Zach, R. (1998). Numbers and functions in Hilbert's finitism. *Taiwanese Journal for Philosophy and History of Science* 10, 33–60.

Zach, R. (2001). Hilbert's finitism: Historical, philosophical and metamathematical perspectives. PhD Dissertation, University of California, Berkeley.

Zach, R. (2015). Hilbert's Program. In E. N. Zalta (Ed.), *The Stanford Encyclopedia of Philosophy.*
https://plato.stanford.edu/archives/sum2015/entries/hilbert-program/.

Conversations with Bill about Functionals and Terms

WILLIAM A. HOWARD

ABSTRACT: Bill's idea of infinite terms and his notion of what is known as "Tait computability" of finite terms have played an important role in the development of proof theory. The main purpose of the following is to provide an account of five conversations in which he discussed these ideas with me. Since four of these conversations took place in the context of Kreisel's seminar at Stanford (summer of 1963), I have provided an account of the origin and philosophical goals of the seminar.

1 Introduction

Gödel's paper (1958) on the functional interpretation of first-order arithmetic by means of primitive recursive functionals of finite type gave rise to various issues and opportunities:

(1.1) There were questions about the nature of the functionals themselves. Gödel described his functionals as computable but he did not give much in the way of technical details. On the one hand, he takes the general notion of computable functional of finite type as immediately intelligible, and he makes a few remarks about a mathematical counterpart, but there are not enough details to determine a unique mathematical notion. On the other hand, he gives a brief description of how the specific functionals needed (i.e., the primitive recursive functionals) are to be introduced by means of defining equations, taken as axioms; so presumably one could compute them by means of substitutions into these axioms in the manner of Skolem (first order) arithmetic.

(1.2) What is the relation between the primitive recursive functionals of finite type and Gentzen's ordinals?

(1.3) There was the possibility of extending Gödel's functional interpretation to formal theories stronger than first-order arithmetic.

During the period 1959–1961, Clifford Spector worked out a far-reaching extension of Gödel's paper. This work was published posthumously in 1962 (see next section).

This is the context in which my conversations with Bill took place. Our conversations were especially focused on two of his ideas which have been particularly influential: first, his theory of terms of infinite length; second, his method for showing the computability of finite terms. I have a clear memory of five of our conversations. Since four of these conversations took place while Bill and I were attending Kreisel's seminar at Stanford in the summer of 1963, I need to say a few words about the seminar. The seminar was centered on the opportunities opened up by Spector's paper of 1962. Hence let me first say a few words about that paper.

2 Spector's Paper

Spector, during his stay at the Institute for Advanced Study, succeeded in extending Gödel's functional interpretation to classical analysis (that is, second-order arithmetic with the obvious comprehension axiom scheme for arbitrary formulas) by means of functionals formed by a scheme which he called *bar recursion of finite type*. In the final stages of writing up the work for publication, Spector died of leukemia. Kreisel did the editorial work required to put the paper in final form (Spector 1962). There is some drama here. The paper was submitted by Kreisel on 9/23/61. In a long footnote on the first page, Kreisel mentions that Spector had mailed a typewritten draft to him on 7/26/61, three days before his death, and he quotes from the accompanying letter giving some details as to how Spector had intended to complete the derivation of the main result. Kreisel was unable to reconstruct Spector's intended derivation, so he provided an alternative approach.

It was Kreisel who had gotten Spector interested in Gödel's functional interpretation in the first place. Spector's participation was fortunate on two accounts: first, because of his outstanding ability; and, second, because he had been a student of Kleene and hence was aware of Kleene's recent formulation of Brouwer's bar theorem by means of an axiom scheme now called *bar induction of type 0*. This eventually appeared in (Kleene-Vesley 1965). In fact, Gödel had already recommended, in the final sentence of his paper (1958), that his functional interpretation be extended so as to treat "the sort of inference that Brouwer used in proving the 'fan theorem' ". Presumaby, by 'fan theorem' he meant the bar theorem, since the fan theorem itself adds no strength; cf. p. 174 of (Kreisel 1987). But it is not hard to see that even the bar theorem is nowhere near as strong as full classical analysis. Spector's breakthrough consisted in generalizing bar induction of type 0 to bar induction of finite type and then showing that the corresponding recursion principle, bar recursion of finite type, suffices for the functional interpretation of full classical analysis. Bar recursion of type σ is a sort of transfinite recursion over a tree whose nodes are functionals of type σ. We shall use N for the type of natural numbers. Hence 'bar recursion of type 0' means: bar recursion of type N.

3 Kreisel's Seminar

The seminar gave rise to a mimeographed report, which became Volume 1 because there were so many ideas to be worked out and written down that Kreisel organized a second volume which came out sometime in 1964. These notes are now pretty scarce. The two volumes have been assembled as an unpublished volume, *Seminar on the Foundations of Analysis*, Stanford University 1963; one copy is available in the Mathematical Sciences Library of Stanford University.

The principal purpose of the seminar was to make a systematic study of the possibility that Spector's scheme of bar recursion of finite type might be given a constructive basis, thereby obtaining a constructive consistency proof of classical analysis. At least, that was the most ambitious goal. More realistically, the goal was to use the impetus provided by (Gödel 1958) and (Spector 1963) in order to provide a constructive foundation for various subsystems of full classical analysis. For this purpose, Kreisel had singled out certain subsystems, the three most promising being those based on the axioms of Δ_1^1 comprehension, Σ_1^1 axiom of choice and Π_1^1 comprehension.

The viewpoint was that of *reductive proof theory* (Feferman 2000). A suitable portion of mathematical practice is formalized by means of a theory L, then this theory is analyzed by means of a constructive theory L*, typically by giving a translation of L into L*. In the present case there is an intermediate step: some part of Brouwer's intuitionistic mathematics is expressed in a formal system B; then L is translated into B, and B is translated into L*. The result is a proof of the consistency of L relative to that of L*.

In a narrow sense, the goal of reductive proof theory is to obtain consistency proofs; but broader foundational results are often achieved. For example, Gödel, in his talk at Yale University, 1941 (Gödel 1995, 189–200), indicates that his functional interpretation provides results about the constructive content of intuitionistic mathematics. Thus, although the ostensible goal of reductive proof theory is consistency proofs, one often has broader goals in mind. This was the spirit of the seminar.

Some of the constructive formal systems are taken as *foundational*; these are the systems to which all the others are reduced (of course, the system used in the metalanguage must also be foundational). One of the tasks of the seminar was to specify suitable foundational systems. Upon what considerations will one's choice of the foundational systems be based? According to Kreisel (seminar notes, page 0.5), these considerations will consist largely of "making oneself aware of the intended sense of such notions as constructivity, i.e., as one crudely says, of the sense implicit in what one has commonly accepted so far."

3.1 Philosophical viewpoint

Following Kreisel's suggestion, we note that constructivism is commonly understood to arise from the following philosophical viewpoint.

Our mathematical ideas, and our mathematical objects, are constructed by us; they do not already exist in some Platonistic realm 'beforehand'. There is no absolute notion of truth. There is no completed infinite (we cannot perform a search through an infinite set).

3.2 Constructive principles

Corresponding to the philosophical viewpoint just described, we have the following commonly accepted constructive principles.

(3.2.1) To say that an object of a certain kind *exists* is to say that we have a procedure for constructing it.

(3.2.2) To have a constructive function f on natural numbers means to have an effective procedure which, when applied to any number n, yields a number $f(n)$. In other words, f is, or corresponds to, an effective procedure.

(3.2.3) To assert a proposition α is to provide a proof of α. This makes up for the absence of the classical notion of truth.

3.3 Mathematical formulation of the informal notions

(i) The notion of effective procedure in 3.2.2 is implemented mathematically by the notion of an algorithm (in the sense of Turing computability).

(ii) A very restricted notion of proof is implemented mathematically by the notion of derivation in a formal system.

3.4 Brouwer's intuitionistic mathematics

The principles 3.2.1–3.2.3 are embodied in Brouwer's intuitionistic mathematics but the problem is that Brouwer's framework of thought, as presented in his papers, differs radically from the framework of ordinary mathematics. Since the 1930s, efforts have been made to close the gap. In fact, this was one of the purposes of Kreisel's seminar.

Already in 1945, Kleene had shown, by means of his theory of recursive realizability:

(3.4.1) Corresponding to every proof of a formula of the form $(\exists y)\phi(y)$ in Heyting arithmetic HA, there is a numeral $\bar{n}$ and a proof, in HA, of $\phi(\bar{n})$.

The correspondence from the given proof to the pair $(\bar{n}, \text{proof of } \phi(\bar{n}))$ is itself given by an algorithm (Turing machine).

By 1963 Kleene had given an interpretation of a substantial portion of Brouwer's work by means of an extension of his original notion of recursive realizability. This eventually appeared in (Kleene-Vesley 1965).

3.5 Criteria for constructive formal systems

For systems C containing existential quantifiers, Kleene's result 3.4.1 supplies the paradigm. Suppose the variable X ranges over objects of a given type. Then we require:

(3.5.1) Corresponding to every proof in C of the of a formula of the form $(\exists X)\phi(X)$, there is a term A and a proof in C of $\phi(A)$.

For a free-variable system such as Gödel's T, where the terms are formulated as lambda terms or combinators, we have a notion of reduction, and the main criterion for the constructiveness of T is simply:

(3.5.2) Corresponding to every closed term d of type N, there is a reduction sequence which reduces d to a numeral, $val(d)$, say, where the correspondence must be given by an algorithm. The relation to 3.2.2 is as follows. Let h be a closed term of type $N \rightarrow N$. Then we get a constructive mapping from numerals $\bar{k}$ to numerals $val((h\bar{k}))$. Thus h represents an algorithm. The algorithm, in the sense of a set of instructions, is given by the lambda-term itself (plus a specification of which redex is to be reduced). The issue is: Does the procedure terminate? (Halting problem for the corresponding Turing machine.)

Perhaps one also wants the normalizability of terms of higher type. This depends on one's attitude toward the equality relation.

3.6 Candidates for constructive formal systems

At the time of the seminar, we were thinking of the following candidates for constructive formal systems.

(3.6.1) Some sort of extension of Gödel's T; e.g., add a convincing definition scheme; or maybe go to transfinite types.

(3.6.2) The system PRA(d) consisting of free variable, primitive recursive arithmetic PRA extended by adding a rule for the introduction of functions by a simple form of transfinite recursion with respect to a given system of ordinal notations less than d, and a corresponding rule of (free variable) transfinite induction. In my view, the transfinite recursion schema (ultimately, the descending chain principle) for the ordinal notations less than d needs to be justified on constructive grounds.

(3.6.3) Ordinal progressions of theories, as long as this is done constructively. Of course, the strength of these theories, at least those developed in the early 1960s (Feferman 1964 and Schütte 1965) is bounded by the ordinal Γ_0.

(3.6.4) An intuitionistic system HA+O. In section III of the notes (page 3.7), Kreisel proposed (essentially) Heyting arithmetic extended by a predicate O such that O(n) says that n is a code of a Kleene-Church constructive ordinal. The predicate O was to be introduced by means of his notion of a *generalized inductive definition* (g.i.d.). His interest in HA+O was due to the possibility that the general theory of (non-iterated) g.i.d.s might be reduced to HA+O by means of a Kleene-style recursive realizability argument.

4 First Conversation

My first conversation with Bill took place in January 1963 at a meeting of the American Mathematical Society in Berkeley. We discussed two topics. First, the ordinal analysis of Gödel's primitive recursive functionals; second, the possibility of extending Gödel's functional interpretation to Schütte's predicative analysis.

Concerning the first topic, we had both decided that it was reasonable to formulate the primitive recursive functionals within the typed lambda calculus. Hence an ordinal analysis would consist of assigning to each term an ordinal (notation) in such a way that a reduction step on a term would lower the corresponding ordinal. Computability (reduction to normal form) would be a consequence of the descending chain principle for ordinals. But finding the required assignment of ordinals is a hard problem, and Bill had found a way around this by exploiting an analogy between normalization in the lambda calculus and cut elimination in Gentzen-style proofs. Just as Schütte had greatly simplified Gentzen's ordinal analysis by introducing infinite proofs, Bill had simplified the ordinal analysis of finite terms by introducing a system of infinite terms. We did not discuss the details of his normalization procedure at that time; but, in outline, the ideas were clear enough just on the basis of the analogy between infinite terms and infinite proofs.

To extend Gödel's functional interpretation to Schütte's predicative analysis, it is necessary to extend the primitive recursive functionals to transfinite type levels. From our conversation, I got the impression that Bill accomplished this by taking a suitably defined sequence of functionals $F_0, \ldots, F_i, \ldots$, where F_i has type α_i, to be another functional, say F, whose type is taken to be the sequence $\alpha_0, \ldots, \alpha_i, \ldots$. Probably I did not hear quite what he was saying, because I see from his paper (Tait 1965) that, from the viewpoint of the functionals that underlie his terms, the concept would be: a *countable set* of functionals $\{F_0, \ldots, F_i, \ldots\}$, where F_i has type α_i, is taken to be a functional whose type is essentially the set $\{\alpha_0, \ldots, \alpha_i, \ldots\}$. Moreover, the types α_i are required to have a special form, essentially $\sigma_i \longrightarrow \tau_i$, where, in addition, one requires the condition:

(4.1) $\sigma_i \neq \tau_i$ for $i \neq j$.

The domain of F consists of all functionals G such that the type of G is in $\{\sigma_0, \ldots, \sigma_i, \ldots\}$, and FG is defined to be $F_k G$, where k is uniquely determined by the condition 4.1. I shall discuss the corresponding theory of infinite terms in section 6.

I told Bill my own ideas about extending the primitive recursive functionals to transfinite type levels. These were based on ideas about mappings from types to functionals. For example, the identity functional of type ω would consist of the functional $(\lambda \sigma \prec \omega).(\lambda X^\sigma).X^\sigma$, where σ ranges over all finite types. To this functional I assigned a type symbol in an ad hoc manner. Of course, today we would use the type symbol $(\forall \sigma \prec \omega)(\sigma \longrightarrow \sigma)$ or something like $\Pi_{(\sigma \prec \omega)}(\sigma \longrightarrow \sigma)$. Since the mappings from types to functionals, and from numbers to types, etc., were defined explicitly by primitive recursion schemata, I did not need to appeal to infinite terms (in fact, the idea of using infinite terms had not occurred to me). I had ideas about using all this to extend Gödel's functional interpretation to Schütte's predicative analysis. It would have been pretty complicated. I had abandoned this project when Spector's paper came out and I saw the possibility of doing something more exciting.

The remaining four conversations took place during the summer of 1963, concurrently with Kreisel's seminar at Stanford.

5 Second Conversation

We were walking in Palo Alto and Bill was explaining his method of proving that the terms for primitive recursive functionals, formulated by means of typed combinators, were computable in the sense that a certain, very specific, sequence of reductions must terminate. This was the method he eventually published in (Tait 1967). The paper is a bit hard to read because two ideas are involved: (1) hereditary computability (Bill used the term *convertiblity*) and (2) intensional equality. They can be separated, and idea #1 has become very influential, especially in its simplest form in which one concentrates on the computability of the terms of type N.

The method is as follows. We are working with closed terms. Computability of a term is defined by induction on its type level. A term of level 0 is computable if it has a reduction sequence ending in a numeral. A term of level $n+1$ is computable if the result of applying that term to a complete list of arguments consisting of computable terms (necessarily of level less than $n+1$) is computable. Computability of the basic elementary combinators follows essentially by definition, and computability of the primitive recursor of a given type is easily provable by induction. But if A and B are computable then so is AB, essentially by definition (we assume that AB is well-formed). Computability of an arbitrary closed term M now follows from the buildup of M from the basic combinators and the primitive recursor. The method is so simple that, when he first told it to me, I did not think it could work.

Concerning the second feature of Bill's computability method, namely, in its more sophisticated form it proves normalizability and hence provides an approach to the question of intensional equality, let me say a few words about the latter.

Gödel (1958) made use of a notion of equality that he called *definitional equality*. Moreover, in his formulation of the free variable theory he used

(5.1) propositional combinations of equations between terms of higher type.

I did not know what to make of this, so I had mainly just ignored it. In our conversation, Bill told me that the notion of intensional equality should be taken seriously. Well, this was an interesting new idea. We wondered how it might be exploited, perhaps by adding a term for an equality predicate between functionals of higher type, but we were not able to come up with anything interesting. In my Buffalo 1970 paper (Howard 1970), in order to be assured that the normalizability of closed terms implies the consistency of T, I used Gödel's formulation of T, but I was not very happy about it because

(5.2) the notion of intensional equality depended on whether the functionals were formulated via combinators or via lambda terms, and, if the latter, what kinds of reductions were involved.

As Troelstra notes on p. 228 of (Gödel 1990), Gödel's functional interpretation of HA can be carried out in a reformulated version of T in which there are no equations between terms of higher type. In fact, the consistency of this version of T follows from computability of terms of level 0.

Spector (Spector 1962) uses a *rule of extensional equality*. Thus the papers of Gödel and Spector raise fundamental issues about the equality relation.

I encountered issues of this kind during my review (Howard 1986) of Martin-Löf's 1984 lectures (Martin-Löf 1984). What is the reduction process being used? In addition, Martin-Löf introduces an equality notion via the pair belonging to a type (where a type is regarded as a proposition, i.e., the species of all proofs of that proposition). Thus here also we have fundamental issues about the equality relation.

6 Third Conversation

This took place in the summer house I rented from Kreisel and was about Bill's theory of infinite terms, which was in the final stages of preparation for publication (Tait 1965).

Just as Schütte first showed how the introduction of infinite proofs provided an easy analysis of Gentzen's consistency proof for PA, then, in subsequent papers, extended this to predicative analysis, one can develop Bill's theory of infinite terms in two stages: (1) infinite terms of finite type (this gives an analysis of the terms of Gödel's T), and (2) extension to infinite terms of transfinite type.

6.1 Infinite terms of finite type

Concerning #1, the basic idea can be regarded as follows. Normalization of simple typed lambda terms (i.e., terms without primitive recursion operators R), is easy: the redexes have a natural rank such that it is easy to pick a redex of highest rank n such that any new redexes produced by reduction of this redex must have lower rank. Thus we get a normalization by double recursion on pairs (n, k), where k is the number of redexes of rank n. When primitive recursion operators R are included, a difficulty is encountered. To reduce a closed term $Rtba$ of type N, where a and b are irreducible, it is necessary to reduce the term t to a numeral $\bar{i}$; maybe $\bar{i}$ turns out to be $\bar{1}$, in which case the next reduction step gives us $b\bar{0}a$, where $b\bar{0}$ may be a redex of rank k (say). The problem is that the reduction of t to a numeral may require the contraction of redexes of rank $j < k$; thus one cannot proceed according to the recursion on pairs (n, k) just described. There is a similar problem if t reduces to numeral greater than $\bar{1}$.

Bill's solution was to represent the operator R as the infinite term $(r_0, r_1, \ldots, r_i, \ldots)$, where r_i is (essentially) $\lambda x.\lambda y.f_i$, and f_i is defined by the recursion $f_0 = y$, $f_{i+1} = xif_i$. By use of the rule III, page 181, the term $Rtba$ reduces in two 'steps' to the term

$$(\lambda x.x, \lambda x.b\bar{0}x, \lambda x.b\bar{1}(b\bar{0}x), \lambda x.b\bar{2}(b\bar{1}(b\bar{0}x)), \ldots)ta$$

One can now proceed by eliminating the redexes of maximum rank from this term. The price for all this is that one must use reduction processes that simultaneously reduce every one of an infinite set of redexes.

Although this approach does not supply a proof of normalization of the terms of higher type in Gödel's T, it does provide an ordinal analysis of these terms in the sense that one gets, in a straightforward way, the computability of the numerical terms of T by means of a reasonably simple form of ordinal recursion within the ordinals less than ϵ_0, hence a characterization of the "provably recursive functions" of PA.

Bill regarded his theory of infinite terms as a pun on Schütte's theory of infinite proofs, but we both felt that there must be something going on at a deeper level. In particular, there must be an exact connection between Gentzen's cut elimination and the normalization of terms of finite type. This was very much on my mind a few years later, when I found such a connection via some work of Curry; but this required an extension of the notion of type (Howard 1980).

6.2 Infinite terms of transfinite type

During our conversation, I had difficulty following the details; but, on looking at the paper, I see that he meant the following. The crucial rule for generating transfinite types is:

(Tp 2) If $(\sigma_0, \tau_0), \ldots, (\sigma_i, \tau_i), \ldots$ is a sequence of ordered pairs of types of rank less than b, where $\sigma_i \neq \sigma_j$ for $i \neq j$, then the set $\rho = \{(\sigma_0, \tau_0), \ldots, (\sigma_i, \tau_i), \ldots\}$ is a type of rank b,

where i ranges over a nonempty finite initial segment of the natural numbers $\mathbb{N}$ or all of $\mathbb{N}$ (we omit the tag that keeps track of the rank of the type).

Corresponding to this rule for generating types, we have the following rule for generating terms:

(Tm 4) If ρ is as in Tp 2 and $t_0, \ldots, t_i, \ldots$ is a sequence of terms, where t_i has type τ_i, then the set $r = \{\lambda x^{\sigma_0}.t_0, \ldots, \lambda x^{\sigma_i}.t_i, \ldots\}$ is a term of type ρ.

Then:

(Tm 6) If r has type $\{(\sigma_0, \tau_0), \ldots, (\sigma_i, \tau_i), \ldots\}$, and s is a term of type σ_k, for some k, then rs is a term of type τ_k.

The condition $\sigma_i \neq \sigma_j$ for $i \neq j$ in Tp 2 assures us that k is unique. Correspondingly to the term formation rules Tm 4 and Tm 6, we have the reduction rule:

(I) If r is as in Tm 4, and s has type σ_k, then $rs \longrightarrow t_k(s/x^{\sigma_k})$.

In addition, there is a term formation rule

(Tm 5) if $r = (r_0, \ldots, r_i, \ldots)$ is a sequence of terms of type τ, then r is a term of type $\{(N, \tau)\}$,

with corresponding reduction rules (II), (III), page 181 of (Tait 1965).

6.3 Narrow constructive version CV

In the narrow constructive version of this theory (page 180), the sets would be given by numerical codes, for example by codes for recursively enumerable sets (footnote 1, page 179); we also need codes for ordered pairs (for Tp 2). Of course, there are various ways to do this, but one way is the following. A code for a recursively enumerable set is any code for a computable function that generates the set. The extensional objects are: sets (case 1) and ordered pairs (case 2). Thus we will have

(6.3.1) an equivalence relation between codes, where two codes are equivalent if they represent the same extensional object,

where 'represent the same extensional object' is meant in a hereditary sense as follows. Let $ext(\alpha)$ be the extensional object represented by the code α. Thus α is a code of a computable function that generates a sequence of codes $\alpha_0, \ldots, \alpha_i, \ldots$, the range of i being specified (case 1); or α is a code of a pair (σ, τ) of codes (case 2). In case 1, $ext(\alpha)$ is the set $\{ext(\alpha_0), \ldots, ext(\alpha_i), \ldots\}$, whereas in case 2, $ext(\alpha)$ is the ordered pair $(ext(\sigma), ext(\tau))$. This defines $ext(\alpha)$ by transfinite recursion on the rank of α.

We will identify a computable function with the sequence it generates (including the case in which the function is given only for a nonempty finite initial segment of N). Also, in this and the next section, we will use 'type' for 'code of a type'. Of course, the terms themselves will now be generated by computable functions given by suitable codes.

In the set-theoretic version, every term has a unique type, as Bill remarks on page 180. In a narrow constructive version, CV, a term will have an equivalence class of types. For the moment, we will take this to be the equivalence relation defined in 6.3.1; but, for a natural alternative, see 6.3.3, below. For a constructive metamathematics, there is some question concerning the mathematical details of how the assignment of equivalence classes of types to terms is to be carried out. I shall use an approach that assigns a unique type to each term. The system CV is formulated as follows.

In line with footnote 1, page 179, Tp 2 is replaced by:

(Tp 2*) If ρ is a sequence $(\sigma_0, \tau_0), \ldots, (\sigma_i, \tau_i), \ldots$ of ordered pairs of types of rank less than b, where σ_i is not equivalent to σ_j if $i \neq j$, then ρ is a type of rank b.

The rule Tm 4 becomes:

(Tm 4*) If ρ is as in Tp 2* and $t_0, \ldots, t_i, \ldots$ is a sequence of terms, where t_i has type τ_i, then the sequence $r = (\lambda x^{\sigma_0}.t_0, \ldots, \lambda x^{\sigma_i}.t_i, \ldots)$ is a term with type ρ.

The rule Tm 5 becomes:

(Tm 5*) If $r = (r_0, \ldots, r_i, \ldots)$ is sequence of terms, where t_0 has type τ, and the type of t_i is equivalent to τ, for all i, then r is a term with type $((N, \tau))$.

The rule Tm 6 is replaced by:

(Tm 6*) Suppose ρ is as in Tp 2*, r is a term with type ρ, and s is a term with a type equivalent to σ_k, for some k. Then rs is a term with type τ_k.

Corresponding to the reduction rule (I), we will have:

(I*) If r and s are as in Tm 4*, Tm 6*, then $rs \longrightarrow t_k(s/x^{\sigma_k})$.

Is the system CV, as just formulated, coherent?

The main item to be checked is:

(6.3.2) if the type of s is equivalent to σ, and t has type τ, then $t(s/x^\sigma)$ is a well-formed term with type equivalent to τ.

But that is easily checked (by transfinite induction on the length of a term).

Useful fact:

If rs is well-formed in CV, and s is replaced by a term $\tilde{s}$ with equivalent type, we get a term $r\tilde{s}$ whose type is the the same as that of rs. Actually, this would be used a proof of 6.3.2.

The equivalence relation between types: We have been working with the equivalence relation 6.3.1; but, since types are now given as codes of sequences, the equivalence relation corresponding to this would be more appropriate. Namely, consider the extensional version of CV in which we work with sequences of types rather than their codes. The extensional objects are now: sequences and ordered pairs. We will have

(6.3.3) an equivalence relation based on types-as-sequences,

where two codes are equivalent if they represent the same extensional object in a hereditary sense similar to that discussed for 6.3.1, above.

The condition σ_i not equivalent to σ_j if $i \neq j$ in Tp 2* might be difficult to handle in a constructive metamathematics.

6.4 Relation to current constructive type theory

What is the relation between Bill's notion of type (the constructive version) and current notions of constructive type theory? One clue is the following. The criterion for the term rs in Tm 6* to be well-formed is: there exists k such that s has type equivalent to σ_k. Hence we can think of r as acting on pairs (k, s) such that s has type equivalent to σ_k. In current type theory, every pair of this kind is given the type $\sum_{(i:N)} \sigma_i$, supposing i to range over numbers, where numbers are represented by terms of type N.

In the present situation, the number i is, conceptually, just a tag, not an object of type N. It is convenient to work with the corresponding numerals. Hence we will introduce a type *Tag* for numerals used in this way. The 'bare' numeral $\bar{0}$ does not come to us with a type. It is assigned a type by

(Tm 2) $\bar{0}$ has type N,

page 179. The result is a *typed term* $\bar{0}^N$. With this understanding of the role of Tm 2, we are free to introduce an additional typed term $\bar{0}^{Tag}$:

(Tm 2#) $\bar{0}^{Tag}$ is a term with type Tag.
In addition to Tm 3, we will have:
(Tm 3#) If i has type Tag, then so does Si.
Summary: A 'bare' numeral $\bar{n}$ becomes a term of type Tag or a term of type N when it is provided with one of these two types.

In accordance with the above, we introduce:
(Tp Σ) To a function $g(i) = \sigma_i$ from tags to types with rank less than b, we associate the type $\alpha = \sum_{(i:Tag)} \sigma_i$ with rank b.
(Tm Σ) Let α be as in Tp Σ. If k is a tag and s is a term of type σ_k, then the typed pair $(k,s)^\alpha$ is a term with type α.

Remark 6.4.1 If $\beta = \sum_{(i:Tag)} \tau_i$ is a type which is different from α, but $\tau_k = \sigma_k$ for k as above, then the typed term $(k,s)^\beta$ is regarded as different from the typed term $(k,s)^\alpha$. In other words, the ordered pair (k,s) is not itself a term; it becomes a term only when assigned a type as just described. Thus we have preserved the property: every term has a unique type (up to equivalence of types).

Corresponding to the rule of pair formation Tm Σ, we include projection operations (...)0 and (...)1 which apply to any term whose type has the form $\sum_{(i:Tag)} \sigma_i$, and reductions
(Proj 0) $(k,s)0 \longrightarrow k$
(Proj 1) $(k,s)1 \longrightarrow s$.
Here, of course, 0 and 1 are of type Tag.

Let CV# be the system obtained by modfiying CV as follows. Add Tm 2#, Tm 3#, Tp Σ, Tm Σ, Proj 0 and Proj 1 as above. Replace Tm 2*, Tm 4*, Tm 6* and (I*) by:

(Tp 2#) If ρ is a sequence $(\sigma_0, \tau_0), \ldots, (\sigma_i, \tau_i), \ldots$ of ordered pairs of types of rank less than b, then ρ is a type of rank $b+1$. (We take the rank to be $b+1$ rather than b in order to accommodate the possibility that type of the term p in Tm 6#, below, may have rank b.)

Thus we no longer require the condition: σ_i not equivalent to σ_j if $i \neq j$.

(Tm 4#) Suppose $\rho = ((\sigma_0, \tau_0), \ldots, (\sigma_i, \tau_i), \ldots)$ is as in Tp 2#. Let $t_0, \ldots, t_i, \ldots$ be a sequence of terms such that t_i has type τ_i, for all i. Then $r = (\lambda x_0^{\sigma_0}.t_0, \ldots, \lambda x_i^{\sigma_i}.t_i, \ldots)$ is a term with type ρ.

(Tm 6#) If r has type ρ, where ρ is as in Tp 2#, and p has type equivalent to $\sum_{(i:Tag)} \sigma_i$, then rp is a term with type $\tau_{first(p)}$.

(I#) If r is as in Tm 4#, and (k,s) has type equivalent to $\sum_{(i:Tag)} \sigma_i$, then $r(k,s) \longrightarrow t_k(s/x_k^{\sigma_k})$.

We need a special type, say $N\tau$, for Tm 5#:
(Tm 5#) If $r = (r_0, \ldots, r_i, \ldots)$ is sequence of terms, where t_0 has type τ, and the type of t_i is equivalent to τ, for all i, then r is a term with type $N\tau$.
(Tm 7#) If r has type $N\tau$ and s has type N, then rs has type τ.

The equivalence relation 6.3.3 is extended to CV# in the obvious way.
I think that the system CV# is probably in line with Bill's intentions, but, in any case, the following exercise may be of interest.

Embedding of CV in CV#: Let t be a term of type σ in CV. We define the image t^* of t by transfinite recursion on the length of t. The term t^* will belong to CV# and will have the same type-code as t. This gives us a one-to-one mapping $t \mapsto t^*$ from CV onto a subset CV* of CV#.

Case 1 If r is as in Tm 4*, r^* is defined to be $(\lambda x_0^{\sigma_0}.t_0^*, \ldots, \lambda x_i^{\sigma_i}.t_i^*, \ldots)$.

Case 2 If $r = (r_0, \ldots, r_i, \ldots)$ is as in Tm 5*, r^* is defined to be $(r_0^*, \ldots, r_i^*, \ldots)$.

Case 3 Suppose r and s are as in Tm 6*. Then the condition σ_i not equivalent to σ_j if $i \neq j$ assures us that the k just mentioned is unique. We define $(rs)^*$ to be $r^*(k, s^*)^\alpha$, where $\alpha = \sum_{(i:Tag)} \sigma_i$.

Case 4 If r and s are as in Tm 7#, we define $(rs)^*$ to be r^*s^*.

Case 5 We define: $\bar{0}^* = \bar{0}$, $N^* = N$ and $Tag^* = Tag$, $(x^\sigma)^* = x^\sigma$.

As mentioned above,

(6.4.2) t^* is the same as t.

But in 6.4.2 there is an implicit assumption that

(6.4.3) t^* is well-defined.

Hence

(6.4.4) We prove 6.4.2 and 6.4.3 simultaneously by transfinite induction on the length of t. For example, suppose rs is well-formed in CV. Consider case 3. Thus s has a type $\tilde{\sigma}_k$ equivalent to σ_k. By transfinite induction assumption, r^* and s^* are well-defined terms of CV#, with the same types as r and s, respectively. Hence r^* has the same type as r, namely, ρ, above; and the type $\tilde{\alpha}$ of (k, s^*) is obtained by replacing σ_k by $\tilde{\sigma}_k$ in the type $\alpha = \sum_{(i:Tag)} \sigma_i$. Thus $\tilde{\alpha}$ is equivalent α. Hence $r^*(k, s^*)$ is a well-defined term of CV#.

Thus we can consider 6.4.2–6.4.3 to be proved.

It is easy to verify, for b, c, d, e in CV:

(6.4.5) $(d(e/y))^* = d^*(e^*/y^*)$,

hence

(6.4.6) $b \longrightarrow c$ iff $b^* \longrightarrow c^*$,

hence

(6.4.7) b red c iff b^* red c^*.

(6.4.8) *Remark* From the viewpoint of current type theory, the term r of Tm 4#, Tm 6# has a Π-type (dependent product type) as follows. According to (I#), the term r acts on a term (k, s) with type equivalent to α to produce a term $t_k(s/x_k^{\sigma_k})$ with type equivalent to τ_k. Thus r, as a term of CV#, has type $\prod_{(u:\alpha)} \tau_{u0}$.

6.5 Summary

I have provided two constructive versions of Bill's system of infinite terms. The version CV is pretty straightforward. To understand CV from the viewpoint of current notions of constructive type theory, I formulated a variant, CV#.

In any constructive version of Bill's theory, the equivalence relation between codes of types plays a crucial role. In particular, in CV, the condition

(6.5.1) σ_i not equivalent to σ_j if $i \neq j$

assures us that the number (or tag) k in the reduction rule (I*), $rs \longrightarrow t_k(s/x^{\sigma_k})$, is well-determined. On the other hand, the condition 6.5.1 plays no role in the system $CV^{\#}$. For $CV^{\#}$ itself, there is a great deal of latitude for the choice of the equivalence relation between codes of types. In fact, we could take the equivalence relation to be the identity relation. This would provide a sort of 'free object' from which various versions of $CV^{\#}$ are obtained by the choice of various equivalence relations. Essentially, the only reason we mentioned, in section 6.4, that the equivalence relation used in CV can be extended to $CV^{\#}$ was to make the embedding go through.

We have not looked at general equivalence relations on the *terms* or, more generally, the codes of terms. Of course, various equivalence relations on terms correspond to various models. In a given model, a term denotes a *functional*. In current type theory, (hence in $CV^{\#}$), the domain of this functional consists of all functionals of a given type; but this is not the case in CV (or the set-theoretic version of Bill's theory).

In any case, it appears to me that $CV^{\#}$, or something similar, might provide an interesting variant of Bill's theory, worth looking into.

7 Fourth Conversation

This took place when I visited Bill in his house in Palo Alto. We returned to the issue of intensional equality. Bill was concerned with the fact that, although normalizaion of terms provided a notion of intensional equality, one has not only problem (5.2) (see second conversation) but also problem (7.1): the terms chosen depend on the particular system of functionals being studied. Was there a more general notion of intensional equality that applied to an arbitrary constructive functional? Eventually Troelstra came up with such a notion: HRO, defined on p. 125 of (Troelstra 1973). This is a nice result; my only reservation is that it depends on a pretty specific notion of constructive functional, based in a direct manner on Turing computability (of functions of lowest type). In 1972 I asked Gödel what he meant by the notions of computable functional and intensional equality. In his reply, he described something that closely resembled HRO.

This conversation also touched on a basic foundational issue: On the one hand, we have systems of terms; on the other hand, we have the abstract objects that these terms are supposed to denote. In what sense do these abstract objects have any reality? Of course, for a person who regards sets as real, there is no problem: the functionals are certain sets. I have always found this viewpoint problematic, and I wondered what Bill's view was. I was in a positivistic mood at the time, so I took the position that only the terms were real. This led to the sort of argument that has no useful outcome.

To elaborate on this a bit: If we don't take the primitive functionals as certain sets, where the domain and counterdomain are *defined concepts*, then the question, 'What is a type?' confronts us; in other words, we don't just apply a given term to every term that comes along; that leads to paradoxes. Curry struggled with this. In his approach, a term belongs to (or has) a type by means of a proof. A term may belong to lots of different types. In contemporary computer science, this is called 'Curry-style typing' as opposed to 'Church-style typing' in which every term carries with it a unique type. I have always favored the Church-style approach but I have also found it worthwhile to keep the Curry-style approach in mind.

8 Fifth Conversation

The seminar was over and I was due to leave soon. The conversation took place on a patio on campus. Bill told me some ideas he had about how one might obtain an ordinal analysis of bar recursion of type 0 (BR_0). He intended to capitalize on the fact that the continuity property necessary for BR_0 is automatically provided by the computability of the terms, where computability of a term containing variables of level one means computability relative to free choice functions. Let a be a closed term of type 2 and f a variable of type $N \longrightarrow N$. Then computability of af in the sense just mentioned means that the tree of unsecured sequences of the functional represented by a is well-founded. Bar recursion of type 0 can be understood as recursion over this tree. He intended to exploit these ideas somehow by means of his theory of infinite terms. I had already been developing ideas of my own about getting an ordinal analysis of BR_0. It was one of those conversations in which, on the one hand, both parties were eager to share their ideas, but, on the other hand, for competitive reasons neither party wanted to say too much.

9 After the Seminar

It would be beyond my ability to give even a brief summary of the work presented in the seminar or in the notes, but let me say a few words about one of the items; namely, Kreisel's notion of a *generalized inductive definition* (Sec. III, vol. 1 of the notes).

In Sec. VI, vol. 2, of the notes, I reformulated Kreisel's theory HA+O (cf. 3.6.4, above) by introducing a new type W for well-founded trees (thought of as representing constructive ordinals), thus obtaining a theory $H(N, W)$ based on two ground types, N and W. Let $T(N, W)$ be the corresponding extension of Gödel's theory T. It is easy to give the Gödel functional interpretation of $H(T, W)$ into $T(N, W)$. In the following year (spring of 1965), I obtained an ordinal analysis of $T(N, W)$ by means of the Bachmann ordinal $\phi(\epsilon_{\Omega+1})0$. For this I used Bill's theory of infinite terms of finite type extended to terms of uncountable length. This eventually appeared in (Howard 1972).

The predicate O is given by the simplest case of Kreisel's generalized inductive definitions (g.i.d.s), but in HA+O it is easy to develop Kreisel's theory of non-iterated positive g.i.d.s. (the intuitionistic theory). Thus the above gives us a proof-theoretic treatment of non-iterated positive g.i.d.s. Jeffrey Zucker extended this to a treatment of general non-iterated g.i.d.s, even for classical logic (Zucker 1973).

G.i.d.s can be iterated, and this iteration can be extended into the transfinite. The theory of iterated g.i.d.s has been a fruitful topic in proof theory. Combined with Gentzen-style ordinal analyses, this has led to consistency proofs for various subsystems of classical analysis (Feferman 2000, 2010). The strongest of these is somewhat weaker than Π_2^1 comprehension. My impression is that, in this work, the metamathematics can be reduced to the system $PRA(d)$ mentioned in 3.6.2, for appropriate d. Of course, if this is to be used as a *constructive foundation*, a constructive justification of the descending chain principle for the ordinal notations less than d would be required.

References

Feferman, S. (1964): Systems of predicative analysis, *Journal of Symbolic Logic* 29, 1–30.

Feferman, S. (2000): Does reductive proof theory have a viable rationale?, *Erkenntnis* 53, 63–96.

Feferman, S. (2010): The proof theory of classical and constructive inductive definitions. A 40 year saga, 1968–2008, in R. Schindler, ed., *Ways of Proof Theory*, Ontos Verlag, Frankfurt, pp. 7–30.

Gödel, K. (1958): Über eine bisher noch nicht benutzte Erweiterung des finiten Standpunktes, *Dialectica* 12, 208–287.

Gödel, K. (1990): *Collected Works*, vol. II, S. Feferman et al., eds., Oxford University Press, New York.

Gödel, K. (1995): *Collected Works*, vol. III, S. Feferman et al., eds., Oxford University Press, New York.

Howard, W. A. (1970): Assignment of ordinals to terms for primitive recursive functionals of finite type, in A. Kino et al., eds., *Intuitionism and Proof Theory*, North Holland, Amsterdam, pp. 443–458.

Howard, W. A. (1972): A system of abstract constructive ordinals, *Journal of Symbolic Logic* 37, 355–374.

Howard, W. A. (1980): The formulae-as-types notion of construction, in J. Seldin and R. Hindley, eds., *To H.B. Curry: Essays on Combinatory Logic, Lambda Calculus and Formalism*, Academic Press, Boston, pp. 479–490.

Kleene, S. C. (1945): On the interpretation of intuitionistic number theory, *Journal of Symbolic Logic* 10, 109–124.

Kleene, S. C. and Vesley, R. E. (1965): *The Foundations of Intuitionistic Mathematics, Especially in Relation to Recursive Functions*, North-Holland, Amsterdam.

Kreisel, G. (1987): Gödel's excursions into intuitionistic logic, in P. Weingartner and L. Schmetterer, eds., *Gödel Remembered*, Bibliopolis, Naples, pp. 65–186.

Martin-Löf, P. (1984): *Intuitionistic Type Theory*, Bibliopolis, Naples.

Schütte, K. (1965): Predicative well-orderings, in J. Crossley and M. Dummett, eds., *Formal systems and recursive functions*, North-Holland, Amsterdam, pp. 280–303.

Spector, C. (1962): Provably recursive functionals of analysis: a consistency proof of analysis by an extension of principles in current intuitionistic mathematics, in J. Dekker, ed., *Recursive Function Theory*, Proceedings of Symposia in Pure Mathematics, vol. 5, American Mathematical Society, Providence, Rhode Island, pp. 1–27.

Tait, W. W. (1965): Infinitely long terms of transfinite type, in *Formal Systems and Recursive Functions*, J. Crossley and M. Dummett, eds., North-Holland, Amsterdam, pp. 176–185.

Tait, W. W. (1967): Intensional interpretations of functionals of finite type I, *Journal of Symbolic Logic* 3, 198–212.

Troelstra, A. S., ed. (1973): *Metamathematical Investigation of Intuitionistic Arithmetic and Analysis*, Springer Lecture Notes in Mathematics, no. 344, Springer, Berlin.

Zucker, J. I. (1973): Iterated inductive definitions, trees and ordinals, in Troelstra (1973), pp. 392–453.

Part II:
Logic and Philosophy of Mathematics

A Proposition is the (Homotopy) Type of its Proofs

STEVE AWODEY[*]

ABSTRACT: We explain and motivate the recently developed, homotopical interpretation of constructive type theory.

> There are, at first blush, two kinds of construction involved: constructions of proofs of some proposition and constructions of objects of some type. But I will argue that, from the point of view of foundations of mathematics, there is no difference between the two notions. A proposition may be regarded as a type of object, namely, the type of its proofs. Conversely, a type A may be regarded as a proposition, namely, the proposition whose proofs are the objects of type A. So a proposition A is true just in case there is an object of type A.
>
> W.W. Tait (1994)

Overview

Homotopy type theory is a new field devoted to a recently discovered connection between Logic and Topology — more specifically, between constructive type theory, which was originally invented as a constructive foundation for

---

[*]Thanks to Ulrik Buchholtz and Michael Shulman for comments on an earlier draft. This research was partially supported by the U.S. Air Force Office of Scientific Research through MURI grant FA9550-15-1-0053. Any opinions, findings and conclusions or recommendations expressed in this material are those of the authors and do not necessarily reflect the views of the AFOSR.

mathematics and now has many applications in the theory of programming languages and formal proof verification, and homotopy theory, a branch of algebraic topology devoted to the study of continuous deformations of geometric spaces and mappings. The basis of homotopy type theory is an interpretation of the system of intensional type theory into abstract homotopy theory. As a result of this interpretation, one can construct new kinds of models of constructive logic and study that system semantically, e.g. proving consistency and independence results. Conversely, constructive type theory can also be used as a formal calculus to reason about abstract homotopy. This is particularly interesting in light of the fact that the type theory used underlies several computational proof assistants, such as Coq and Agda; this allows one to use those systems to reason formally about homotopy theory and fully verify the correctness of definitions and proofs using these computer proof systems. Potentially, this could provide a useful tool for mathematicians working in fields like homotopy theory and higher category theory. Finally, new logical principles and constructions based on homotopical and higher categorical intuitions can be added to the system, providing a way to formalize many classical spaces and sophisticated mathematical constructions. Examples include the so-called *higher inductive types* and the *univalence axiom* of Voevodsky.

More broadly, *univalent foundations* is an ambitious new program for foundations of mathematics, proposed by Voevodsky, which is based roughly on homotopy type theory and intended to capture a very broad range of mathematics (I am reluctant to use the phrase "All of Mathematics", but there is nothing in particular that could not, in principle, be done). The new univalence axiom, which roughly speaking implies that isomorphic structures can be identified, and the general point of view that it promotes sharpen the expressiveness of the system and make it more powerful, so that new concepts can be isolated and new constructions can be carried out, and others that were previously ill-behaved (such as quotients) can be better controlled. The system is not only more expressive and powerful than previous type- and set-theoretic systems of foundations; it also has two further, distinct novelties: it is still amenable to computer formalizations, and it captures a conception of mathematics that is distinctly "structural". These two seemingly unrelated aspects, one practical, the other philosophical, are in fact connected in a rather subtle way. The structural character of the system, which the univalence axiom requires and indeed strengthens, permits the use of a new "synthetic" style of foundational axiomatics which is quite different from conventional axiomatic foundations. One might call the conventional, set-theoretic style of foundations an "analytic" (or perhaps "bottom-up") approach, which "analyses" mathematical objects into constituent *material* (e.g. sets or numbers), or at least constructs appropriate "surrogate objects" from such material — think of real numbers as Dedekind cuts of rationals. By contrast, the "synthetic" (or "top-down") approach permitted by univalent foundations is based on describing the fundamental *structure* of mathematical objects in terms of their universal properties, which in type theory are given by rules of inference determining directly how the new objects map to and from all other ones. This fundamental shift in foundational methodology has the practical effect of simplifying and shortening many proofs by taking advantage of a more axiomatic approach, as opposed to the more laborious analytic constructions.[1] Indeed, in a relatively short time, a large amount of classical mathematics has already been developed in this new system: basic homotopy theory, category theory, real analysis, the cumulative hierarchy of set theory, and many other topics. The proofs of some very sophisticated, high-level theorems have now been fully formalized and verified by computer proof assistants — a foundational achievement that would be very difficult to match using conventional, "analytic" style foundational methods.

Indeed, this combination of a synthetic foundational methodology and a powerful computational implementation has the potential to give new life, and a new twist, to the old idea of reducing mathematics to a purely formal calculus. Explicit formalizations that were once too tedious or complicated to be done by hand can now be accomplished in practice with a combination of synthetic methods and computer assistance. This new formal reduction of mathematics raises again the epistemological question of whether, and in what sense, the type-theoretic basis of the formal system is purely "logical", and what this means about mathematics and the nature of a priori knowledge. That is a question of significant philosophical interest, but it is perhaps better pursued independently, once the mathematical issues related to the formalization itself are more settled.

1 Type theory

In its current form, constructive type theory is the result of contributions made by several different people, working both independently and in collaboration. Without wanting to give an exhaustive history (for one such, see Kamareddine et. al. (2004)), it may be said that essential early contributions were made by H. Curry, W. Howard, F.W. Lawvere, P. Martin-Löf, D.S. Scott, and W.W. Tait.

Informally, the basic system consists of the following ingredients:

- **Types:** $X, Y, \ldots, A \times B, A \to B, \ldots$, including both primitive types and type-forming operations, which construct new types from given ones, such as the product type $A \times B$ and the function type $A \to B$.

- **Terms:** $a : A$, $b : B$, ..., including variables $x : A$ for all types, primitive terms $b : B$, and term-forming operations like $\langle a, b \rangle : A \times B$ and $\lambda x.b(x) : A \to B$ associated to the type-forming operations.

---

[1] In a related context, it has been said that such an approach has "all the advantages of theft over honest toil", but the issue of how to *justify* the rules for new constructions can be separated from that of their expedience.

One essential novelty is the use of so-called *dependent types*, which are regarded as "parametrized" types or *type families indexed over a type*.

- **Dependent Types**: $x : A \vdash B(x)$ means that $B(x)$ is a type for each $x : A$, and thus it can be thought of as a function from A to types. Moreover, one can have iterated dependencies, such as:

    $x : A \vdash B(x)$
    $x : A, y : B(x) \vdash C(x, y)$
    $x : A, y : B(x), z : C(x, y) \vdash D(x, y, z)$
    etc.

- **Dependent Type Constructors**: There are special type constructors for dependent types, such as the sum $\sum_{x:A} B(x)$ and product $\prod_{x:A} B(x)$ operations. Associated to these are term constructors that act on dependent terms $x : A \vdash b(x) : B(x)$, such as $\lambda x.b(x) : \prod_{x:A} B(x)$.

- **Equations**: As in an algebraic theory, there are then equations $s = t : A$ between terms of the same type, such as $\bigl(\lambda x.b(x)\bigr)(a) = b(a) : B(a)$.

The entire system of constructive type theory is a formal calculus of such typed terms and equations, usually presented as a deductive system by formal rules of inference. For one modern presentation, see the appendix to Univalent Foundations Project (2013). This style of type theory is somewhat different from the Frege-Russell style systems of which it is a descendant. It was originally intended as a foundation for *constructive* mathematics, and it has a distinctly "predicative" character — for instance, it is usually regarded as open-ended with respect to the addition of new type- and term-forming operations, such as universes, so that one does not make use of the notion of "all types" in the way that set-theory admits statements about "all sets" via its first-order logical formulation. Type theory is now used widely in the theory of programming languages and as the basis of computerized proof systems, in virtue of its good computational properties.

Propositions as types

The system of type theory has a curious dual interpretation:

- On the one hand, there is the interpretation as **mathematical** objects: the types are some sort of constructive "sets", and the terms are the "elements" of these sets, which are being built up according to the stated rules of construction.

- But there is also a second, **logical** interpretation: the types are "propositions" about mathematical objects, and their terms are "proofs" of the corresponding propositions, which are being derived in a deductive system.

A Proposition is the (Homotopy) Type of its Proofs 57

This is known as the *Curry-Howard correspondence*, and it can be displayed as follows:

0	1	$A+B$	$A \times B$	$A \to B$	$\sum_{x:A} B(x)$	$\prod_{x:A} B(x)$
$\bot$	$\top$	$A \vee B$	$A \wedge B$	$A \Rightarrow B$	$\exists_{x:A} B(x)$	$\forall_{x:A} B(x)$

For instance, regarded as propositions, A and B have a conjunction $A \wedge B$, a proof of which corresponds to a pair of proofs a of A and b of B (via the $\wedge$-introduction and elimination rules), and so the terms of $A \wedge B$, regarded as a type, are just pairs $\langle a, b \rangle : A \times B$ where $a : A$ and $b : B$. Similarly, a proof of the implication $A \Rightarrow B$ is a function f that, when applied to a proof $a : A$ returns a proof $f(a) : B$ (*modus ponens*), and so $f : A \to B$. The interpretation of the existential quantifer $\exists_{x:A} B(x)$ mixes the two points of view: a proof of $\exists_{x:A} B(x)$ consists of a term $a : A$ and a proof $b : B(a)$; so in particular, when it can be proved, one always has an instance a of an existential statement. In classical logic, by contrast, one can use "proof by contradiction" to establish an existential statement without knowing an instance of it, but this is not possible here. This gives the system a distinctly constructive character (which can be specified in terms of certain good proof-theoretic properties). This is one reason it is useful for computational applications.

Identity types

Under the logical interpretation above we now have:

- **propositional logic**: $0, 1, A+B, A \times B, A \to B$,
- **predicate logic**: $B(x), C(x,y)$, with the **quantifiers** $\prod$ and $\sum$.

It would therefore be natural to add a primitive relation representing equality of terms $x = y$ as a *type*. On the logical side, this would represent the proposition "x is identical to y". But what would it be *mathematically*? How are we to continue the above table:

0	1	$A+B$	$A \times B$	$A \to B$	$\sum_{x:A} B(x)$	$\prod_{x:A} B(x)$	?
$\bot$	$\top$	$A \vee B$	$A \wedge B$	$A \Rightarrow B$	$\exists_{x:A} B(x)$	$\forall_{x:A} B(x)$	x = y

We shall add to the system a new, primitive type of *identity* between any terms $a, b : A$ of the same type A:

$$\mathsf{Id}_A(a, b).$$

The *mathematical* interpretation of this identity type is what leads to the homotopical interpretation of type theory. Before we can explain that, however, we must first consider the rules for the identity types.

The **introduction** rule says that $a : A$ is always identical to itself:

$$\mathsf{r}(a) : \mathsf{Id}_A(a, a)$$

The **elimination** rule is a form of what may be called "Lawvere's Law":[2]

$$\frac{c : \mathsf{Id}_A(a, b) \qquad x : A \vdash d(x) : R(x, x, \mathsf{r}(x))}{\mathsf{J}(a, b, c, d) : R(a, b, c)}$$

That may look a bit forbidding when seen for the first time. Schematically, it is saying something like:

$$a = b \ \& \ R(x, x) \ \Rightarrow \ R(a, b).$$

Omitting the proof terms, this characterizes identity by saying that it is the *least* (or better: *initial*) reflexive relation.

The rules for identity types are such that if a and b are *syntactically equal* as terms, $a = b : A$, then they are also *identical* in the sense that there is a term $p : \mathsf{Id}_A(a, b)$. But the converse is not true: distinct terms $a \neq b$ may still be propositionally identical $p : \mathsf{Id}_A(a, b)$. This is a kind of *intensionality* in the system, in that terms that are identified by the propositions of the system may nonetheless remain distinct syntactically, e.g. different polynomial expressions may determine the same function. Allowing such syntactic distinctions to remain (rather than including a "reflection rule" of the form $p : \mathsf{Id}_A(a, b) \Rightarrow a = b$, as is done in "extensional type theory"), gives the system its good computational and proof-theoretic properties. It also gives rise to a structure of great combinatorial complexity.

Although only the syntactically equal terms $a = b : A$ are fully interchangeable everywhere, propositionally identical ones $p : \mathsf{Id}_A(a, b)$ are still interchangeable *salva veritate* in the following sense: assume we are given a type family $x : A \vdash B(x)$ (regarded, if you like, as a "predicate" on A), an identity $p : \mathsf{Id}_A(a, b)$ in A, and a term $u : B(a)$ (a "proof of $B(a)$"). Then consider the following derivation, using the identity rules.

$$\cfrac{u : B(a) \qquad \cfrac{p : \mathsf{Id}_A(a, b) \qquad \cfrac{\cfrac{x : A \vdash B(x)}{x : A, y : B(x) \vdash y : B(x)}}{\cfrac{x : A \vdash \lambda y.y : B(x) \to B(x)}{p_* : B(a) \to B(b)}}}{p_* u : B(b)}}$$

Here $p_* = \mathsf{J}(a, b, p, \lambda y.y)$. The resulting term $p_* u : B(b)$ (which is a derived "proof of $B(b)$") is called the *transport* of u along p. Logically, this just says

$$a = b \ \& \ B(a) \ \Rightarrow \ B(b),$$

---

[2]See Lawvere (1970) for a closely related principle.

i.e. that a type family over A must respect the identity relation on A. As we shall see below, the homotopy interpretation provides a different view of transport; namely, it corresponds to the familiar *lifting property* used in the definition of a "fibration of spaces":

$$\begin{array}{ccc} B & u \dashrightarrow p_*u & \quad (4.1) \\ \downarrow & & \\ A & a \xrightarrow{\ p\ } b & \end{array}$$

2 The homotopy interpretation

Given any terms $a, b : A$, we can form the identity type $\mathsf{Id}_A(a,b)$ and then consider its terms, if any, say $p, q : \mathsf{Id}_A(a,b)$. Logically, p and q are "proofs" that a and b are identical, or more abstractly, "reasons" or "evidence" that this is so. Can p and q be different? It was once thought that such identity proofs might themselves always be identical, in the sense that there should always be some $\alpha : \mathsf{Id}_{\mathsf{Id}_A(a,b)}(p,q)$; however, as it turns out, this need not be so. Indeed, there may be many distinct (i.e. non-identical) terms of an identity type, or none at all. Understanding the structure of such iterated identity types is one result of the homotopical interpretation.

Suppose we have terms of ascending identity types:

$$a, \ b : A$$
$$p, \ q : \mathsf{Id}_A(a,b)$$
$$\alpha, \ \beta : \mathsf{Id}_{\mathsf{Id}_A(a,b)}(p,q)$$
$$\ldots : \mathsf{Id}_{\mathsf{Id}_{\mathsf{Id}_\ldots}}(\ldots)$$

Then we can consider the following informal interpretation:

$$\begin{array}{rcl} \text{Types} & \rightsquigarrow & \text{Topological spaces} \\ \text{Terms} & \rightsquigarrow & \text{Continuous maps} \\ a : A & \rightsquigarrow & \text{Points } a \in A \\ p : \mathsf{Id}_A(a,b) & \rightsquigarrow & \text{Paths } p \text{ from } a \text{ to } b \\ \alpha : \mathsf{Id}_{\mathsf{Id}_A(a,b)}(p,q) & \rightsquigarrow & \text{Homotopies } \alpha \text{ from } p \text{ to } q \\ & \vdots & \end{array}$$

So for instance A may be a space with points a and b, and then an identity term $p : \mathsf{Id}_A(a,b)$ is interpreted as a path in A from a to b, i.e. a continuous function $p : [0,1] \to A$ with $p0 = a$ and $p1 = b$. If $q : \mathsf{Id}_A(a,b)$ is another such path from a to b, a higher identity term $\alpha : \mathsf{Id}_{\mathsf{Id}_A(a,b)}(p,q)$ is then interpreted as a homotopy from p to q, i.e. a "continuous deformation" of p into q, described formally as a continuous function $\alpha : [0,1] \times [0,1] \to A$ with the expected behavior on the boundary of the square $[0,1] \times [0,1]$. Higher identity terms are likewise interpreted as higher homotopies.

Note that, depending on the choice of space A and points $a, b \in A$ and paths p, q, it may be that there are no homotopies from p to q because, for example, those paths may go around a hole in A in two different ways, so that there is no continuous way to deform one into the other. Or there may be many different homotopies between them, for instance wrapping different numbers of times around the surface of a ball. Depending on the space, this can become quite a complicated structure of paths, deformations, higher-dimensional deformations, etc. — indeed, the investigation of this structure is what homotopy theory is all about.

One could say that the basic idea of the homotopy interpretation is just to extend the well-known topological interpretation of the *simply-typed* λ-calculus Awodey (2000), Awodey and Butz (2000) (which interprets types as spaces and terms as continuous functions) to the *dependently typed* λ-calculus with Id-types. The essential new idea is then simply this:

> An identity term $p : \mathsf{Id}_A(a, b)$ is a path in the space A from the point a to the point b.

Everything else essentially follows from this one idea: the dependent types $x : A \vdash B(x)$ are then forced by the rules of the type theory to be interpreted as *fibrations*, in the topological sense, since one can show from the rules for identity types that the associated map $B \to A$ of spaces must have the *lifting property* indicated in diagram (4.1) above (a slightly more intricate example shows that one can "lift" not only the endpoint, but also the entire path, and even a homotopy). The total Id-types $\sum_{x,y:A} \mathsf{Id}_A(x, y)$ are naturally interpreted as *path spaces* A^I, and the maps $f, g : A \to B$ that are identical as terms of function type $A \to B$ are just those that are *homotopic* $f \sim g$.

The homotopy interpretation was first proposed by the present author and worked out formally (with a student) in terms of Quillen model categories—a modern, axiomatic setting for abstract homotopy theory that encompasses not only the classical homotopy theory of spaces and their combinatorial models like simplicial sets, but also other, more exotic notions of homotopy (see Awodey and Warren (2009)). The interpretation was shown to be *complete* in the logical sense by Gambino and Garner (Gambino and Garner (2008)).[3] These results show that intensional type theory can in a certain sense be regarded as a "logic of homotopy", in that the system can be faithfully represented homotopically, and then used to reason formally about spaces, continuous maps, homotopies, and so on. The next thing one might ask is, how much general homotopy theory

---

[3] There is a technical question related to the selection of path objects and diagonal fillers as interpretations of Id_A-types and elimination J-terms in a "coherent" way, i.e. respecting substitution of terms for variables; various solutions have been given, including Warren (2008), van den Berg and Garner (2012), Voevodsky (2009), Lumsdaine and Warren (2015), Awodey (2018).

The fundamental groupoid of a type

Like path spaces in topology, identity types endow each type with the structure of a *groupoid*: a category in which every arrow has an inverse.

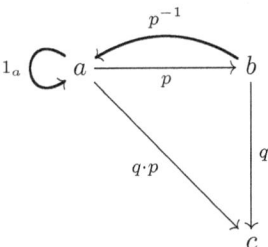

The familiar *laws of identity*, namely reflexivity, symmetry, and transitivity are provable in type theory, and their proof terms therefore act on identity terms, providing the *groupoid operations* of unit, inverse, and composition:

$$r : \mathsf{Id}(a, a) \qquad \text{reflexivity}$$
$$s : \mathsf{Id}(a, b) \to \mathsf{Id}(b, a) \qquad \text{symmetry}$$
$$t : \mathsf{Id}(a, b) \times \mathsf{Id}(b, c) \to \mathsf{Id}(a, c) \qquad \text{transitivity}$$

The *groupoid laws* of units, inverses, and associativity also hold "up to homotopy", i.e. up to the existence of a higher identity term. This means that instead of e.g. $p^{-1} \cdot p = 1_a$, we have a higher identity term:

$$\alpha : \mathsf{Id}_{\mathsf{Id}}\left(p^{-1} \cdot p, 1_a\right)$$

as indicated in:

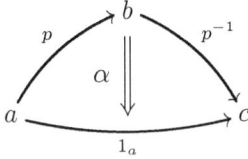

Indeed, this is just the same situation that one encounters in defining the fundamental group of a space in classical homotopy theory, where one shows e.g. that composition of paths is associative up to homotopy by reparametrization of the composites. In fact, in virtue of the homotopy interpretation, the classical case is really just an instance of the more general, type theoretic one.

Inspired by this occurence of type theoretic groupoids, Hofmann and Streicher (Hofmann and Streicher (1998)) discovered an interpretation of the entire system of type theory into the category of all groupoids, which was a precursor of the homotopy interpretation. It was used, for instance, to establish the above mentioned fact that identity types may have elements that are not themselves identical.

The identity structure of a general type may actually be much richer than that of just a groupoid; as in homotopy theory, there may be non-trivial higher identities, representing higher homotopies between homotopies, and this structure may go on to higher and higher identities without ever becoming degenerate.

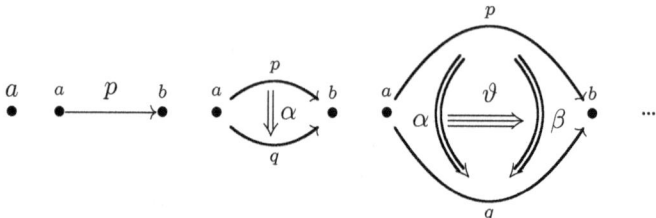

The resulting structure is that of an ω-*groupoid*, which is something that has also appeared elsewhere in mathematics — twice! As already mentioned, such "infinite-dimensional groupoids" also occur in homotopy theory, where the fundamental ω-groupoid of a space is an algebraic invariant that respects the homotopy type (according to Grothendieck's famous "homotopy hypothesis" these groupoids contain all the essential information of the space up to homotopy); but also in category theory, one has considered the idea of an ω-*category*, with not only objects and arrows between them, but also 2-arrows between arrows, 3-arrows between 2-arrows, and so on. It is indeed remarkable that the same notion has now appeared again in logic, as exactly the structure of iterated identity in type theory.[4]

Homotopy levels

One of the most useful new discoveries is that the system of all types is naturally stratified into "homotopy levels" by a hierarchy of definable conditions.[5]

At the lowest level are those types that are *contractible* in the following sense.

$$X \text{ is contractible} =_{\text{def}} \sum_{x:X} \prod_{y:X} \mathsf{Id}_X(x,y).$$

---

[4] See Lumsdaine (2010) and van den Berg and Garner (2011) for details.

[5] This concept is due to Voevodsky, cf. Voevodsky (2009). Also see Univalent Foundations Project (2013), ch. 7.

Under the logical reading, this condition says that X is a "singleton", in that there is an element $x : X$ such that everything $y : X$ is identical to it. So roughly, these are the types that have just one element, up to homotopy.

The next level consists of the *propositions*, defined as those types whose identity types are always contractible,

$$X \text{ is a proposition} =_{\text{def}} \prod_{x,y:X} \mathsf{Contr}(\mathsf{Id}_X(x,y)).$$

It is not hard to see that such types are contractible *if* they are inhabited—thus they are like "truth values", either false (i.e. empty) or true (i.e. contractible), and then essentially uniquely so. In other words, the elements of a proposition contain no further information, other than the mere inhabitation of the proposition, which we interpret to mean that it holds.

At the next level are the *sets*, which are types whose identity relation is always a proposition:

$$X \text{ is a set} =_{\text{def}} \prod_{x,y:X} \mathsf{Prop}(\mathsf{Id}_X(x,y))$$

These types have the familiar, set-like behavior that the identity proofs, when they exist, are unique (again "up to homotopy").

Next come the types whose identity types are sets, which may be called *groupoids*, because they are like the algebraic groupoids just discussed:

$$X \text{ is a groupoid} =_{\text{def}} \prod_{x,y:X} \mathsf{Set}(\mathsf{Id}_X(x,y))$$

These types may have distinct identity proofs between elements, but all higher identity proofs are degenerate.

The general pattern is now clear:

$$X \text{ has homotopy level } n+1 =_{\text{def}} \prod_{x,y:X} \mathsf{H}_n(\mathsf{Id}_X(x,y))$$

Thus the types X of homotopy level $n+1$ (for which we write $\mathsf{H}_{n+1}(X)$) are the types whose identity relation is of homotopy level n; these types correspond to the higher-dimensional groupoids of category theory, when we think of identity terms as higher-dimensional arrows. To start the numbering, we may set the contractible types to be level 0.

The homotopy level of a type is the height at which the tower of iterated identity types becomes degenerate; under the homotopy interpretation this corresponds (up to a shift in numbering) to the notion of a space being a *homotopy n-type*, which is usually defined as the greatest n such that the n-th homotopy group is non-trivial. In each case, it is a measure of the complexity of the type/space — in the former case in terms of higher identities, and in the latter in terms of higher homotopies.

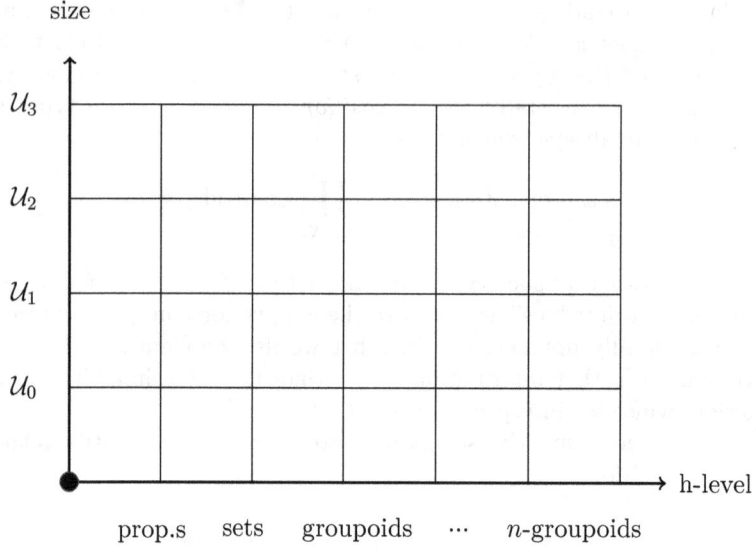

The 2D hierarchy of types

The recognition that types have these different degrees of complexity allows for a more refined version of the propositions-as-types idea, according to which only those types that are "propositions" in the sense of the homotopy levels are read as bare *assertions*, while others are regarded more discriminately as *structured objects* of various kinds. Accordingly, a type family $x : A \vdash B(x)$ such that all values $B(x)$ are propositions can be regarded as a simple "predicate" (or a "relation" depending on the arity), while a family of sets, groupoids, etc. is viewed more accurately as a structure on A.

The stratification of types by homotopy levels gives us a new view of the mathematical universe, which is now seen to be arranged not only into the familiar, one-dimensional hierarchy of *size*, determined by a system of universes $\mathcal{U}_0, \mathcal{U}_1, \mathcal{U}_2, \ldots$, but also into a hierarchy of *homotopy levels*, which form a second dimension independent of the first (see Fig. 1).

3 Higher inductive types

The recognition and use of the notion of homotopy level of a type has made the entire system of type theory more expressive and powerful, for example by allowing greater control over the introduction of new type constructions. One such construction that was formerly problematic but is now better behaved is the construction of the *quotient type* $A/\sim$ of a type A by an equivalence relation $x, y : A \vdash x \sim y$. When $x \sim y$ is known to be a proposition for all $x, y : A$, then the quotient $A/\sim$ will be a set, and the introduction and elimination rules can be determined without difficulty. Such "set quotients" can be constructed, roughly speaking, as equivalence classes Voevodsky (2009); or they can be introduced axiomatically Univalent Foundations Project (2013), essentially by stating rules that say that the identity type of $A/\sim$ is a *relation* (i.e. a family of propositions) that is freely generated by the equivalence relation $x \sim y$.

The latter, axiomatic approach is a special case of the very powerful construction method of *higher inductive types*, which are a systematic way of introducing new types with stipulated points, paths, higher paths, etc.. In order to explain this further, let us first recall how type theory deals with *ordinary* inductive types, like the natural numbers. The natural numbers $\mathbb{N}$ can be implemented as an inductive type via rules that may be represented schematically as:

$$\mathbb{N} := \begin{cases} 0 : \mathbb{N} \\ s : \mathbb{N} \to \mathbb{N} \end{cases}$$

The terms 0 and s are the *introduction rules* for this type. The recursion property of $\mathbb{N}$ is captured by an *elimination rule*:

$$\frac{a : X \qquad f : X \to X}{\mathsf{rec}(a, f) : \mathbb{N} \to X}$$

which says that given any structure of the same kind as $\mathbb{N}$, there is a map $\mathsf{rec}(a, f)$ to it from $\mathbb{N}$, which furthermore preserves the structure, as stated by the following *computation rules*:

$$\mathsf{rec}(a, f)(0) = a,$$
$$\mathsf{rec}(a, f)(sn) = f(\mathsf{rec}(a, f)(n)).$$

The map $\mathsf{rec}(a, f) : \mathbb{N} \to X$ is actually required to be the *unique* one satisfying the computation rules, a condition that can be ensured either with a further computation rule or by reformulating the elimination rule as a more general *induction principle* rather than a recursion principle (cf. Awodey et. al. (2012)).

In more algebraic terms, one would say that $(\mathbb{N}, 0, s)$ is the *free structure* of this kind. We remark that it can be shown on the basis of these rules, and without further assumptions, that $\mathbb{N}$ *is a set* in the sense of the hierarchy of homotopy levels.

The circle

We now want to use the same method of specifying a new type by introduction and elimination rules (which amount to specifying the mappings to and from other types), but now with generating data that may include also elements of identity types, in addition to elements of the type itself and operations on it. A simple example is the following.

The homotopical circle $\mathbb{S}$ can be given as an inductive type involving one "base point" and one "higher-dimensional" generator:

$$\mathbb{S} := \begin{cases} \mathsf{base} : \mathbb{S} \\ \mathsf{loop} : \mathsf{Id}_{\mathbb{S}}(\mathsf{base}, \mathsf{base}) \end{cases}$$

The element $\mathsf{loop} : \mathsf{Id}_{\mathbb{S}}(\mathsf{base}, \mathsf{base})$ can therefore be regarded as a "loop" at the basepoint $\mathsf{base} : \mathbb{S}$, i.e. a path that starts and ends at base. The corresponding recursion property of $\mathbb{S}$ is then given by the following elimination rule,

$$\frac{a : X \qquad p : \mathsf{Id}_X(a, a)}{\mathsf{rec}(a, p) : \mathbb{S} \to X}$$

with computation rules,

$$\mathsf{rec}(a, p)(\mathsf{base}) = a,$$
$$\mathsf{rec}(a, p)_!(\mathsf{loop}) = p.$$

There is an obvious analogy to the rules for $\mathbb{N}$.[6] The map $\mathsf{rec}(a, p) : \mathbb{S} \to X$ is then moreover required to be unique up to homotopy, which again is achieved either with additional computation rules or a generalized elimination rule in the form of "circle induction" rather than "circle recursion" (see Univalent Foundations Project (2013), Sojakova (2014)).

Conceptually, these rules suffice to make the structure $(\mathbb{S}, \mathsf{base}, \mathsf{loop})$ into the "free type with a point and a loop". To see that it actually behaves as it should to be the homotopical circle, one can verify that it has the correct homotopy groups (cf. Licata and Shulman (2013)):

Theorem 1 (Shulman 2011). *The type-theoretic circle $\mathbb{S}$ has the following homotopy groups:*

$$\pi_n(\mathbb{S}) = \begin{cases} \mathbb{Z}, & \text{if } n = 1, \\ 0, & \text{if } n \neq 1. \end{cases}$$

The homotopy groups $\pi_n(X, x)$ for any type X and basepoint $x : X$ can be defined as usual in terms of loops at x in X, i.e. identity elements $\mathsf{Id}_X(x, x)$, "modulo homotopy", i.e. modulo higher identities. The proof of the above theorem can be given entirely within the system of type theory, and it combines methods from classical homotopy theory with ones from constructive type theory in a novel way, using Voevodsky's univalence axiom (a sketch is given in Section 5 below). The entire development has been fully formalized in Licata and Shulman (2013).

---

[6]A map $f : A \to B$ induces a map on identities, taking each $p : \mathsf{Id}_A(a, b)$ to a term in $\mathsf{Id}_A(fa, fb)$ which we here write $f_! p$ (see Univalent Foundations Project (2013), ch. 2).

The interval

The homotopical interval $\mathbb{I}$ is also a higher inductive type, this time generated by the basic data:

$$\mathbb{I} := \begin{cases} 0, 1 : \mathbb{I} \\ \mathsf{path} : \mathsf{Id}_{\mathbb{I}}(0, 1) \end{cases}$$

Thus $\mathsf{path} : \mathsf{Id}_{\mathbb{I}}(0, 1)$ represents a path from 0 to 1 in $\mathbb{I}$. The elimination and computation rules are analogous to those for the circle, but now with separate endpoints 0 and 1. So given any path $p : \mathsf{Id}_X(a, b)$ between points a and b in any type X, there is a unique (up to homotopy) map $\mathbb{I} \to X$ taking 0 to a, 1 to b, and path to p. This specification makes the structure $(\mathbb{I}, 0, 1, \mathsf{path})$ the "free type with a path".

In terms of this example, we can plainly compare the methodology behind the use of higher inductive types in homotopy type theory with the conventional approach of classical topology:

> In classical topology, we start with the *interval* and use it to define the notion of a *path*.
> In homotopy type theory, we start with the notion of a *path*, and use it to define the *interval*.

The notion of a *path*, recall, is a primitive one in our system, namely a term of identity type. In terms of these, one can then determine the interval $\mathbb{I}$ via its mappings, rather than the other way around.

Constructing higher inductive types

The higher inductive types mentioned so far were introduced *axiomatically*, by stating their basic rules. Can we instead *construct* them, similarly to the way that quotients can also be constructed from equivalence classes? One possible way to do this is by what is sometimes called an "impredicative encoding", which is a construction that involves a quantification over "all types".

Consider first some related examples that are not *higher* inductive types, but which can be determined by such impredicative encodings. First let p and q be propositions and consider:

$$p \vee q =_{\mathrm{def}} \forall_x \big[(p \Rightarrow x) \wedge (q \Rightarrow x) \Rightarrow x\big]$$

where the quantifier $\forall_x$ is over *all propositions* x (in a given universe). Among propositions, this type has the correct behavior to be the disjunction of p and q.

Next let A and B be sets, and consider:

$$A + B =_{\mathrm{def}} \prod_X \big[(A \to X) \times (B \to X) \to X\big]$$

where the product $\prod_X$ is over *all sets* X (again, in a given universe). In order for this to actually be the coproduct among all sets, this specification requires an additional *coherence condition* saying that the transformations

$$\alpha_X : ((A \to X) \times (B \to X)) \longrightarrow X$$

are *natural in* X, in a straightforward sense that we will not spell out here.

The same general idea can be used for the interval and the circle: for a type X, define the *path space* $I(X)$ and the *(free) loop space* $L(X)$ by:

$$I(X) =_{\text{def}} \sum_{x,y:X} \mathsf{Id}_X(x,y)$$

$$L(X) =_{\text{def}} \sum_{x:X} \mathsf{Id}_X(x,x)$$

Morally, we expect that $I(X) = \mathbb{I} \to X$ and $L(X) = \mathbb{S} \to X$, so as a first approximation we set:

$$\mathbb{I} =_{\text{def}} \prod_X \big[I(X) \to X\big]$$

$$\mathbb{S} =_{\text{def}} \prod_X \big[L(X) \to X\big]$$

where the product $\prod_X$ is now over all *groupoids* X (types of homotopy level 3, in a given universe). In order to get the correct elimination rules for these types, we again add a further *coherence condition*, now involving higher-order naturality, which again we will not spell out here.[7]

The possibility of a "logical construction of the circle" and similar constructions of some other higher inductive types are current work in progress. At present they require either a general assumption of "impredicativity", or more specialized "resizing rules", or some other device to handle the shift in universes involved in the quantification over "all types".

Many basic spaces and constructions can be introduced directly as higher inductive types. These include, for example:

- higher spheres $\mathbb{S}^n$, mapping cylinders, tori, cell complexes,
- suspensions ΣA, homotopy pushouts,
- truncations, such as connected components $\pi_0(A)$ and "bracket" types $[A]$ (cf. Awodey and Bauer (2004)),
- (higher) homotopy groups π_n, Eilenberg-MacLane spaces $K(G,n)$, Postnikov systems,

---

[7] See Awodey (2015). The displayed formula for the circle was first considered by M. Shulman.

- a Quillen model structure on the system of all types,
- quotients by equivalence relations and more general quotients,
- free algebras, algebras presented by generators an relations,
- the real numbers, the surreal numbers,
- the cumulative hierarchy of Zermelo-Fraenkel sets.

The use of higher inductive types is a topic that is curently under very active investigation (see e.g. Lumsdaine (2011)).

4 Univalence

Voevodsky has proposed a new foundational axiom to be added to type theory: the *univalence axiom*. It is motivated by the homotopy interpretation and makes precise the informal mathematical practice of "identifying" isomorphic objects. Especially when combined with higher inductive types, this new axiom is a powerful addition to the system. Although it is formally incompatible with the naive interpretation of type theory according to which all types are sets, it is provably consistent with the homotopical interpretation Kapulkin et. al. (2012). Its status as a constructive principle is still unsettled, however, and that question is the focus of much current research.

Isomorphism, equivalence, and invariance

In type theory, the notion of a *type isomorphism* $A \cong B$ is definable as usual; namely, the statement

there are $f : A \to B$ and $g : B \to A$ with $gf(x) = x$ and $fg(y) = y$

is formalized by the type of isomorphisms,

$$\mathrm{Iso}(A, B) =_{\mathrm{def}} \sum_{f:A\to B} \sum_{g:B\to A} \left(\prod_{x:A} \mathsf{Id}_A(gf(x), x) \times \prod_{y:B} \mathsf{Id}_B(fg(y), y) \right).$$

Under the logical reading, this type expresses exactly the preceding informal statement. The types A and B are isomorphic just if this type is inhabited by a term, which is then exactly an isomorphism between A and B. Here we see the propositions-as-types idea at work: a proof of the proposition $A \cong B$ is the same thing as a term of the type $\mathrm{Iso}(A, B)$, namely, an isomorphism.

There is also a more refined notion of *equivalence* of types $A \simeq B$ which adds a further "coherence" condition relating the identity terms of $\mathsf{Id}_A(gf(x), x)$ and $\mathsf{Id}_B(fg(y), y)$ via f and g (see Univalent Foundations Project (2013), chapter 4). Since every isomorphism can be "promoted" to an equivalence, the latter condition is no "stronger" logically; nonetheless, it is worth the extra trouble to consider, because being an equivalence $f : A \simeq B$ is always a propositional condition, whereas being an isomorphism $f : \text{Iso}(A, B)$ need not be one. Under the homotopy interpretation, the type $A \simeq B$ consists of the *homotopy equivalences* of spaces. The notion of type equivalence also subsumes *categorical equivalence* (for groupoids), *isomorphism* (for sets), and *logical equivalence* (for propositions).

Now, it is an important fact about type theory that all "definable properties" $P(X)$ of types X (formally, any type expression with a type variable X) can be shown to respect type equivalence, in the sense that $A \simeq B$ and $P(A)$ imply $P(B)$; indeed, if $A \simeq B$ then $P(A) \simeq P(B)$. Briefly, we may say that all type-theoretic properties and concepts are *invariant*.[8] It therefore follows that equivalent types $A \simeq B$ are *indiscernable* within the system. Thus it is natural to ask how equivalence is related to the *identity* of the types A and B.

The univalence axiom

To reason internally about identity of types A and B, we need to add to the basic system a *type universe* $\mathcal{U}$, with an identity type,

$$\text{Id}_\mathcal{U}(A, B)$$

expressing the relation of identity of types. The usual rules for identity then imply that identity of types implies their equivalence (because equivalence is reflexive), and so there is a comparison map,

$$\text{Id}_\mathcal{U}(A, B) \to (A \simeq B).$$

The *univalence axiom* asserts that this map is itself an equivalence:

$$\text{Id}_\mathcal{U}(A, B) \simeq (A \simeq B) \tag{UA}$$

So UA can be read: *identity is equivalent to equivalence*. It internally identifies those types that are equivalent, and therefore indiscernable. Indeed, since UA is an equivalence, there is a map coming back:

$$\text{Id}_\mathcal{U}(A, B) \longleftarrow (A \simeq B).$$

Regarded as a map of types, we write this as $\text{ua} : (A \simeq B) \to \text{Id}_\mathcal{U}(A, B)$, which maps equivalences of types to identities between them. Read logically, ua is a proof of the statement that *equivalent types are identical*; thus in particular, isomorphic sets, groups, etc., are also identified.

Note that in the extended system with a universe $\mathcal{U}$, the univalence axiom is just what is needed to maintain the above-mentioned *invariance* of all "properties" $P(X)$:

$$A \simeq B \text{ implies } P(A) \simeq P(B),$$

---

[8]This of course does not hold in set theory. For example, consider the sets $\{\emptyset\}$ and $\{\{\emptyset\}\}$, which are isomorphic but are distinguished by the property $P(X) = (\exists x, y)\, x \in y \in X$.

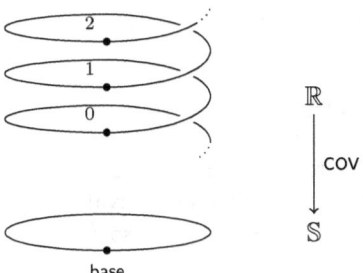

The winding map in classical topology

for we can take $P(X) = \mathsf{Id}_{\mathcal{U}}(A, X)$ to see that equivalent types must be identified.

Note also that UA implies that $\mathcal{U}$, in particular, is not a set: for there are two distinct isomorphisms $2 \cong 2$, and these therefore correspond by UA to two distinct identity terms in $\mathsf{Id}_{\mathcal{U}}(2, 2)$.

Finally, we mention that the computational character of UA is still an open question. The system of type theory without it has some desirable properties, like the so-called "strong normalization" of terms, which implies the decidability of the syntactic equality relation of terms $a = b : A$. Adding a new axiom like univalence is likely to disrupt this property, but does it completely destroy the constructive character of the system? This is one of the open questions currently under active investigation.

5 Synthetic reasoning

In homotopy type theory, what is called the "synthetic" style of reasoning involves making use of the *primitive geometric element* introduced by taking the notion of a *path* as basic, rather than reducing it to maps from the real interval $[0, 1]$. This method is especially powerful in combination with the univalence axiom. By way of example, let us sketch the proof of the above-mentioned theorem that the fundamental group of the circle $\mathbb{S}$ is the integers $\mathbb{Z}$.

Computing $\pi_1 \mathbb{S}$

To compute the fundamental group of the circle $\mathbb{S}$, just as in the classical proof, we shall make use of the "universal cover" (see Fig. 4.2). As a covering space and therefore a fibration, the universal cover will be a dependent type over $\mathbb{S}$, which, in the presence of a universe $\mathcal{U}$, is simply a map

$$\mathsf{cov} : \mathbb{S} \longrightarrow \mathcal{U}.$$

We can define such a type family using the recursion property of the circle; indeed, we just need the following data:

- a point $A : \mathcal{U}$
- a loop $p : \mathsf{Id}_{\mathcal{U}}(A, A)$

For the point A we shall take the integers $\mathbb{Z}$. By univalence, to give a loop $p : \mathsf{Id}_{\mathcal{U}}(A, A)$ in $\mathcal{U}$, it suffices to give an equivalence $q : \mathbb{Z} \simeq \mathbb{Z}$. But since $\mathbb{Z}$ is a set, equivalences are just isomorphisms; so for q we can take the successor function $\mathsf{succ} : \mathbb{Z} \cong \mathbb{Z}$.

Definition (Universal cover of $\mathbb{S}$). The type family $\mathsf{cov} : \mathbb{S} \longrightarrow \mathcal{U}$ is given by circle recursion with

$$\mathsf{cov}(\mathsf{base}) = \mathbb{Z},$$
$$\mathsf{cov}(\mathsf{loop}) = \mathsf{ua}(\mathsf{succ}).$$

As in classical homotopy theory, we then use the universal cover to define the "winding number" of any path $p : \mathsf{Id}_{\mathbb{S}}(\mathsf{base}, \mathsf{base})$ by $\mathsf{wind}(p) = p_*(0)$, where p_* is the transport operation along p. This gives a map from the type $\Omega(\mathbb{S})$ of (based) loops in $\mathbb{S}$ to the integers,

$$\mathsf{wind} : \Omega(\mathbb{S}) \longrightarrow \mathbb{Z},$$

which can be shown to be inverse to the map $\mathbb{Z} \longrightarrow \Omega(\mathbb{S})$ defined by composing loop with itself a given number of times i,

$$i \mapsto \mathsf{loop}^i.$$

This proof can be formalized in a very efficient way, and the result is no longer than a conventional, "unformal" mathematical proof (see Licata and Shulman (2013)). This is a real advance over the traditional "analytic" style of formalization, which would require defining the circle as a subspace of the Euclidean plane $\mathbb{R}^2$, defining homotopies via continuous maps from the unit interval $[0, 1]$, using reparametrizations of paths to define their composition, defining the real numbers $\mathbb{R}$ and the winding map via trigonometric functions, and so on. To be sure, those are worthwhile mathematical objects and constructions in their own right! But by avoiding the need for them in this case, the synthetic approach seems to be somehow closer to the real "essence" of the homotopical fact being proved.

The cumulative hierarchy of sets

As a final example of synthetic reasoning in the full system of homotopy type theory with higher inductive types and univalence, we consider a somewhat "experimental" construction which gives the cumulative hierarchy of sets; see Univalent Foundations Project (2013) for the details.

Given a universe $\mathcal{U}$, we make the *cumulative hierarchy* V of sets in $\mathcal{U}$ as a higher inductive type with the following generating constructors (we shall write $x =_V y$ for $\mathsf{Id}_V(x,y)$ for easier comparison with more familiar treatments):

1. For any type $A : \mathcal{U}$ and any map $f : A \to V$, there is a "set",

$$\mathsf{set}(A, f) : V.$$

We think of $\mathsf{set}(A, f)$ as the image of A under f, i.e. the classical set

$$\{f(a) \mid a \in A\}.$$

2. For all $A : \mathcal{U}$ and $f : A \to V$ and $B : \mathcal{U}$ and $g : B \to V$ such that

$$\big(\forall a : A\, \exists b : B\ f(a) =_V g(b)\big) \wedge \big(\forall b : B\, \exists a : A\ f(a) =_V g(b)\big),$$

we put in a path in V from $\mathsf{set}(A, f)$ to $\mathsf{set}(B, g)$.

3. The "set-truncation" constructor: for all $x, y : V$ and all $p, q : x =_V y$, we add a (higher) path from p to q.

In (2) we used the "logical" notation $\exists$ and $\forall$, etc., to indicate that we are working with the "propositional truncations" of the corresponding type theoretical operations Σ and Π (cf. Univalent Foundations Project (2013) ch. 3).

Next, the membership relation $x \in y$ is defined for elements of V by

$$(x \in \mathsf{set}(A, f)) =_{\mathsf{def}} (\exists a : A.\ x =_V f(a)).$$

One can then show entirely within the system that the resulting structure $(V, \in)$ satisfies *almost* all of the axioms of Aczel's constructive set theory CZF Aczel (1978) (e.g. Strong Collection is missing).

Finally, assuming the usual axiom of choice just for those types that are *sets* in the sense of the homotopy levels, it then follows that $(V, \in)$ is a model of the full system of ZFC set theory (cf. Univalent Foundations Project (2013), ch. 10.5). The proofs of these results make essential use of the univalence axiom and have been fully formalized in the Coq proof assistant (cf. Ledent (2014)).

The system just mentioned is an interesting hybrid of classical set theory and constructive type theory that not only contains a model of ZFC but also many types of higher homotopy level that do not behave like classical sets. This provides a new, more refined view of one possible relationship between classical and constructive foundations. Whereas constructive foundations are usually regarded as incompatible with classical logic, in this system the classical sets form a subsystem of constructive type theory consisting of certain objects distinguished by a natural, intrinsic, structural property—namely that their identity relation is always a proposition.

References

P. Aczel. The type theoretic interpretation of constructive set theory. In A. MacIntyre, L. Pacholski, and J. Paris, editors, *Logic Colloquium '77*, volume 96 of *Studies in Logic and the Foundations of Mathematics*, pages 55–66. North-Holland, Amsterdam, 1978.

S. Awodey. Topological representation of the lambda-calculus. *Mathematical Structures in Computer Science*, 1(10):81–96, 2000.

S. Awodey. Natural models of homotopy type theory. *Mathematical Structures in Computer Science*, 28(2):241–286, 2018.

S. Awodey and A. Bauer. Propositions as [types]. *Journal of Logic and Computation*, 14(4):447–471, 2004.

S. Awodey and C. Butz. Topological completeness for higher-order logic. *Journal of Symbolic Logic*, 3(65):1168–1182, 2000.

S. Awodey, J. Frey, and S. Speight. Impredicative encodings of (higher) inductive types. In *Proceedings of the 33rd Annual IEEE/ACM Symposium on Logic in Computer Science (LICS 2018)*. IEEE Computer Science, 2018.

S. Awodey, N. Gambino, and K. Sojakova. Inductive types in Homotopy Type Theory. In *Proceedings of the 27th IEEE/ACM Symposium on Logic in Computer Science (LICS 2012)*, pages 95–104. IEEE Computer Society, 2012.

S. Awodey and M. Warren. Homotopy-theoretic models of identity types. *Mathematical Proceedings of the Cambridge Philosophical Society*, 146(1):45–55, 2009.

N. Gambino and R. Garner. The identity type weak factorisation system. *Theoretical Computer Science*, 409(3):94–109, 2008.

M. Hofmann and T. Streicher. The groupoid interpretation of type theory. In *Twenty-five years of constructive type theory 1995*, volume 36 of *Oxford Logic Guides*, pages 83–111. Oxford Univ. Press, 1998.

F. D. Kamareddine, T. Laan, and R. P. Nederpelt. *A modern perspective on type theory: from its origins until today*. Springer, 2004.

C. Kapulkin, P. Lumsdaine, and V. Voevodsky. The simplicial model of univalent foundations. arXiv:1211.2851v1, 2012.

F. W. Lawvere. Equality in hyperdoctrines and comprehension schema as an adjoint functor. In A. Heller, editor, *Proceedings of the AMS Symposium on Pure Mathematics*, volume XVII, pages 1–14, 1970.

J. Ledent. The cumulative hierarchy of sets, 2014. Post on the Homotopy Type Theory blog.

D. Licata and M. Shulman. Calculating the fundamental group of the circle in Homotopy Type Theory. In *Logic in Computer Science (LICS 2013)*, pages 223–232. IEEE Computer Society, 2013.

P. Lumsdaine. Weak ω-categories from intensional type theory. *Logical Methods in Computer Science*, 6:1–19, 2010.

P. Lumsdaine. Higher inductive types: a tour of the menagerie, 2011. Post on the Homotopy Type Theory blog.

P. Lumsdaine and M. Warren. The local universes model: An overlooked coherence theorem. *ACM Transactions on Computational Logic*, 2015.

K. Sojakova. Higher inductive types as homotopy-initial algebras. Technical Report CMU-CS-14-101, Carnegie Mellon University, 2014. Available at http://reports-archive.adm.cs.cmu.edu/.

W. W. Tait. The law of excluded middle and the axiom of choice. In A. George, editor, *Mathematics and Mind*, pages 45–70. Oxford University Press, 1994.

The Univalent Foundations Program, Institute for Advanced Study. *Homotopy Type Theory - Univalent Foundations of Mathematics*. Univalent Foundations Project, 2013.

B. van den Berg and R. Garner. Types are weak ω-groupoids. *Journal of the London Mathematical Society*, 102(2):370–394, 2011.

B. van den Berg and R. Garner. Topological and simplicial models of identity types. *ACM Transactions on Computational Logic*, 13(1), 2012.

V. Voevodsky. Notes on type systems, 2009. Available from the author's web page.

M. Warren. *Homotopy-theoretic aspects of constructive type theory*. PhD thesis, Carnegie Mellon University, 2008.

Reflections on Reflection in a Multiverse

Geoffrey Hellman

ABSTRACT: After explaining the challenge of introducing higher-order reflection principles into modal-structural interpretations of set theories, we motivate and describe two methods. The first runs into difficulties, both informal, regarding motivation, and formal, regarding consistency. After reviewing structuralist motivation for the Axiom of Replacement (answering Boolos), we then argue, analogously, for directly positing the possibility of second-order Reflection, noting its (relative) consistency and emphasizing the avoidance of "ultimate infinities". This approach naturally generalizes to a more thoroughgoing model-relativity, in line with a "height-potentialist" as opposed to "universalist" view. We conclude with some comparisons with Tait's "Constructing Cardinals from Below", which has helped us navigate these treacherous waters.

1 Introduction and Background

Questions of justification of mathematical axioms can arise within mathematics proper, but they naturally occupy center stage in foundational-philosophical investigations. When it comes to axioms of set theory, such questions take on added significance for at least two reasons: first, of course, set theory is itself seen as a foundational framework capable of incorporating and reconstructing all of ordinary mathematics; and, second, set theory, pursued as a branch of (extraordinary) mathematics in its own right, is actively involved in pushing the boundaries of known mathematics ever upward and outward, proposing new axioms in the process. Zermelo (1930) viewed this as "creative progress", often in tension with a tendency to seek and claim "all-embracing completeness". As I have understood this, his advice was to resist the latter in the interests of the former. Articulating a framework that could accomplish this was a principle motivation for developing modal-structuralism, and this will be developed further in what follows.

These issues come to the fore when one seeks to motivate and/or justify second- and higher-order reflection principles.[1] Usually, reference is made to "the universe" of all sets or to an "absolutely infinite" totality, echoing Cantor's language of "inconsistent multiplicities", courting the very "all-embracing completeness" that Zermelo sought to avoid. The idea of "reflection on the universe" is supposed to be that its vastness is *indescribable*, in the sense that any condition we can formulate in set-theoretic language, if true in the whole universe V of (pure) sets, is also true when interpreted over a proper initial segment V_λ thereof.[2] Now such motivating language is anathema to a modal-structural (MS) interpretation, since that is framed so as to resolve the set-theoretic paradoxes by blocking the recognition of any maximal universe or world of set-like objects. It thus initially seemed that we would have to live without higher-order reflection principles.

That, however, is unwelcome news since second-order reflection, R^2 (especially when extended by higher-order parameters), is a powerfully unifying principle, implying—in the presence of, e.g., Zermelo set theory less Infinity but with Choice, Z^2_- C— Infinity, Replacement^2, Inaccessible and Mahlo cardinals of many levels, through weakly compact and higher levels of Indescribable cardinals, significantly beyond what was established in MWON.[3]

---

[1] Let $\phi(x_1,\ldots,x_n)$ be a first-order formula (without abstraction terms) with just the x_i free, let V_β denote the sets of ranks $< \beta$, and let $\phi^{V_\beta}(x_1,\ldots,x_n)$ be the result of restricting all quantifiers in ϕ to V_β: Then the instances of first-order Reflection scheme are of the form,

$$\forall \alpha \exists \beta > \alpha \forall x_1,\ldots,x_n \in V_\beta[\phi(x_1,\ldots,x_n) \leftrightarrow \phi^{V_\beta}(x_1,\ldots,x_n)] \qquad (\text{Rf}_1)$$

This is standardly read: "The true situation (in the universe of sets) is reflected in arbitrarily high levels of the cumulative hierarchy". These instances are derivable as theorems of ZFC. In the other direction, first-order Reflection together with Extensionality, Choice, and Separation yield the Axiom of Infinity and (the instances of) Replacement. See Drake (1974) and references therein.

Second- and higher-order reflection principles are independent of impredicative second-order ZF^{2}C, implying significant large-cardinal extensions. More on this will be discussed below.

[2] Here we deploy Zermelo's (1930) quasi-categoricity results, characterizing models of his (and Fraenkel's) second-order axioms, ZF^{2}C, up to isomorphism by fixing two parameters, the cardinality of urelements and the ordinal-cardinal height of a domain as a strongly inaccessible (regular strong limit) cardinal . Zermelo's key assumption was that any such domain (model) has a proper extension (also a model). Further he assumed that power sets are maximal or "full". For pure set theory, no urelements are present beyond the null set. Then, in modern notation, the models can be represented up to isomorphism as $\langle V_\lambda, \in \upharpoonright V_\lambda \rangle$. Further, for any V_λ and V_κ, $\kappa > \lambda$, strongly inaccessible cardinals, V_κ is an elementary end-extension of V_λ (sets of V_λ occur identically in V_κ, and quantified formulas are preserved in passing to the extension). Below, we refer to structures of this form–including ranges of 2nd and higher-order variables–simply as V_λ, etc.

[3] Tait (2005), following Baumgartner and others, spells this out in detail, obtaining stationary sets of indescribables by certain classes of second-order formulas with higher-order parameters. More will be said about this below.

In what follows, we will describe two ways of incorporating second-order reflection within the MS framework. While neither of these presupposes any absolutely infinite totalities, the first—based on what we call an "extendability interpretation"—does draw on a vestige thereof in its appeal to "all possible models" or "the possibilities of satisfying the axioms". Furthermore, as has recently been shown,[4] a satisfactory principle has yet to be precisely stated as the first attempts lead to inconsistency. A second, more direct method, however, improves on the first in both respects: not even a vestige of absolute infinity is involved, even at the level of motivation; and the proposed axiom scheme is clearly consistent relative to its ordinary set-theoretic counterpart, e.g. $Z^2_-C + R^2$.

2 Reflection on Extendability Interpretation

The core MS axiom on which the extendability interpretation is based is a modalized version of Zermelo's (1930) principle that any model of ZF^2C has a proper extension to a more inclusive model (indeed a proper end-extension in light of quasi-categoricity, as noted above):

$$\Box \forall M \Diamond \exists M'[M' \succ_e M], \qquad \text{(EP)}$$

where the variables range over standard models of Z^2C and where $\succ_e$ means "is a proper end-extension of".[5]

Building on this, the proposed interpretation uses a translation scheme devised by Putnam (1967) applicable to "unbounded" set-theoretic sentences, i.e. with unrestricted first- or second-order quantifiers. One thereby treats unrestricted universal (respectively, existential) set quantifiers not as ranging over elements of a fixed, maximal universe, but as ranging over any items of any (respectively, some) model of the set-theory we're considering that there might be. More precisely, we relativize such quantifiers to the domain of any (respectively, some) possible extension of any model "already assumed", as illustrated by, say, the Axiom of Inaccessibles:

$$\forall \alpha \exists \beta > \alpha[Inac(\beta)]. \qquad \text{(AI)}$$

For related results as well as a demonstration that inconsistency results if third-order universal quantifiers are allowed in the reflecting formulas, see Koellner (2009).

[4] By Roberts (forthcoming).

[5] Zermelo's unmodalized principle runs up against 2nd-order logical comprehension: the class of all models or all inaccessible cardinals, etc., would then exist as inextendable totalities, contrary to what is intended. The modal EP does not succumb to the analogue of this: properly understood, there is no reason to countenance any totality or even plurality of "all possible models, etc."; it makes sense to speak of what exists or would exist under a hypothesis, but not of "merely possible objects", *possibilia, simpliciter*. This theme will recur below, in the final section.

The Putnam-translate of this, call it AI_{PT} is

$$AI_{PT} = \Box \forall M, \alpha \in |M| \; \Diamond \exists M', \beta \in |M'| \, [M' \succ M \; \& \qquad (AI_{PT})$$
$$\beta > \alpha \; \& \; \text{Inac}(\beta)].^6$$

In this manner, all unrestricted quantifiers of the original sentence become restricted or relativized to models in the modal translate, as desired. Only the modal quantifiers over models are unrelativized, which is as it should be. It is straightforward to generalize this pattern to sentences with any finite number of unrestricted quantifiers,[7] and to allow unrestricted second-order quantifiers as well.[8]

Consider now the following motivating principle for "reflection on possibilities":

> *The mathematical possibilities of ever larger structures (of appropriate type) are so vast as to be "indescribable": whatever condition we attempt to lay down to characterize that vastness fails in the following sense: if indeed it is accurate regarding the possibilities of mathematical structures, it is also accurate regarding a mere segment of them, where such a segment can be taken as a single structure (of the relevant type).*

Now, without reifying "possibilities" or "worlds", we can apply the Putnam translation scheme to formulate a MS reflection scheme as follows: Given a first- or second-order set theoretic sentence S with unrestricted quantifiers, take its Putnam translate, S_{PT}, to express the antecedent of an instance of MS reflection on S, i.e. S_{PT} replaces, by a kind of approximation to "absolute infinity", the condition in ordinary reflection principles that expresses that S "holds in the set-theoretic universe". It does this in a manner appropriate to MS in that in considers what holds in arbitrary extensions of set-theoretic models (as possibilities). The conclusion, then, of such a reflection principle simply asserts that it is possible that S itself holds in some set-theoretic model, i.e. that, with respect to what S says about sets, it fails to discriminate between arbitrary extensions of models and some fixed model. Summarizing, then, the form of MS reflection is this:

$$S_{PT} \to \Diamond \exists M [\text{``} M \text{ models } T \; \& \; S\text{''}], \qquad (MSR)$$

where T stands for a set theory (or a precursor) already "secured" (i.e. such that the possible existence of a model is already accepted), and where "models

---

[7]One first transforms the sentence to prenex form and then proceeds from left to right, as illustrated in the example.

[8]Relativizing a universal second-order monadic quantifier, $\forall X[\ldots]$ to a domain D means writing $\forall X[X \subseteq D \to \ldots]$; a second-order monadic existential, $\exists X[\ldots]$, goes over to $\exists X[X \subseteq D \; \& \; [\ldots]$. ($\subseteq$ is defined in the obvious way via second-order or plural predication.) Generalization to polyadic second-order quantifiers is straightforward.

T & S" abbreviates the result of writing out the axioms of T & S with quantifiers relativized to the domain $|M|$ of M (officially treated with a plural variable), replacing occurrences of $\in$ with a term referring to the binary relation playing the role of $\in$ in M.

Applying this to our example, the axiom of inaccessibles, AI, we see that AI_{PT} holds for $T = ZF^2C$, as it follows from the quasi-categoricity of second-order models of ZF^2C that the height of any is a strongly inaccessible cardinal, and by extendability, (EP), that cardinal appears as a "set" in any proper extension satisfying the definition of such a cardinal. Thus, this instance of MSR yields the possibility of AI's holding in a single possible model. The height of such will then be a fixed point in the enumeration of strongly inaccessibles, i.e., a strongly inaccessible κ with κ-many strongly inaccessibles beneath it, i.e. a *hyperinaccessible*. Evidently the combination of MSR and EP has climbing potential.[9]

As it stands, however, we cannot lay down *all* instances of MSR on pain of contradiction. This is because there are cases of set-theoretic sentences *refutable* in standard systems of set theory whose Putnam-translates nevertheless are true (given basic principles of MS, e.g. EP). For an instructive example,[10] consider the sentence expressing that every class is coextensive with some set, i.e. $\forall X \exists y \forall z [X(z) \leftrightarrow z \in y]$. Obviously this leads to contradiction via Russell's paradox, or Cantor's, i.e. instantiate on X with the class of all z such that $z \notin z$ to obtain Russell's paradox, with the class of z such that $z = z$ to obtain Cantor's. However, the Putnam translate of this sentence looks like this:

$$\Box \forall M, X \subseteq |M| \Diamond \exists M' \succ M \exists y \in |M'| \Box \forall M'', z \in |M''| [X(z) \leftrightarrow z \in y],$$

which follows easily from the fact that in any extension M' of M, say of height κ, the κth rank exists and has the domain $|M|$ of M occurring as a set. Thus, this instance of MSR would have it that possibly there is a set-theoretic model of the original S, which is impossible. One obvious solution is to restrict MSR explicitly to sentences S that are consistent with the accepted set theory, T.

Unfortunately, however, even with this restriction, inconsistent instances of MSR arise.[11] Moreover, no further, well motivated restriction is readily available that would restore consistency while allowing that MSR would still

---

[9]Arguments like the one for inaccessibles can be given for small large cardinals through the indescribables, that is, cardinals λ such that V_λ itself satisfies the axiom scheme

$$\forall X [\phi(X) \rightarrow \exists \beta \phi^\beta (X \cap R(\beta))],$$

where $\phi^\beta(Y)$ is the result of restricting the 1st- and 2nd-order quantified variables in $\phi(Y)$ to the cumulative ranks V_β and $V_{\beta+1}$, repsectively. Indeed, the Putnam-translate of this scheme itself is implied by (EP), so that MSR implies the modal-existence of a model thereof.

[10]Pointed out to me by Øystein Linnebo in correspondence.

[11]See Roberts (forthcoming), who supplies an example applying the Gödel-Rosser first incompleteness theorem.

apply to second-order sentences.[12] It is worth taking a different tack in order to bring higher-order reflection into MS set theory.

3 Interlude: Motivating/Justifying Replacement[13]

As a way into this approach, consider how the Replacement Axiom can be motivated from the standpoint, say, of Zermelo set theory. Two sorts of considerations stand out, the first, well known *mathematical* motivation, the second more *philosophical-foundational*. Now we take it as uncontroversial that Replacement *is* well-motivated mathematically. In Zermelo set theory with Choice (ZC), it cannot even be proved that the von Neumann ordinal $\omega + \omega$ exists! More generally, ordinal arithmetic requires suitable closure under its operations. Replacement, of course, delivers that, as we get general forms of transfinite induction and recursion. Furthermore, we get the theorem of Hartogs, that, given any infinite set S of ordinals, there is a least ordinal *cardinally* greater than any member of S; and then (especially with Choice) a useful (if seriously incomplete) theory of transfinite cardinal arithmetic can be constructed. Moreover, Replacement unifies a wide variety of applications, those just mentioned and many more, in a principled way: any objects of a universe of discourse not more numerous than those of a "set" (object of the universe, possible domain of a function) should qualify as "elements of a set" (co-domain of the function).

When it comes to *philosophical-foundational* justification of Replacement, however, matters are considerably more vexed. Consider Boolos: In spelling out the "iterative conception", although he builds in an ω'th level (but why?), he writes, "The axioms of replacement do not seem to us to follow from the iterative conception." (1971)[14] "Ok, but so what?" you may ask. Do they (or does it) follow from *anything* we have reason to "accept"? Apparently not: Regarding the minimal fixed point, $\kappa = \aleph_\kappa$, of Cantor's aleph function, he writes: "I am... doubtful that anything could be provided that should be called a reason and would settle the question." (2000). Replacement + Z guarantees $\aleph_\kappa$ (via transfinite induction). On Boolos' account, the iterative story provides

---

[12] As Roberts demonstrates, the Putnam translates of first-order (non-modal, set-theoretic) sentences preserve truth-values. Cf. Hellman (1989), Ch. 2, Appendix.

[13] For a comprehensive treatment of the Axiom(s) of Replacement in mathematical and historical contexts, see Kanamori (2012), to which this section is intended as a friendly amendment.

[14] It has never been clear to me why an ω'th level is accepted as part of "the iterative conception", while Replacement or other conditions on the extent of infinite levels are not. Once "iterative" is understood as "with respect to ordinally indexed levels", the extent of levels or ordinals in principle available comes to the fore as an intrinsic matter that the approach must treat. (The fact that Replacement, like other principles, including Reflection, can also be motivated by considerations of "limitation of size", is not in itself a reason to detach it from an "iterative conception". Different sources of motivation can overlap.) Boolos' stance, that Replacement requires motivation "beyond the iterative conception", would seem equally applicable to *any* axiom guaranteeing infinite levels.

a reason for accepting Z, so Replacement is cast into doubt. In the end, Boolos' position seems to be that there is a standoff between set theory as practiced and common sense. For he asks, "Do you really think there are as many sets as that?" Maybe there simply aren't "that many things in existence". And then "set theory is not true". To the retort that going set theory requires κ (and much more), he responds ironically that legend requires dragons![15]

Well, in pursuit of dragons, let us compare now how a structuralist (Hilbert-inspired) view of mathematical axioms approaches the matter. First, consider a non-modal, platonist version such as Shapiro's "*ante rem*" structuralism (1997). In this framework, *coherence* of an axiom system in pure mathematics suffices for existence of the objects posited (i.e. as places in a structure). There are multiple, coherent but pairwise conflicting set theories. None is "the true one" or true *simpliciter*. Rather, each is *true of* its *ante rem* realization. The Frege-Boolos type questions, "Are there really self-membered sets?", "Are there really κ-many sets?", etc., are either each answered, "Yes, in such-and-such *ante rem* structures, but no, not in thus-and-so ones..."; or the questions may be rejected as meaningless, as would be a question such as, "Is Tom taller?" (without further context).

Next, consider the modal-structuralist response. This is broadly along similar lines, with "logico-mathematical possibility" playing the role of "coherence". But not only are questions of mathematical existence relativized to given or hypothetical domains of models, since logical possibility is all that matters for pure mathematics, Boolos' musings about whether there are "in fact" as many sets (better, objects)[16] as some $\kappa = \aleph_\kappa$ are seen as beside the point. What matters is merely that there *could*, logically speaking, be so many things. Furthermore, there is a point to be made about "burden of proof". Both *ante rem* structuralism and modal structuralism take conceptual coherence as sufficient for "existence" in pure mathematics. Such a standard can be seen as a natural fallback—in the wake of Gödel's incompleteness theorems—of Hilbert's view that, for pure mathematics, *consistency* is the most that can be demanded of its postulate systems, *not actual, literal truth*. Seen in this light, skepticism regarding an existence axiom such as Replacement should be (re)construed as expressing doubt about conceptual coherence and suspicion that there is a logical impossibility lurking in the proposed axiom (in relation to other axioms or assumptions already operative). In the specific case of Replacement, reflect on what such a suspicion amounts to: that, regardless of how rich a structure of ordinals and sets we may conceive, some set is so large that it can be "spread out" among ordinals (via a function from the set to ordinals) so high that it

---

[15] One may wonder, in connection with higher infinities, why common sense should matter. Shouldn't a "natural ontological attitude" as Fine (1986) has described in connection with theoretical physics apply in mathematics as well, deferring such questions to experts in the discipline? (Cf. "naturalist" themes of Maddy 1997.)

[16] NB: officially, MS interpretations eliminate predicate and relation constants such as 'set' and 'is a member of' in favor of variables, just as in the cases of the number systems or structures for abstract algebra, etc.

would be *impossible* coherently to conceive of ordinals that are strictly larger than all of them (in the range of the function). What grounds are there for thinking that there is such a conceptual limitation? Thus, the shift from actual to possible existence carries with it a shift in the burden of proof: appeals to the limitations of "common sense" or even "common mathematical experience" are nugatory. Rather, the default position would seem to be that of a set-theoretic liberal who tolerates any posit whatever as a possibility unless some reason can be given for at least suspecting a conceptual conflict with other assumptions or inferences therefrom. Rather than the question's being, "Why should we believe there really are infinite sets as large as *that?*", it becomes, "Why should we think there is any logical or conceptual problem with positing as *merely possible* enough things for there to be ordinals *that* large?"

Along with this shift of burden of proof comes the freedom of choice of the investigator to posit structures of interest, in the cases at hand, of "large" structures in the relevant sense. For Replacement, that means that a universe of discourse for set theory should be incomparably larger (by the Cantorian standard of (non)existence of bijections) than any of its (set-like) objects. The same sort of considerations, we suggest, should apply to Reflection principles.

4 Possibility of Reflection

Whereas we motivated our first strategy leading to MSR by appeal to the vastness of the possibilities of set-theoretic structures, here we adjust and simplify that language as follows:

> *We are interested in studying (higher-order) models $\mathcal{M}$ for $Z^2_- C$ so vast as to be indescribable in the following sense: if a (second-order) condition Φ holds in $\mathcal{M}$, its relativisation **also** holds in a proper substructure $\mathcal{M}'$ of $\mathcal{M}$.*

Without loss, we can take $\mathcal{M}$ to be of the form V_κ and $\mathcal{M}'$ to be of the form V_α with $\alpha < \kappa$. In effect, we are postulating the possibility of $\mathcal{M}$'s satisfying the second-order reflection schema,

$$\forall X[\Phi(X) \to \exists \beta(\Phi^\beta(X \cap V_\beta)], \tag{Rf^2}$$

where 'X' is a second-order variable to be assigned a subclass or plurality of the domain of $\mathcal{M}$. Note here the difference this makes as compared with the standard, fixed maximal universe view. There the variable 'X' would purport to range over *inextendable* proper classes in an absolute sense; whereas here it is relativized to a hypothetical model $\mathcal{M}$. The EP of course applies to this, and in any proper extension, any class or plurality over the domain of $\mathcal{M}$, including that domain itself, would be coextensive with a "set" i.e. first-order object. In this manner, we reap the benefits of Rf^2 without any hint of "absolute infinity". It is straightforward to show that the postulate,

$$\Diamond \exists X, Y [Z^2_- \wedge Rf^2]^X_{\in/Y} \tag{Poss Rf^2}$$

(where 'Y' is a two-place relation variable), together with the EP holds in an S-5 Kripke structure on the assumption that $Z_-^2 + Rf^2$ is consistent, where this is carried out, for instance, in ZF^1C + the Axiom of Inaccessibles (AI).[17]

Still stronger reflection principles can be brought within the purview of the MS framework by postulating the possibility of the first Erdös cardinal, $\kappa = \kappa(\omega)$, defined by the condition, $\kappa \to (\omega)_2^{<\omega}$, which abbreviates the statement that κ is the least cardinal such that any partition of the finite subsets of κ into two pieces has a countable "homogeneous subset", a countable subset s of κ such that for each finite n, the n-membered subsets of s are mapped to the same piece (which may depend on n).

Theorem (Koellner): Assume there exists $\kappa = \kappa(\omega)$; then there is a δ such that V_δ satisfies $\Gamma_n^{(2)}$-reflection for all $n < \omega$, where Γ_n^2 is the class of formula of the form,

$$\forall X_1^{(2)} \exists Y_1^{(k_1)} \ldots \forall X_n^{(2)} \exists_n^{(k_n)} \varphi[X_1^{(2)}, Y_1^{(k_1)}, \ldots, X_n^{(2)}, Y_n^{(k_n)}, A^{(l_1)}, \ldots, A^{(l_m)}]$$

where φ lacks 2nd or higher-order quantifiers and the superscript k's and l's range over natural numbers.[18]

The proof uses results of Silver on models with indiscernibles; but it makes essential use of the fact that the initial universal quantifiers are only 2nd-order. Indeed, Koellner, in the same paper, proves that, when even a single 3rd-order initial universal quantifier is allowed (with just a single 4th-order parameter), a counterexample to the corresponding reflection principle arises. Thus, $\kappa(\omega)$ emerges as a barrier, and reflection principles seem insufficient to take us further, beyond compatibility with the axiom of constructibility ("$V = L$").[19]

Still, the positive theorem is significant in implying the existence of many indescribable cardinals specified with higher-order parameters. Furthermore, when combined with earlier well-known consequences of 2nd-order reflection, a great unification of axioms of infinity (below $\kappa(\omega)$) is achieved, including the initial huge step in passing from the finite to the countably infinite, and the second major step in passing, via Replacement[2] and the EP, to the first uncountable strongly inaccessible.

---

[17]The proof (in first-order ZFC + AI) runs thus: We assume an axiomatic system of second-order logic for proving consequences of Z_-^2 + Rf^2. By hypothesis, Z_-^2 + Rf^2 is consistent, so by Henkin-style logical completeness for axiomatic second-order theories, there is a (set-theoretic) model (not necessarily standard), M, of Z_-^2 + Rf^2. Take that to be the "actual world" of a Kripke frame for S-5 along with the other worlds taken as the natural models, V_κ, κ strongly inaccessible (by AI), with accessibility taken as elementary end-extension (i.e. preserving satisfaction in M) or its converse. So EP is satisfied, with all structures satisfying ZF^2C and (at least) that Rf^2 is satisfiable (as by hypothesis it holds in the initial model).

[18]These are formulas of a language of finite orders (over the language of set theory) which are "positive" in a technical sense (due to Tait 2005, see Koellner 2009, §3) to avoid known counterexamples to higher-order reflection principles (Reinhardt 1974).

[19]For example, the partition cardinal $\kappa(\omega_1)$ is inconsistent with $V = L$. See e.g. Drake (1974), Theorem 4.8, p. 253.

In this manner, then, the MS framework incorporates second-order reflection principles much as standard set theory does, and does so consistently, relative to the latter, but with "the whole cumulative hierarchy" replaced by (the possibility of) a sufficiently rich, standard transitive model of Z^2_-C that itself can have indefinitely many proper extensions. That seems rather a gain than a loss.

5 Revising MS Interpretation of Standard Set Theory

Reflecting on the above sections, we are naturally led to modify the original MS interpretation of unbounded set-theoretic sentences, which used the Putnam extendability translation. It now seems, however, in light of Roberts' results, that that translation works well (faithfully, in a precise sense[20]) for first-order sentences but is unreliable for second-order. Instead, we propose that *all unbounded sentences be treated as implicitly bounded*, viz. treating V as some possible V_κ, κ *inaccessible*. Then all quantifiers, first- and second-order, are relativized respectively to the domain of such a structure or to subclasses or pluralities of the domain (where this includes n-ary relations via a set-theoretic pairing function). But then, contrary to orthodoxy, "the universe" is only provisional, as it has potentially indefinitely many proper extensions, as the EP implies.

A proponent of the fixed, maximal universe view may object that this distorts the theory, especially the first-order theory, which doesn't even allow reference to a domain of quantification as an object. Perhaps this should be conceded. But the real question is whether anything worth saving, especially anything of mathematical significance, is lost on the re-interpretation. Certainly the MS program requires the move to second-order formulations so that quasi-categoricity holds.[21] Furthermore, to avoid metaphysical questions over "the nature of *sets* or *classes*",[22] we prefer the machinery of plural quantifiers instead of class quantifiers along with the elimination of relation constants via quantifiable variables. It would be surprising indeed if these moves implied any mathematical limitation *vis-à-vis* standard theories. But the question will not be pursued further in this paper.

6 Comparison with Tait's "Large Cardinals from Below"

In this insightful and instructive paper, Tait begins with Cantor's way of introducing the totality Ω of (transfinite) ordinals via the principle,

---

[20]See Roberts (forthcoming), also Hellman (1989), Ch. 2, Appendix.

[21]Furthermore, second- but not first-order ZFC is finitely axiomatizable, enabling the hypothetical component of MS interpretation, that is, enabling formulating conditionals with the conjunction of the axioms (relativized to any given structure) as antecendent and the (relativized) statement itself as consequent. See my (1989), Ch's 1 and 2.

[22]E.g. whether modal language even makes sense in combination with such purely mathematical terms, or whether it should be avoided, much as tensed language is.

If an initial segment Σ of Ω is a set, then it has a least strict upper bound $S(\Sigma)$ as a member of Ω. (CP)

As he then says, this presupposes the "set/non-set" distinction, since Ω itself cannot be a set or a transfinite ordinal on pain of contradicting well-foundedness. Cantor's distinction between "consistent" vs. "inconsistent" multiplicities only labels the problem but provides no solution. Tait then proposes that (CP) be replaced with a hierarchy of limited principles formulated in terms of conditions C, in place of 'is a set', that are mathematically precise. These take the form

If an initial segment Σ of Ω_C satisfies the condition C, then it has a least strict upper bound $S(\Sigma) \in \Omega_C$; (TP)

such a condition C is calleld an "existence condition". Here Ω_C is a limit ordinal which itself *can* consistently have a strict upper bound, $S(\Omega_C)$, with the consequence that Ω_C cannot satisfy condition C, lest $S(\Omega_C) \in \Omega_C$ and then $S(\Omega_C) < S(\Omega_C)$; so that introducing $S(\Omega_C)$ depends on another, more inclusive existence condition.

Tait then proceeds to develop a series of increasingly powerful set theories answering to the prior notion of "logical set" formalized in second-order logic, presupposing the principle of "Separation", that elements x of a *given* set Y (satisfying any condition whatever) form a set, a subset of Y, so that one presupposes passing from Y to the totality of *all* subsets of Y, i.e. to $\mathcal{P}(Y)$, the power set of Y. This is combined with the idea of iterating this operation through ordinally indexed levels of a cumulative hierarchy, where the ordinals are taken as the von Neumann ordinals, defined internally in set theory itself (autonomous rather than externally given somehow). Then whether an initial segment Σ has a strict upper bound so that the iteration of power sets can continue past the ordinals of Σ should depend only on Σ, "or better, of the (provisional) universe of sets obtained by iterating the power set operation through Σ". This is the sense in which the theories to be developed generate large cardinals "from below".

The connection with second-order Reflection principles is this: Suppose we're given a second-order formula $\Phi(X)$ (with just the second-order variable X free) and we want to introduce an ordinal Ω_C such that the domain V_{Ω_C} satisfies Φ-reflection, i.e.

$$\forall X \subseteq V_{\Omega_C}[V_{\Omega_C} \models \Phi(X) \rightarrow \exists \alpha \Phi^\alpha(X \cap V_\alpha)],$$

where the superscript α indicates relativization of quantifers in Φ to V_α. Then we take the existence condition C on initial segments $A \subseteq \Omega_C$ simply to be the negation of this instance of Rf^2 relative to A, viz.

$$\exists X \subseteq V_A[V_A \models \Phi(X) \land \forall \alpha \in A \neg \Phi^\alpha(X \cap V_\alpha)]$$

and apply the relativized Cantorian principle TP to it, implying by well-foundedness of ordinals that C does *not* hold of Ω_C, i.e. that V_{Ω_C} does satisfy Φ-reflection. The remainder of Tait's paper explores in detail the consequences of such methods to obtain many—indeed stationary sets of—indescribable cardinals.[23]

It should be clear that the results Tait establishes are available in the MS framework by direct postulation as described above for Rf^2. As far as mathematics is concerned, there would seem to be little to choose between these approaches. Both methods lead to the same small large cardinals, and both are limited by Koellner's results ruling out reflection at the level of third-order. But isn't there a significant philosophical difference? Tait's approach works in the modal-free language of Morse-Kelley set theory, and the reflection scheme Rf^2 explicitly quantifies over arbitrary (proper) classes, avoided in the MS framework, which relativizes Rf^2 to a hypothetical standard model of Z^2_-C. The latter is consistent with the extendability principle, EP, whereas Morse-Kelley directly violates it by admitting second-order constants like Ω and V itself. Still, Tait has this to say about Zermelo's stance in his (1930):

> Zermelo takes as his starting point [his axioms of second-order ZFC less the Axiom of Infinity]. He notes that it is not a categorical theory and so does not itself determine a structure. Rather its various categorical (and consistent) extensions determine structures [e.g. domains such as V_ω or V_λ with λ strongly inaccessible, etc.] But, contrary to a common understanding of set theory as a theory about 'the' universe of sets, *there is no such universe: the notion of 'all ordinals' or 'all sets' lacks rigorous mathematical sense. Or rather it has only a relative sense: all sets or ordinals of this or that categorically determined domain. More generally there is no absolute notion of proper class.* ... If we take set theory itself to have a [uniquely preferred, maximal] model, then its universe is indeed the 'class of all sets' in the absolute sense and paradoxes arise if we do not make the absolute distinction, *which otherwise has no foundation,* between sets and proper classes. (P. 11) (My emphases and language in square brackets.)

As far as it goes, this is all friendly to the MS approach, although neither Zermelo nor Tait attempts to reconstruct set theory in a modal language. However, second-order class comprehension has somehow to be restricted, as straight ZF^2C implies the existence of Ω, V, etc. But how is this to be done? If

---

[23] As Koellner points out, however, the method sketched can be applied so widely that the existence of cardinals satisfying any Φ you like follows too easily, simply by taking a condition C on segments A to be $V_A \not\models$ "There is a Φ-cardinal", and then applying the relative Cantorian TP to it. Thus, justification of the TP depends on the choice of condition C, and, as he puts it, " it is far from immediate" that "in the case of conditions C that give rise to reflection principles the resulting instances of the relativized Cantorian principle [TP] are intrinsically justified".

modal logic is available, one also has available natural restrictions that block monstrosities such as "the class or plurality of *all possible* models, ordinals, cardinals, etc.". These derive from our "actualist" understanding of modal discourse, viz. that collections contain only things which coexist, or they *would, in given hypothetical circumstances,* contain only things that *would then coexist;* and likewise *a fortiori* with regard to "any things" in the sense of plural quantifiers. (We summarize this by saying that collections and pluralities are "world-bound", but that does not involve any reification of worlds or *possibilia* as objects.) It literally *makes no sense* to speak of "all possible set-like objects" or "all possible ordinals", etc. Thus the modal analogues of proper classes simply do not arise. But in the non-modal, extensional language of set theory proper, no such restrictions on second-order comprehension are available, and proper classes or their plural counterparts, e.g. "all the ordinals, etc.", come with the territory. Perhaps the idea is to get by with our informal understanding and simply ignore such implications. In any case, Tait's position is clearly *not* to retreat to first-order ZFC with its ambient but ineffable, all-encompassing universe; and that much, along with the above quoted remarks, sounds a clear note of agreement on which to conclude these reflections.

References

Boolos, G. (1971): "The Iterative Conception of Set", *The Journal of Philosophy* 68, 215–232.

Boolos, G. (1998): "Must We Believe in Set Theory?", in *Logic, Logic, and Logic* (Harvard University Press), pp. 120–132.

Drake, F. (1974): *Set Theory: An Introduction to Large Cardinals* (North Holland).

Fine, A. (1986): *The Shaky Game: Einstein Realism and the Quantum Theory* (University of Chicago Press).

Hellman, G. (1989): *Mathematics without Numbers: Towards a Modal-Structural Interpretation* (Oxford University Press).

Hellman, G. (1996): "Structuralism without Structures", *Philosophia Mathematica* (3) 4, 100–123.

Kanamori, A. (2012): "In Praise of Replacement", *Bulletin of Symbolic Logic* 18, 46–90.

Koellner, P. (2009): "On Reflection Principles", *Annals of Pure and Applied Logic*, 157:2–3, 206–219.

Maddy, P. (1997): *Naturalism in Mathematics* (Oxford University Press).

Putnam, H. (1967): "Mathematics without Foundations", *The Journal of Philosophy* 64, 5–22.

Roberts, S. (forthcoming): "Modal Structuralism and Reflection", *Review of Symbolic Logic*.

Shapiro, S. (1997): *Philosophy of Mathematics: Structure and Ontology* (Oxford University Press).

Tait, W. W. (2005): "Constructing Cardinals from Below", in *The Provenance of Pure Reason* (Oxford University Press), Chapter 6, pp. 133–154.

Zermelo, E. (1930): "Über Grenzzahlen und Mengenbereiche: Neue Untersuchungen über die Grundlagen der Mengenlehre", *Fundamenta Mathematicae* 16, 29–47.

Concepts versus Objects

Charles Parsons[*]

ABSTRACT: Some writers on mathematics, in particular Solomon Feferman, Daniel Isaacson, and Donald A. Martin, have argued that concepts rather than objects are what mathematics is primarily about. Without deploying a sophisticated analysis of 'about' (which these writers also do not do), we contest this view. The language of mathematics, whether regimented as in formalized theories or in informal usage, contains reference to objects. The notion of concept in application to mathematics is probably less well understood than the notion of object. Furthermore, when mathematical theories are regimented as first-order theories, it is questionable that they involve reference to concepts.

The paper discusses the views of Isaacson and Martin. Some of their arguments involve questions or puzzles about reference to particular mathematical objects. We agree that such reference is not as present as reference to particular objects is in everyday discourse and some science, and the literature on structuralism shows that they are not as definite in their individuation as one might expect. Martin argues from categoricity results that if a proposition like the continuum hypothesis lacks a truth value, then the full universe of sets does not exist. We maintain that it does not follow that there are no sets, since his argument does not imply that first-order set theories do not have models.

---

[*]It is a great pleasure to write in honor of my friend Bill Tait, whom I have known since 1957 and who has ever since been a valued discussion partner on many matters concerning the foundations and philosophy of mathematics.

An earlier version of this paper was presented to a Very Informal Gathering of logicians at UCLA in honor of Tony Martin, 31 January 2015. Thanks to Martin for comments on that version. An intermediate version was presented to the conference Representation and Axiomatization, IHPST, Paris, 19 March 2015. Thanks to Marco Panza and Alberto Naibo for that invitation and to Dagfinn Føllesdal and the editor for comments on still later versions.

The present essay proposes to deal with the claim that mathematics is primarily about concepts and thus only secondarily, if at all, about objects. I will discuss some writings of two people who have made statements amounting to this claim, Daniel Isaacson and D. A. Martin. Martin has been quite explicit about it: "I share with Kurt Gödel and Solomon Feferman the view that mathematical concepts, not mathematical objects, are what mathematics is about."[1] In the later of two essays bearing on this theme, Isaacson speaks of structures rather than concepts, but the issues he raises are at least very closely related. Concepts are more prominent in the earlier paper.[2]

I

This claim that concerns us contains three elements that could give rise to problems: 'concept', 'object', and 'about'. Each of these notions has been studied by philosophers, although 'about' has been a stepchild. The writers who have advanced the thesis with which I am concerned seem prepared to use the term 'about' in an informal and rather naïve way. I will follow them in this.[3]

The manner in which I have looked at this question is roughly as follows: Keeping to the informal use of 'about', it should be obvious that mathematics is in some way about objects, because reference to objects is embodied in singular terms and first-order quantifiers. Although this view as I understand it owes much to Quine and, more distantly, Frege, I don't think that any alternative to viewing the informal language of mathematics in that way has gone anywhere.[4] However, one might then view Martin and Isaacson as offering skeptical arguments or puzzles about mathematical objects.

The question whether, and in what sense, mathematics is about concepts is not so straightforward. In the first instance, it depends on whether the language of mathematics contains reference to concepts. Assuming that it does, one then has to ask what is the significance of that fact.

At this point one has to distinguish between the informal language of mathematics and the regimented versions reflected in formal axiomatic theories. So

---

[1] Donald A. Martin, "Completeness or incompleteness of basic mathematical concepts," forthcoming in P. Koellner and C. Parsons (eds.), *Exploring the Frontiers of Incompleteness*. Also relevant is his "Gödel's conceptual realism," The Bulletin of Symbolic Logic 11 (2005), 207–224.

[2] "Mathematical intuition and objectivity," in Alexander George (ed.), *Mathematics and Mind* (New York and Oxford: Oxford University Press, 1994), pp. 118–140; also "The reality of mathematics and the case of set theory," in Z. Novák and A. Simonyi (eds.), *Truth, Reference, and Realism* (Budapest: Central European University Press, 2010), pp. 1–75.

[3] I have not so far determined whether the systematic studies of 'about' that exist are relevant to the issues discussed here. The most thorough known to me is Stephen Yablo, *Aboutness* (Princeton University Press, 2014).

[4] Clearly, this general attitude underlies the treatment of issues about mathematical objects in my *Mathematical Thought and its Objects* (Cambridge University Press, 2008). (This work will be cited as MTO.) My general outlook on the notion of object is presented in chapter 1 of that work.

by 'informal' I mean at this point "not regimented or formalized." I emphasize the case of regimented language, because it is more tractable. The most standard formalizations of mathematical theories are first-order, so that as already noted the language involves reference to objects. There are, of course, many second-order theories discussed by logicians. In the mathematical context, however, it is typical to take the second-order variables to range over sets, subsets of the domain of the first-order language. But on this understanding, these theories can still be expressed in the first-order language of set theory, so that the second-order variables simply range over the power set of the domain of the first-order variables. I will not pursue here the difficult question of how to understand the second-order language when the first-order quantifiers range over "all sets."[5]

We can take the first-order language of set theory as the appropriate example of regimented mathematical language. Usually it has the primitive predicate of membership and no primitive singular terms. However, because even minimal axioms imply the existence of some sets, we can presume that it involves reference to mathematical objects, namely sets. Statements by the two authors I mentioned at the outset seem to call this into question, but I will postpone consideration of their views for the present. However, to forestall misunderstanding I should emphasize some aspects of how I use the phrase "reference to objects." The language used in a given context may involve reference to objects even if it contains no singular terms and so no direct means of referring to individual objects. Definite descriptions are almost always available as a more indirect means of reference, but I do not claim that even that form of reference to individual objects is essential to mathematics as such, even though it certainly occurs, maybe most visibly in number theory. I should also note that I use the term 'reference' in the way usual in semantics, in which a statement involving reference to objects will not be true unless objects satisfying certain conditions exist. The term is sometimes used in a more intensional way, according to which a speaker might refer to, say, unicorns without believing or presupposing that unicorns exist.

A more urgent question is whether the first-order language of set theory makes reference to concepts. The vehicle of such reference would most likely be the predicates of the language. Two eminent logician-philosophers, Frege and Gödel, have developed views according to which predicates as such designate concepts. I will mention some differences between their views, but most important is something that they have in common. That is that for them concepts are not essentially mental. They are objective in a sense stronger than being shared by different minds. In this respect, Frege departed from a tradi-

---

[5]The reader of MTO will discern a "first-order bias" on a number of issues. I have tried not to presuppose that here. Although I think second-order logic is fine in its place and, unlike Quine, have no objection to its being called "logic," I am skeptical of the claims made for it by many writers. Cf. also my "Some consequences of the entanglement of logic and mathematics," in Michael Frauchiger (ed.), *Reference, Rationality, and Phenomenology: Themes from Føllesdal* (Frankfurt: Ontos Verlag, 2013), pp. 153–178.

tion that can be traced back at least to Kant, according to which concepts are a species of representation and are thus contents of the mind. In Kant's view, they may still be objective, but where this is so it is by virtue of the fact that the individual mind has a basic structure that is common to all human minds and in some respects also still common to some purely hypothetical non-human minds.

Frege's departure from this tradition is clearest where he introduces his mature views in writings of the 1890s. The notion of concept is explained by a mathematical analogy. Concepts are characterized as functions, in roughly the mathematical sense, which can take only two values, the True and the False, the truth-values now familiar from elementary logic.[6] If we ignore his disastrous Basic Law V (which he anyway abandoned in his late writings), Frege's logic is a version of second-order logic, which is familiar today. But he takes the second-order variables to range over functions, and in many important contexts these will be concepts. Predicates, even logically complex ones, have concepts as their reference.[7]

Frege's view of predicates is on the model of his view of function expressions, which are singular terms hollowed out by the removal of the expressions for their arguments. A predicate is then a sentence similarly hollowed out. Because of the empty argument places, Frege calls such expressions "unsaturated" and holds that this unsaturated character is inherited by the functions and concepts that they designate. They are thus fundamentally different from objects. Interesting as this view is, it disappears from later logic, although Russell's idea of the "ambiguity" of a propositional function has some kinship with it.

Frege did not single out the first-order part of his logic. Given his view of predicates as designating concepts, it would probably have seemed perverse to him to renounce variables and quantifiers for concepts. Frege's logic incorporates in effect full (impredicative) second-order logic, so that in its natural reading, it embodies a realistic view of concepts.

Gödel did not articulate a theory of meaning and reference on a level with Frege's, but on the other hand he worked early on with first-order logic. However, he expresses a realistic view of concepts in his first published philosophical essay:

> Classes and concepts may, however, also be conceived as real objects, namely classes as "pluralities of things" or as structures consisting of a plurality of things and concepts as the properties and relations of things existing independently of our definitions and constructions. It seems to me that the assumption of such objects is

---

[6] See e.g. Frege, *Funktion und Begriff* (Jena: Pohle, 1891), p. 15.

[7] I favor 'reference' as a rendering of Frege's *Bedeutung*, although some translations render it as 'meaning', which is of course closer to the ordinary meaning of the German word.

quite as legitimate as the assumption of physical bodies and there is quite as much reason to believe in their existence.[8]

Note that Gödel, in contrast to Frege, thinks of concepts as a species of objects. This is the view of his published writings and other essays written in the 1950s and 60s. He may have come to question this in the 1970s, where the reports of Hao Wang are almost the only evidence we have of his views.[9] This evidence does not indicate what manner of entity they are if they are not objects or call into question the attribution to him of a realistic view of concepts.[10]

However that may be, Gödel earlier made clear that even rather elementary predications involve reference to concepts, so that he would give an affirmative answer to our question whether in the first-order language of set theory, predicates designate concepts.[11]

Frege thought of concepts as extensional, although strictly speaking identity is only a relation of objects. Gödel at one point allowed that concepts might be extensional,[12] but in the conversations with Wang in the 1970s he clearly no longer holds that view.[13] The remarks attributed by Wang to Gödel do not make clear why he changed his view. The fact that he refers to them as intensions (remark 8.6.1) suggests that reflection on Carnap may have played a role. Also, he seems to have made an effort to sort out the difference of the three notions set, concept, and class. (On this point see the Appendix.)

In "Gödel's conceptual realism" (2005) Martin evidently shares the conception of concepts as essentially non-mental, and he is sympathetic toward Gödel's view. Where he gets off is at Gödel's version of mathematical intuition, of which "perception of concepts" is an integral part. However, I will not be discussing this issue. In the later "Completeness or incompleteness" paper he is more reserved. He mentions the view of Solomon Feferman that mathematical concepts are "human creations," which he does not accept. He goes on to remark:

---

[8] "Russell's mathematical logic" (1944), in *Collected Works, volume II: Publications 1938–1974*, Solomon Feferman et al. (eds.), (New York: Oxford University Press, 1990), p. 128. Gödel's use of the term 'class' in this essay can give rise to confusion. He follows Russell in using it for entities that would generally be called sets. Russell's theory of types did not have room for the distinction between sets and (proper) classes characteristic of later set theory.

[9] On this matter see the Appendix.

[10] On the other hand, Gödel's idea of perception of concepts is difficult enough to understand, and surely the difficulty is added to if concepts are considered not objects, while no real story is told about what ontological category they do belong to. However, neither Isaacson nor Martin embraces the idea of perception of concepts.

[11] I document this in "Platonism and mathematical intuition in Kurt Gödel's thought," *The Bulletin of Symbolic Logic* 1 (1995), 44–74, §4. Key texts are in version III of "Is mathematics syntax of language?" in *Collected Works, volume III: Unpublished Essays and Lectures*, Solomon Feferman et al. (eds.) (New York: Oxford University Press, 1995).

[12] "Russell's mathematical logic" (1944), p. 129.

[13] Hao Wang, *A Logical Journey: From Gödel to Philosophy* (Cambridge, Mass.: MIT Press, 1996), pp. 274–75. This work will be cited as LJ. Remark 8.6.10 on p. 275 explicitly repudiates the statement in the Russell paper just cited.

> I would say I stand between the two [Feferman and Gödel], but in fact I don't have anything to say about the ontology of mathematical concepts. I do think it is more correct to say that they were discovered than that they were created by us. (p. 1.)

I have noted that both Frege's and Gödel's conceptions of concepts depart from a tradition that goes back at least as far as Kant. I will mention a different conception of concepts, whose departure from Kant is more subtle. This might be called the psychologist's conception. According to it concepts are mental representations and would naturally be taken to belong to individual minds.[14] Thus the psychologist Susan Carey writes, "Concepts are mental representations."[15] They may, however, be intersubjective, in that they might be shared by different subjects, possibly even common to all human beings. One direction in which this conception could be developed is into a naturalized version of the Kantian conception. Something more or less like that is common in the philosophical literature.

Intermediate between a view of concepts as in no way mental and as belonging to the realm of reference is one that regards them as objective but as belonging to the realm of sense. That might be a version of the psychologist's view, but it does not have to be. Frege says little about the senses of predicates, but in general he does not think of senses as "in the mind" and distinguishes them sharply from what he calls ideas (*Vorstellungen*), which belong to the individual subject.[16] But conceptions of concepts along these lines differ about how much of an intrinsic connection they have to the mind.

I have noted that Martin represents Solomon Feferman as agreeing that mathematics is primarily about concepts but as holding that mathematical concepts are "human creations."[17] His conception could be of one of the types just sketched. His view harmonizes with his skepticism about higher set theory and sympathy for constructivism. It is intended as philosophical backing for those views, which differ sharply from Martin's. However, I will not pursue Feferman's views further. His view implies a less realist view of mathematics in general, which raises difficult questions quite different from the ones that concern me here.

I have not canvassed at all the difficulties that these different conceptions of concepts might be subject to. But their existence might be a warning sign about

---

[14] Evidently the psychologist's conception differs from Kant's in that an inquiry into concepts is an empirical inquiry.

[15] *The Origin of Concepts* (Oxford University Press, 2009), p. 2. She goes on to say on the next page that representations refer. Since among the possible referents she includes properties, it is likely that at least some concepts refer to properties. Some conceptions of properties would bring them close to concepts on a realistic view.

[16] Note that Frege's term *Vorstellung* is the same as the one Kant uses as a generic term, usually translated 'representation'. The genus includes concepts.

So far as I know, Frege does not address the question whether senses are "created by us." It is unlikely that he would have viewed the idea favorably.

[17] See for example his "Logic, mathematics, and conceptual structuralism," in Penelope Rush (ed.), *The Metaphysics of Logic* (Cambridge University Press, 2014), pp. 72–92.

the thesis that mathematics is primarily about concepts rather than objects. Is it really the case that the notion of concept, in application to mathematics, is better understood than the notion of object? However, my remarks only note a warning sign and do not yet offer an argument.

With respect to regimented languages like the first-order language of set theory, the view that predicates designate concepts in any of the senses I have mentioned has to compete with a more nominalist view, which distinguishes sharply between predicates and names and denies that predicates stand in a naming-like relation to anything like concepts in the sense of Frege or Gödel or more garden-variety entities like properties and relations. Such a view can be found in the writings of Quine. I have argued elsewhere that it is found in quite early writings of his and that he never gave it up, maintaining it after he did give up the project of a more full-blooded nominalism that rejects mathematical objects and most other abstract objects.[18] The meaning of predicates is in the first instance explained by saying what objects they are true or false of. Concepts, if they enter in at all, would belong to reflection on mathematics and mathematical language rather than to the content of mathematics itself.

A semantical account even of first-order language typically assigns entities to predicates, in the most straightforward case classes and relations. But this typically arises when we want to generalize, either over languages or over interpretations of a particular language. If we accept an inductive definition of truth, satisfaction and truth for a single language can be introduced without any added ontology, provided only that the object language already has finite sequences of elements of the domain. This possibility is exploited in Donald Davidson's well-known approach to meaning. This fact makes it possible for set theorists to understand talk of proper classes as being reducible to talk about predicates of the language of set theory and sequences of sets. In particular, variables for classes in the object language are for most purposes not essential, although of course they are sometimes useful. Gödel-Bernays set theory NBG can be interpreted so as to be compatible with the view of classes just alluded to, but of course theories allowing impredicative specification of classes are also studied and applied.

What of second-order languages? In the mathematical case, the second-order variables typically range over subsets of the domain of the first-order variables. But that presupposes that the latter domain is a set. So if we take ourselves to understand a language where the first-order variables range over absolutely all sets, then it is not obvious how to understand a second-order such language. Although Gödel did not put it that way in his writings, that would have been a motive for him to appeal to concepts in his objective, realist sense. Other ideas have been proposed, either regarding classes as another kind of "set-like" entities, taking at least monadic second-order variables as ranging over pluralities (the "classes as many" of the early Russell), or adopting George Boolos's idea of motivating monadic second-order logic by the plural of natural

---

[18]See "Quine's nominalism," *American Philosophical Quarterly* 48 (2011), 213–228.

language and (in my view unpersuasively) taking this as implying that second-order quantification so understood introduces no new ontology.[19] Occasionally writers invoke Frege's original conception of concept.

The comments above have been focused on regimented languages, the languages of formalized mathematics and in particular of set theory. Without having made a systematic study, I think that in informal, unformalized mathematical discourse reference to concepts is common, of course not limited at all to the rather basic concepts that are emphasized by Martin and Isaacson.[20] I doubt very much that a study of this usage would lead us very far in determining what conception of concepts is most relevant.

II

Thus far I have said nothing about the considerations that my two authors offer in favor of the primacy of concepts (or structures) over objects. I will begin with Isaacson, because in places he invokes rather well known philosophical arguments. I should state at the outset that Isaacson defends a structuralist view of mathematics, and I have pursued that project as well. It is unfortunate that "Mathematical intuition and objectivity" was not brought into the debate about structuralism. I must reproach myself about this, but I was not alone.[21] A difference between us is that I have tried to develop a structuralist view of mathematical *objects*. In "Mathematical intuition and objectivity," Isaacson is concerned to reject what he calls "objects platonism," which he describes as the view

> ... that mathematics is about particular mathematical objects in the way that physical discourse is about particular physical objects, and that the objectivity of mathematical statements is explained by the existence of those mathematical objects that the statement is about. (p. 121)

I don't think he has intended to apply that label to me. To the extent that reference to objects in mathematics resembles that in physics, that is due to the fact that physics is permeated by mathematics. It does not much resemble reference to objects in common-sense contexts, where the objects involved are at the scale of ordinary perception. I wonder if anyone would say that the existence of certain mathematical objects is sufficient to explain the objectivity

---

[19] Gödel's essays say little about proper classes. But he did in conversations with Hao Wang suggest a view of them as a kind of logical fictions. See the Appendix.

[20] A cursory look at some mathematical texts in my own library turned up fewer references to concepts than I expected.

[21] I have less excuse than others, since I attended the conference where it was first presented and read the published version when it appeared. It does not fit comfortably into my preferred classification of structuralist views into eliminative and non-eliminative. See MTO, pp. 51–52.

of mathematical statements. However, it should be a truism that for the (presumably objective) truth of some statements, the existence of objects satisfying certain conditions is a necessary condition.

Concerning mathematical objects, Isaacson makes some rather extreme remarks:

> Nevertheless these arguments [of Paul Benacerraf] do show that even if mathematical objects exist, their existence can play no role in answering our motivating questions. One might, by Ockham's razor, conclude that they do not exist.[22]

> It might be claimed that the account I have just given does not succeed in avoiding the primacy of objects. After all, my condition for a thought to be mathematical was that it be invariant with respect to change of objects. In saying this I may appear to have slipped into object Platonism. Such appearance reflects a source of the natural tendency toward object Platonism, namely that mathematical concepts are concepts of objects. The paradoxical sounding truth of the matter is that mathematics is about objects, but at the same time there are no mathematical objects.[23]

Isaacson cites with approval two well-known papers by Paul Benacerraf, "What numbers could not be"[24] and "Mathematical truth."[25] The first concludes that numbers are not objects. Although what Benacerraf dramatized is the different alternatives for construing numbers as sets, the general phenomenon is what could be called the elusiveness of mathematical objects. If you try to state what object you are referring to when mentioning a particular number or set, it seems that you can at best give an answer internal to some "home" structure[26]; in some cases (e.g. when the relevant structure has nontrivial automorphisms), you cannot even do that; you can at best state relations to other objects in the structure.[27] Although Benacerraf concludes that numbers are not objects, he does not mean that numbers are entities of some other sort. Rather, he suggests briefly a program of eliminating reference to mathematical objects, which has been carried further by others. His second paper argues that if we understand mathematical language as involving reference to objects, then we

---

[22]"The reality of mathematics" (2010), p. 2.
[23]"Mathematical intuition and objectivity "(1994), p. 127.
[24]*Philosophical Review* 74 (1965), 47–73.
[25]*The Journal of Philosophy* 70 (1973), 661–679. Both papers are reprinted in P. Benacerraf and H. Putnam (eds.), *Philosophy of Mathematics: Selected Readings*, 2d ed. (Cambridge University Press, 1983).
[26]The Arabic numeral Isaacson deploys in the remark quoted below serves roughly to indicate how many iterations of the successor operation lead to that number from zero.
[27]This situation has been offered by some writers as an objection to noneliminative structuralism. I defend my own version of that view in "Structuralism and metaphysics," *Philosophical Quarterly* 54 (2004), 56–77, and "Some objections to structuralism," *Al-Mukhatabat*, no. 10 (2014), 69–96.

have no understanding of how we can have cognitive access to them. Much ink has been spilled about this issue, and ways of understanding it have been proposed that are quite different from Benacerraf's original one.[28]

However, other remarks in the two papers make me doubt that the above quotations, particularly the second, express Isaacson's considered view. The most telling may be the following, occurring just after he has said that he is not a nominalist:

> I accept, for example, that works of fiction exist as abstract entities. So why not also numbers and sets as abstract entities? The point is that an individual natural number, e.g. 90745943887872000, exists only through its determination by the categorical theory of the structure of the natural numbers. It has no independent existence. ("The reality of mathematics" (2010), p. 37)

One could paraphrase this remark as saying that individual natural numbers exist only relative to the structure of the natural numbers. They would be an instance of a view of mathematical existence expressed by Paul Bernays more than sixty years ago in saying that they have relative existence (*bezogene Existenz*).[29]

I would take Isaacson's real intention to be not to reject mathematical objects altogether or even to say that whether they exist does not matter, but rather to insist on the priority of structures to the objects that reside in them.[30] This makes intelligible his denial that structures are mathematical objects, since on the view in question, if a structure is a mathematical object, it seems there must be a "superstructure" that is its home. In some cases, such as the number systems, candidates to be such a superstructure are at hand, such as systems of sets or the entire universe of sets. However, there is an inclination to view the natural numbers as more autonomous. However that may be, in the case of the full universe of sets (assuming we regard it as well-defined and do not take the "potentialist" view advanced by many writers[31]), there is no "higher" structure to be the home of the universe as a structure.

---

[28] My own comments on the subject have been brief. But see MTO, pp. xiv-xv.

[29] Paul Bernays, "Mathematische Existenz und Widerspruchsfreiheit" (1950), in his *Abhandlungen zur Philosophie der Mathematik* (Darmstadt: Wissenschaftliche Buchgesellschaft, 1976), esp. p. 99. I discuss Bernays' point of view in §6 of "Paul Bernays' later philosophy of mathematics" (2008), in *Philosophy of Mathematics in the Twentieth Century* (Cambridge, Mass.: Harvard University Press, 2014). Isaacson's stress on categoricity does not occur in Bernays' discussion.

[30] Isaacson could develop his view by adopting a Carnapian distinction between internal and external questions of existence, but he does not mention this possibility. There are hints of such a view in the paper by Bernays cited in the last note; see "Paul Bernays' later philosophy" (2008), p. 87.

[31] One of them is my own earlier self; see Essays 8 through 11 of *Mathematics in Philosophy* (Ithaca, N. Y.: Cornell University Press, 1983). I am still inclined in that direction but am less sure about it now.

Concepts versus Objects

At one point Isaacson suggests that structures are not objects at all. In general, I have difficulty understanding claims to the effect that certain entities are "not objects": if not, what else are they? Isaacson does not say that there are no structures or that talk of structures is a *façon de parler*. If he takes seriously Benacerraf's problem about cognitive access to mathematical objects, why is there not a similar problem about cognitive access to structures? Since it seems to me that structures play for him a role close to the one that concepts play for Martin, possibly he means that structures are akin to concepts as conceived by Frege. But I think the work is done by the weaker claim that structures are not mathematical objects.

At this point, I think we can explain the apparent rejection of mathematical objects in the statements of Isaacson quoted above. It seems to be driven by his particular take on the structuralist intuition. He writes:

> The compelling and immediate reason for rejecting the idea that mathematics is about particular objects is that for any mathematical theory the domain of objects which that theory is taken to be about can always be replaced by a domain consisting of different objects so long as the second domain has a structure isomorphic to that of the first.[32]

This may seem to fly in the face of the fact that mathematical theorems often mention particular objects. Fermat's last theorem, depending on how it is formulated, mentions the number 2 or the number 3. The continuum hypothesis mentions two particular cardinals, $\aleph_1$ and $2 \exp \aleph_0$, asserting that they are identical. Examples can be cited from other areas of mathematics.

I believe that Isaacson would not be moved by this observation. Take the case of Fermat's last theorem. Truth of such a statement is preserved by isomorphism with another "copy" of the natural numbers. Given such an isomorphism, other objects may play the role of 2 and 3, and different functions will play the role of addition and exponentiation. If we suppose that the isomorphism maps our original natural numbers onto a structure containing entirely different objects, the roles of 2 and 3, and of the x, y, z, and n of a given inequation $x^n + y^n \neq z^n$, will be played by different objects. So, he will say, the theorem is not about particular objects.

There is no doubt that truth of statements in the language of a structure is invariant under isomorphism of that structure with another. But the point seems to presuppose that one carries out a mathematical investigation in a single structure that is fixed for the investigation. In actual investigations outside of elementary mathematics, one may begin with one structure and call in others, facts about which are significant for the problem at hand. An outstanding example is the fact that before Fermat's Last Theorem was finally proved, it had been shown that it was implied by the truth of a certain conjecture in

---

[32] "Mathematical intuition and objectivity" (1994), p. 123.

algebraic geometry. Pragmatically at least, it is natural to treat the natural numbers as a basic structure (i.e. not constructed from anything else). This property would not be preserved by arbitrary isomorphic copies of the structure. Some, for example, where the underlying objects are the even numbers, would be dependent on the natural numbers.

Taking the natural numbers as a basic structure, it has been argued that the natural numbers are *sui generis* objects. An alternative that I prefer has the consequence that the reference of expressions for numbers is context-dependent.[33] It seems clear that on either view there is singular reference to natural numbers, even if they are not "definite objects" as Frege demanded. So I don't think these views imply that theorems containing mathematical singular terms are not in any way about particular objects.

It is true that some mathematical investigations can be carried out in such a way that even the sort of reference to individual objects is avoided that is characteristic of talk of natural numbers, rational numbers, and to a limited extent of real numbers. Consider, for example, theories about finite groups. We may treat groups as individual objects, as for example the "four group" with elements 1, a, b, c, an Abelian group with with identity 1, where aa = bb = cc = 1, ab = c, ac = b, and bc = a. But that this is an individual object is a fiction, since we have not said what 1, a, b, and c are, and for most purposes we do not need to. The same is surely true of more sophisticated examples, such as the "sporadic groups" mentioned in the theorem classifying the finite simple groups.[34]

III

Let me now turn to Martin's view. He is mainly concerned with concepts of basic mathematical structures such as the natural numbers, the natural numbers and their subsets, and the universe of sets.[35] These are examples of what Isaacson calls particular structures and also of what I call basic structures.[36] The main contrast is with concepts of kinds of structures, such as groups, fields, and metric spaces.[37] But surely the other concepts deployed in reasoning about the basic structures belong in the picture as well.

In more than one of his writings Martin has expressed doubt about views about mathematics that he calls "object-based." I'm not sure exactly what he

---

[33] See MTO pp. 102–05.

[34] See Martin W. Liebeck, "The classification of finite simple groups," in Timothy Gowers (ed.), *The Princeton Companion to Mathematics* (Princeton University Press, 2008), pp. 687–691.

[35] "Completeness or incompleteness" (forthcoming), p. 2.

[36] Isaacson, "The reality of mathematics" (2010), pp. 1–2. On basic structures, see my "Structuralism and metaphysics" (2004), p. 69. Isaacson's and my concepts are not the same. The pure sets of rank less than, say, $\omega + 23$, are a particular structure but would not be treated by anyone as a basic structure. But what I call basic structures are what Isaacson has mainly in view.

[37] Isaacson (ibid.) somewhat misleadingly calls these general structures.

means by that. He does not show sympathy for nominalism, although he does suggest that pure mathematics can in a way do without objects.

I think Martin's embrace of the thesis that concepts in mathematics are prior to objects is, like Isaacson's, based on difficulties he finds with mathematical objects. In "Completeness or incompleteness," the primary difficulty is based on a Gödelian idea, that some statements are implied by a basic concept, while others are not:

> A fundamental question about a basic concept is that of which statements are *implied by the concept* — would have to be true in any structure that instantiated the concept. I think of this as a kind of necessity, which I would characterize as a logical necessity. I regard the question of whether a statement is implied by a basic concept to be a meaningful one whether or not there are any structures that instantiate the concept. (p. 3)

He does not seem clearly committed to the view that the concept can be axiomatized in such a way that any statement implied by the concept follows logically (in a logic typically used in mathematical reasoning) from the axioms, even if one allows that the axioms will not be recursively enumerable. As he notes (ibid.), the notion of being implied by a concept is derived from Gödel, and he maintains that Gödel does not treat the notion as epistemic.

Martin states that the concept of natural number is "first-order complete: it determines truth values for all sentences of first-order arithmetic" (p. 4). He is inclined to the stronger view that the concept of natural number is "*fully determinate*: it is determined, in full detail, what a structure instantiating it would be like" (p. 6). He is not convinced that this property implies instantiation but admits that he does not have an argument against the claim that any fully determinate concept is instantiated. I'm not sure what the force is of describing the necessity as logical. Thinking of it as metaphysical would, I would guess, make it empty if the concept is not instantiated.

Martin's claim of the priority of concepts and at least partial skepticism about mathematical objects rests primarily on a puzzle he poses about the concept of the universe of sets. On the basis of an argument originally advanced by Zermelo, Martin argues that at least up to a certain rank, the concept of the universe is categorical.[38] Martin endorses an extension of the argument that would make the concept fully categorical, but that is not needed for the puzzle. Let us make the weak assumption that categoricity obtains for $V_{\omega+2}$. Then any models that, in Martin's language, "meet the concept of set" up to that point will agree with respect to third-order arithmetic. The universe of

---

[38] Ernst Zermelo, "Über Grenzzahlen und Mengenbereiche," *Fundamenta Mathematicae* 16 (1930), 29–47. The argument considered here occurs already in "Gödel's conceptual realism" (2005), and Martin discusses categoricity in set theory already in "Multiple universes of sets and indeterminate truth-values," *Topoi* 20 (2001), 5–16.

sets is up to that point unique. But then it follows that models will agree with respect to the continuum hypothesis.

Martin insists that he does not have a firm view about whether CH has a determinate truth-value. But this argument seems to imply that it does. Considerations of the kind Martin advances were voiced in the early years after Cohen's independence proof, and some writers of that time, notably G. Kreisel, took them to imply that CH must have a determinate truth value. Martin does not accept that conclusion, but his reason is that he questions a premise that was accepted in the earlier debate: in Martin's language, that the concept of the universe of sets is instantiated, or even that what is in effect the concept of a standard model of third-order arithmetic is instantiated. He holds that it is not essential to pure mathematics that the concept of the universe of sets should be instantiated. He has no objection to the claim that the concept of natural number (in effect, of a standard model of arithmetic) is instantiated, but very likely his view is that the truth of that thesis is also not essential for pure mathematics.

But let us return to the set-theoretic case. Instantiation of the concept of the universe of sets, as Martin understands it, is a strong condition, that there should exist a system of objects (which we can certainly call sets) that constitute a maximal universe, in particular containing the full power set of each set, and having an absolute infinity of ranks. The second condition is evidently not at issue when we restrict to third-order arithmetic. We should be clear about what non-instantiation does not imply: it does not imply that there are no sets. It clearly does not imply that a weak concept such as that of the hereditarily finite sets is not instantiated. That it is would follow from the assumption, which he does not contest, that the concept of natural number is instantiated. It doesn't even imply that there are no models of first-order set theories, even with very large cardinals, since it is well known that they can differ with respect to CH. I think this is an important point. The idea of a universe of sets with the maximality properties that Martin emphasizes could be viewed as a transcendental Idea that is part of many set theorists' conception of what their subject is about. But what they work with on a day-to-day basis are first-order theories, ZF and its extensions by various further axioms, especially AC and various large cardinal axioms. Extending these theories by going to second-order logic gives no deductive power not given by an additional large cardinal assumption.

Many have questioned whether the categoricity theorems, either of arithmetic or set theory, can be taken at face value.[39] About the case of the natural numbers, I have written at some length, but I don't plan to repeat that here.[40] It should now be well known that the power of the logic needed to prove these categoricity results is much less than casual accounts would lead one to believe. Simpson and Yokoyama have proved that Dedekind's categoricity theorem for

---

[39] I leave out of account the categoricity of the theory of real numbers or of Euclidean space of a fixed number of dimensions, since they would not raise any new issues.

[40] MTO §§48–49.

the natural numbers is provable in WKL_0 and is in fact equivalent to it over the base theory RCA_0.[41] WKL_0 is well known to be conservative over primitive recursive arithmetic. Thus some of the worries one might have coming from the entanglement of second-order logic with set theory are mitigated. I don't know of a "reverse mathematics" analysis of Zermelo's theorem about set theory. Isaacson points out that Shepherdson proved long ago that Zermelo's theorem is provable in Gödel-Bernays set theory, which is conservative over ZF.[42] This can almost certainly be sharpened, as is at least suggested by a sketch given by Martin, as well as by the observation by Jouko Väänänen that if one has ZF with two $\in$-relations, where separation and replacement for each one holds for formulae containing the other, one can prove the two systems of sets to be isomorphic, a purely first-order result.[43] However, this depends on the fact that the universe is the same for both and the membership relations are expressed by given predicates.[44]

Not everyone is persuaded by these considerations, and I think the issue is not primarily how much logic is presupposed in categoricity proofs. Rather it is that whatever the logic, it is possible to step back from it and view the proof as internal to some framework that could be otherwise. In the number-theoretic setting, the proof would concern two hypothetical sequences of natural numbers and argue that they are isomorphic. But reflection on the language in which it is conducted brings out that the language in which the argument is conducted is subject to non-standard interpretations. On such an interpretation, the two sequences of the original argument are still isomorphic, but they may from the point of view of the metalanguage be non-standard.

I don't want to express a verdict about this issue. I described Martin as posing a puzzle. About a hypothesis like CH, the set theorist has three options, none of which is entirely appealing. He must admit that either CH must have a truth-value, even though we may not have any means of deciding it, or he must decline to take the categoricity arguments as proving what they seem to prove, or he must, with Martin, say that the full concept of the universe of sets may not be instantiated.

Martin himself does not use the term 'puzzle' or the stronger term 'paradox'. It appears that he is reasonably comfortable with the third horn of the trilemma just stated. He embraces it in the following remark:

---

[41] Stephen G. Simpson and Keita Yokoyama, "Reverse mathematics and Peano categoricity," *Annals of Pure and Applied Logic* 164 (2012), 229–243. Thanks to Peter Koellner for drawing this paper to my attention.

[42] Isaacson, "The reality of mathematics" (2010), p. 12; J. C. Shepherdson, "Inner models of set theory – part II," *The Journal of Symbolic Logic* 17 (1952), 225–237.

[43] "Multiple universes of sets," §3; Jouko Väänänen, "Second order logic or set theory?" *The Bulletin of Symbolic Logic* 18 (2012), 91–121, p. 104.

[44] The same assumption (for the natural numbers) was made in the argument of §49 of MTO. That suited the scenario I was describing, but it did not strictly speaking give a categoricity result.

> I have already explained that I accept uniqueness [of the universe of sets] but am unconvinced about existence. Indeed, I am unsure about existence *because* I am sure about uniqueness. I am convinced that we do not know whether the Continuum Hypothesis has a truth-value. If the concept of set is instantiated, then uniqueness implies that all instances give the same truth-value. Hence if we are sure of both instantiation and uniqueness, then we seem to be sure that there is a truth-value. For me, this is a *reductio*.[45]

He says that it is not essential to pure mathematics that its concepts be instantiated. However, in correspondence he writes that he is not as comfortable with the third horn as I took him to be. He writes:

> I do worry very much—though my paper doesn't make this clear—about the possibility that sentences like CH might have no truth value. If that's the case, then there is something seriously defective about the concept of set. For me it means that neither the concept or any refinement of it could be fully determinate.[46]

I am somewhat puzzled by the fact that Martin seems not to entertain seriously the possibility the CH may have a truth value even though we may have no way of determining what it is and it may even be beyond our capacity to do that. As noted above, Martin is unfriendly toward what he calls "object-based" accounts of mathematical concepts. What I have called the elusiveness of mathematical objects seems to bother him, and I think he is not persuaded by what philosophers (myself included) have proposed to accommodate the data while keeping something close to the traditional role of such objects.

There are obvious reasons why mathematical objects will not go away. Concepts are first of all concepts of objects. Leaving aside my own reservations about entities that are not objects, Martin does not even attempt to dispense with the *conception* of mathematical objects. Clearly, what it would be for, say, the concept of the natural number sequence to be instantiated is for there to be a sequence of *objects* and operations on them obeying the principles of arithmetic. Something similar is true for the universe of sets or lesser systems of sets such as $V_{\omega+2}$. Even the hypothesis that the concept of $V_{\omega+2}$ is *not* instantiated involved the conception of mathematical objects, in that it asserts that objects satisfying certain conditions do *not* exist.

I don't think Martin would disagree with anything in the last paragraph. But now, what can we say about mathematical truth if the concept of the immediately relevant structure, say $V_{\omega+2}$, is not instantiated? What Martin says about this may be incomplete, but at least a very central class of truths are those that are implied by the concept. See the remark quoted at the

---

[45] "Gödel's conceptual realism" (2005), p. 221, italics in the text. However, I am not sure whether 'reductio' was italicized for emphasis or just because it is a Latin word.

[46] Communication of 13 February 2015.

beginning of this section. The remark that the necessity involved is "a logical necessity" might suggest that it would be conferred by some formal logic. But typically a basic mathematical concept will be axiomatized, and the role of formal logic is to underwrite the derivation of truths from the axioms. If the axioms themselves are to be implied by the concept, it has to be in a more informal sense. This is an instance of what the logical empiricists called analyticity. Martin is much influenced on this point by Gödel, whose differences with logical empiricism are well known to be great. However, Gödel developed a conception of analyticity that would be compatible with his own conceptual realism.[47]

However, Martin says that the notion of being implied by a concept is not epistemic, and it could obtain for a proposition even if it is unknowable. But it is apparently meant to include all truths about the kind of structure the concept is a concept of.

In the case of the natural numbers, this raises a question, perhaps about what it means to say that a proposition is implied by a concept. I think Martin agrees that part of a characterization of the natural numbers is naturally given by open-ended schemata of induction, recursion, or both, where, for example, anything we recognize as a well-defined predicate can give rise to an instance of induction. However, once we get beyond first-order arithmetic, concepts such as real or complex number, or of certain sets, will figure in inductions in the proofs of statements in the language of first-order arithmetic. In many important cases it turns out that the result is in the end provable in PA, but this is not always so. Does Martin want to say that these statements are still implied by the concept of natural number? In the last section of "Completeness or incompleteness," Martin argues for an affirmative answer. Since a version of the paper containing this argument came to me only near the end of my work on the present paper, I will not undertake to go into it. I will only note that it seems to presuppose the full determinateness of the concept of natural number.

This issue will arise for other number systems and for geometries, if we take them to be basic. But because the further assumptions that might be needed for theorems in the language of set theory are also stated in the language of set theory, it does not arise in the same way for set theory. There it arises for cases about which Gödel says that it has not been made clear that they are implied by the concept of set. That raises questions about extrinsic evidence, about which Martin wrote an important paper some years ago and about which he comments further in "Completeness and incompleteness."

---

[47] That his conception occupied a distinctive place in differences about the analyticity of mathematics was pointed out by Hao Wang, "Two commandments of analytic empiricism," *The Journal of Philosophy* 82 (1985), 449–462. I discuss Gödel's conceptions of analyticity in "Analyticity for realists," in Juliette Kennedy (ed.), *Interpreting Gödel: Critical Essays* (Cambridge University Press, 2014), pp. 131–150. In correspondence, Martin has disputed my interpretation there of implication by a concept as more epistemic than he will accept. I have not yet responded to this criticism.

However, that matter takes us away from the question about mathematical objects. I think the upshot of my discussion is that Martin is quite happy to operate with a conceptual scheme in which mathematical objects have a place. I don't see how he could do otherwise without recasting mathematical language in a way that I think we don't have a real idea of. But he thinks that mathematics, or at least pure mathematics, can do without supposing that they *exist*. I would maintain that he is asking too much of existence. I fear that, without wishing to do so, he will put himself into the company of "fictionalists," who maintain that reference to mathematical objects is like reference to persons, places, and other entities in works of fiction. I don't believe he wants to go there, but I am not sure what alternative he envisages.

IV

This discussion is less conclusive about the question with which we started than one could reasonably wish. However, I think there are things that can be said. One is that mathematical objects cannot be made to go away. They may be elusive; they may not be as "definite" as some writers on these issues expect. There are eliminative programs I have not discussed, but they have not survived well the scrutiny exercised on nominalist and eliminative structuralist programs.[48] Neither of our authors aims to pursue such a program.

But now what of concepts? In modern mathematics, objects come in structures, and in a rough and ready way the basic structures and types of structures are given by concepts. Conceptions of concept are diverse, and what we have canvassed here leaves a lot of the territory uncovered. The most I am in a position to claim in this essay is that a clear case has not been made for the priority of concepts over objects or that mathematics can do without the existence of objects.

Appendix: Concepts and objects in the late Gödel

As remarked above, that concepts are objects is the view of Gödel's published writings and other texts (such as the Gibbs Lecture) that belong to the same period. It was probably his most considered view. But from the testimony of Hao Wang, in LJ, it appears that in his last years he called this into question.

In remark 7.3.12 of LJ, Gödel is reported to say that concepts are not objects. In another remark, 8.2.4, he is reported to say that "objects are in space or close to space," something that he would surely not have affirmed about concepts. However, I have not found other remarks that clearly bear on this question.

---

[48]On eliminative structuralism, see MTO §§10–13, 15–17; on nominalism, see John P. Burgess and Gideon Rosen, *A Subject with no Object: Strategies for Nominalist Reconstruction of Mathematics* (Oxford: Clarendon Press, 1997).

In a text that he called Fragment Q, Wang undertook to sort out the distinction of sets, classes, and concepts as Gödel understood them.[49] In that text, Gödel's view of classes is clearly represented as being that they are not objects, because talk of classes is a *façon de parler* for certain (presumably extensional) statements about concepts. So one could say that they are not entities in their own right but rather logical fictions. However, on the question whether concepts are objects Wang is silent.

I think we can conclude that Gödel did not reaffirm his earlier view that concepts are objects but is at the very least not emphatic in his repudiation of it. Whether he was uncertain about the question or simply thought that the issue was not especially important, I do not know.

References

Benacerraf, Paul (1965): "What numbers could not be", *Philosophical Review* 74, 47–73. Reprinted in P. Benacerraf and H. Putnam (eds.), *Philosophy of Mathematics: Selected Readings*, 2d ed. (Cambridge University Press, 1983).

Benacerraf, Paul (1973): "Mathematical truth", *The Journal of Philosophy* 70, 661–679. Reprinted in P. Benacerraf and H. Putnam (eds.), *Philosophy of Mathematics: Selected Readings*, 2d ed. (Cambridge University Press, 1983).

Bernays, Paul (1950): "Mathematische Existenz und Widerspruchsfreiheit", in his *Abhandlungen zur Philosophie der Mathematik* (Darmstadt: Wissenschaftliche Buchgesellschaft, 1976).

Burgess, John P. and Rosen, Gideon (1997): *A Subject with no Object: Strategies for Nominalist Reconstruction of Mathematics* (Oxford: Clarendon Press).

Carey, Susan (2009): *The Origin of Concepts* (Oxford University Press).

Feferman, Solomon (2014): "Logic, mathematics, and conceptual structuralism", in Penelope Rush (ed.), *The Metaphysics of Logic* (Cambridge University Press), pp. 72–92.

Frege, Gottlob (1891): *Funktion und Begriff* (Jena: Pohle).

Gödel, Kurt (1944): "Russell's mathematical logic", reprinted in Solomon Feferman et al. (eds.), *Collected Works, volume II: Publications 1938–1974* (New York: Oxford University Press, 1990).

Gödel, Kurt (1953): "Is mathematics syntax of language?", reprinted in Solomon Feferman et al. (eds.), *Collected Works, volume III: Unpublished Essays and Lectures* (New York: Oxford University Press, 1995).

---

[49]Published as "Sets and concepts, on the basis of discussions with Gödel," edited with introduction and notes by Charles Parsons, in C. Parsons and M. Link (eds.), *Hao Wang, Logician and Philosopher* (London: College Publications, 2011), pp. 79–118.

Isaacson, Daniel (1994): "Mathematical intuition and objectivity", in Alexander George (ed.), *Mathematics and Mind* (New York and Oxford: Oxford University Press), pp. 118–140.

Isaacson, Daniel (2010): "The reality of mathematics and the case of set theory", in Z. Novák and A. Simonyi (eds.), *Truth, Reference, and Realism* (Budapest: Central European University Press), pp. 1–75.

Liebeck, Martin W. (2008): "The classification of finite simple groups", in Timothy Gowers (ed.), *The Princeton Companion to Mathematics* (Princeton University Press), pp. 687–691.

Martin, Donald A. (2001): "Multiple universes of sets and indeterminate truth-values", *Topoi* 20, 5–16.

Martin, Donald A. (2005): "Gödel's conceptual realism", *The Bulletin of Symbolic Logic* 11, 207–224.

Martin, Donald A. (forthcoming): "Completeness or incompleteness of basic mathematical concepts", in P. Koellner and C. Parsons (eds.), *Exploring the Frontiers of Incompleteness*.

Parsons, Charles (1983): *Mathematics in Philosophy* (Ithaca, N.Y.: Cornell University Press).

Parsons, Charles (1995): "Platonism and mathematical intuition in Kurt Gödel's thought", *The Bulletin of Symbolic Logic* 1, 44–74.

Parsons, Charles (2004): "Structuralism and metaphysics", *Philosophical Quarterly* 54, 56–77.

Parsons, Charles (2008): *Mathematical Thought and its Objects* (Cambridge University Press).

Parsons, Charles (2008): "Paul Bernays' later philosophy of mathematics", reprinted in *Philosophy of Mathematics in the Twentieth Century* (Cambridge, Mass.: Harvard University Press, 2014).

Parsons, Charles (2011): "Quine's nominalism", *American Philosophical Quarterly* 48, 213–228.

Parsons, Charles (2013): "Some consequences of the entanglement of logic and mathematics", in Michael Frauchiger (ed.), *Reference, Rationality, and Phenomenology: Themes from Føllesdal* (Frankfurt: Ontos Verlag), pp. 153–178.

Parsons, Charles (2014): "Analyticity for realists", in Juliette Kennedy (ed.), *Interpreting Gödel: Critical Essays* (Cambridge University Press), pp. 131–150.

Parsons, Charles (2014): "Some objections to structuralism", *Al-Mukhatabat*, no. 10, 69–96.

Sheperdson, J.C. (1952): "Inner models for set theory – part II", *The Journal of Symbolic Logic* 17, 225–237.

Simpson, Stephen G. and Yokoyama, Keita (2012): "Reverse mathematics and Peano categoricity", *Annals of Pure and Applied Logic* 164, 229–243.

Väänänen, Jouko (2012): "Second order logic or set theory?", *The Bulletin of Symbolic Logic* 18, 91–121.

Wang, Hao (1985): "Two commandments of analytic empiricism", *The Journal of Philosophy* 82, 449–462.

Wang, Hao (1996): *A Logical Journey: From Gödel to Philosophy* (Cambridge, Mass.: MIT Press).

Wang, Hao (2011): "Sets and concepts, on the basis of discussions with Gödel", edited with introduction and notes by Charles Parsons, in C. Parsons and M. Link (eds.), *Hao Wang, Logician and Philosopher* (London: College Publications), pp. 79–118.

Yablo, Stephen (2014): *Aboutness* (Princeton University Press).

Zermelo, Ernst (1930): "Über Grenzzahlen und Mengenbereiche". *Fundamenta Mathematicae* 16, 29–47.

On Reconstructing Dedekind Abstraction Logically

Erich H. Reck*

ABSTRACT: While Richard Dedekind's technical contributions to the foundations of mathematics have been widely praised from the beginning, his remarks about "abstraction" and "free creation", as tied to his structuralist views, have received a more mixed reaction. Often they have been either ignored as irrelevant or dismissed as crudely psychologistic, while occasionally it has been suggested to interpret, or reconstruct, them "logically". But what exactly could the latter amount to? In this essay, four ways of interpreting "Dedekind abstraction" from a logical point of view are explored, called the "neo-Russellian", "neo-Hilbertian", "neo-Fregean", and "neo-Cantorian" reconstruction, respectively. Distinguishing them is meant as a contribution to Dedekind scholarship, but also to current philosophical debates about structuralism.

Richard Dedekind made many technical contributions to the foundations of mathematics, most of which have been widely accepted and praised. Some of his more informal, philosophical remarks in this context have, however, been received more critically. His remarks about "abstraction" and "free creation", as tied to his structuralist views about the natural and real numbers, constitute a main example. They have often been either dismissed as irrelevant or attacked as a crude form of psychologism. In Dedekind's defense, it has been suggested to interpret these remarks not in a problematic psychological but in a "logical

---

*As my former dissertation advisor and mentor, Bill Tait has had a strong influence on my interests, my research, and my career. I am grateful for his support and inspiration, which continues until today. I am also pleased to be able to contribute a paper on a topic he addressed himself in his writings, even if only tangentially, namely "Dedekind abstraction". Further references will be provided as we go along.

sense", or at least, as having a defensible "logical core", as W. W. Tait put it. But what does that amount to if elaborated more, i.e., how exactly should "Dedekind abstraction" be understood from a logical point of view? This is the core issue to be addressed in the present paper.[1]

The paper is structured as follows. In its first section, some crucial passages from Dedekind's writings, especially from his well-known essay, *Was sind und was sollen die Zahlen?* (1888), will be introduced. In section two, we will turn to several influential criticisms of these passages, by Michael Dummett, Bertrand Russell, and Gottlob Frege, together with some initial, informal defenses of Dedekind against them. In the third section of the paper, four specific and relatively detailed proposals for how to reconstruct Dedekind abstraction logically and more formally will be introduced, called the "neo-Russellian", "neo-Hilbertian", "neo-Fregean", and "neo-Cantorian" reconstructions, respectively, for reasons that will become apparent along the way. This will be followed, in section four, by a further comparative discussion, both about the strengths and weaknesses of these four options and about their Dedekindian credentials. A brief summary and conclusion will round off the paper.

1 Dedekind's Crucial Remarks

There are two passages in Dedekind's *Was sind und was sollen die Zahlen?* that are crucial for present purposes, to be found in Sections 73 and 134 of that text. Before quoting them, let me provide some background. Dedekind's 1888 essay is meant to provide a novel account of the natural numbers $\mathbb{N}$, developed within the framework of a general theory of sets and functions (or "mappings"). Two main ingredients in it are his definitions of what it means for a set to be infinite (Definition 64) and simply infinite (Definitions 71), respectively. As is well known, the latter amounts to a characterization of the natural numbers in terms of (an early version of) the second-order Dedekind-Peano axioms. In between these two definitions, Dedekind presents, as a third main ingredient, a (rather controversial) argument for the existence of an infinite set (Theorem 66).[2] Assuming that result, the existence of a simply infinite subset N then follows straightforwardly (Theorem 72).

The first passage crucial for us occurs as the next step. Dedekind writes:

> 73. Definition. If in the consideration of a simply infinite system N set in order by a mapping Φ we entirely neglect the special character of the elements, simply retaining their distinguishability and

---

[1] Towards the end of the second section, I will mention an alternative, more "pragmatic" defense of Dedekind's remarks about "abstraction" too. Yet another possible defense, suggested to me by Benis Sinaceur, consists of interpreting Dedekind in a broadly "epistemological" sense. I cannot address that alternative in the present paper, including its relation to a "logical reading", but hope to do so in a future publication.

[2] Dedekind's argument for his Theorem 66 involves an appeal to "the totality S of all things which can be objects of my thought" (Dedekind 1963, p. 64). I will leave aside what is problematic about it here; cf. Reck (2003, 2013) and Klev (2018) for details.

> taking into account only the relations to one another in which they are placed by the order-setting mapping Φ, then these elements are called *natural numbers* or *ordinal numbers* or simply *numbers*, and the base-element 1 is called the *base-number* of the *number-series N*. With reference to this freeing the elements from every other content (abstraction) we are justified in calling numbers a free creation of the human mind (Dedekind 1863, p. 68; original emphasis, translation modified slightly).

Note here, in particular, Dedekind's use of the terms "abstraction" and "free creation". It is these two terms that call for further clarification (including disambiguation, as our later discussion will make evident).

The second passage crucial for our purposes occurs later in Dedekind's 1888 essay. Here the context is the following: At this point, Dedekind has established that any two simple infinities are isomorphic to each other (Theorem 132), also that any set that can be mapped 1-1 onto a simple infinity is itself simply infinite (Theorem 133). Then he adds:

> 134. Remark. By the two preceding theorems (132), (133) all simply infinite systems form a class in the sense of (34) [an equivalence class]. At the same time, with reference to (71), (73) it is clear that every theorem regarding numbers, i.e., regarding the elements n of the infinite system N set in order by the mapping Φ, and indeed every theorem in which we leave entirely out of consideration the special character of the elements n and discuss only such notions as arise from the arrangement Φ, possesses perfectly general validity for every other simply infinite system Ω set in order by a mapping θ and its elements ν, and that the passage from N to Ω (e.g., also the translation of an arithmetic theorem from one language into another) is effected by the mapping ψ considered in (132), (133), which changes every element n of N into an element ν of Ω, i.e., into $\psi(n)$. [...] By these remarks, as I believe, the definition of the notion of numbers given in (73) is fully justified. [...] (*ibid.*, pp. 95–96, translation modified slightly).

Particularly relevant in this passage is Dedekind's observation (without proof) that any arithmetic theorem about N "possesses perfectly general validity" also for every other simple infinity, via the corresponding isomorphism ψ, and that this fact "fully justifies" his earlier introduction of "the notion of numbers".

A natural and fairly literal (though not uncontroversial) reading of these two passages would seem to be the following: After first establishing some basic theorems about the notion of simple infinity, including its satisfiability (the existence of systems falling under it) and its categoricity (every two systems falling under it are isomorphic), Dedekind introduces, by an act of "free creation", a special such system worthy of the label "the natural numbers".

And he does so by "abstraction" from whatever non-arithmetic properties the elements of the initial simple infinity have, i.e., properties going beyond those definable in terms of the function Φ. Later this procedure is justified further by noting that its result is invariant under the initial choice of a simply infinite system, since all such systems are isomorphic which implies that any arithmetic theorem that holds for one also holds for any other.

According to this reading of Dedekind, the system of "the natural numbers" that is introduced via "abstraction" is different from whatever simple infinity he started with (although it is isomorphic to it). It is a separate, *sui generis* simple infinity. Why and how so? Unlike the elements contained in other simple infinities, its elements are not determined, i.e. characterized in their nature, by non-arithmetic properties; instead, they are determined, purely and fully, by the relevant "structural" properties (cf. Reck 2003). It is this aspect that makes the resulting position a kind of "structuralism", or more specifically, a version of "non-eliminative structuralism".[3] (Again, this reading is not uncontroversial; and it is in need of disambiguation.)

Such an initial reading of Dedekind finds reinforcement if we consider how he proceeds in his other foundational essay, *Stetigkeit und irrationale Zahlen* (1872), with respect to the real numbers $\mathbb{R}$. In that context too, he first introduces a central notion: that of a continuous ordered field. He argues that this notion is satisfiable, i.e., that complete ordered fields exist (the system of all cuts on the rational numbers, endowed with corresponding operations and an ordering, is his main example). And then he introduces "the real numbers" once again by "creation" (Dedekind 1963, p. 15). Admittedly, Dedekind does not use the word of "abstraction" in this earlier essay; nor does he formulate a categoricity theorem for complete ordered fields yet (although it can be added, as is well known). On the other hand, he is more explicit in this earlier text that the mathematical objects introduced along such lines are *sui generis*, since, unlike the elements in other continuous ordered fields, they have no "foreign properties". For example, the real numbers do not have elements in a set-theoretic sense, as Dedekind cuts do.[4] Furthermore, in a letter from 1888—the year in which *Was sind und was sollen die Zahlen?* was published—he makes clear that the two cases are meant to be parallel, including with respect to the issue of "creation" (Dedekind 1888b).[5]

2 Standard Criticisms and Initial Defenses

As mentioned at the beginning of this essay, Dedekind's remarks about "abstraction" and "free creation" were not received as positively as his corresponding technical results. Indeed, an interpretation of Dedekind that was dominant

---

[3]Cf. Parsons (1990) for this terminology; and see Reck & Price (2000) for more.

[4]On this last point, see also Dedekind (1876), among others.

[5]In an aside in his 1872 essay, Dedekind adds that the negative and fractional numbers should be taken to "have been created by the human mind" as well (Dedekind 1963, p. 4). Hence this amounts to a general theme; cf. Reck (2003) for more.

in English-speaking philosophy for decades, and is still influential today, is to accept our initial reading above while giving it a twist that makes it evidently problematic. A particularly explicit instance of it can be found in Michael Dummett's writings. According to Dummett, Dedekind's position involved the view that "the mind could, by this means [i.e., abstraction], create an object or system of objects lacking the features abstracted from, but not possessing any others in their place" (Dummett 1991, p. 50). Why is this problematic? Because, as Dummett argues, it makes mathematics hopelessly subjective; i.e., its objects exist then only in people's minds or subjective consciousness. This amounts to a crude and problematic form of psychologism.

Dummett is explicit that it is this psychologistic, subjectivist aspect to which he is objecting primarily. Appealing to the authority of Gottlob Frege, whom he ranks far above Dedekind as a philosopher of mathematics, he writes: "Frege devoted a lengthy section of *Grundlagen* [*der Arithmetik*], sections 29–44, to a detailed and conclusive critique of this misbegotten theory" (*ibid.*); and the particular sections of *Grundlagen* invoked contain Frege's criticism of subjectivist views about mathematics, especially views according to which numbers only exist in people's minds. Another authority appealed to in this context is Bertrand Russell. Thus Dummett writes: "[Dedekind] believed that the magical operation of abstraction can provide us with specific objects having only structural properties. Russell did not understand that belief because, very rightly, he had no faith in abstraction thus understood" (p. 52). On the basis of such considerations Dedekind's position is dismissed as "mystical structuralism" by Dummett, while he takes "eliminative" versions of structuralism more seriously, including calling them "hard nosed".

Is Dummett justified in treating Frege and Russell as his allies in this context, and more specifically, in rejecting Dedekind's position as a crude form of psychologism? There is no doubt that these two thinkers are critical of Dedekind's appeal to "abstraction". But on closer inspection this is not (or not primarily) because of its supposed psychologistic character.[6] Russell's main objection is to the alleged structural nature of mathematical objects that he took Dedekind to endorse. As he puts it memorably:

> [I]t is impossible that the ordinals should be, as Dedekind suggests, nothing but the terms of such relations as constitute a progression. If they are to be anything at all, they must be intrinsically something; they must differ from other entities as points from instants, or colours from sounds. (Russell 1903, p. 249)

Basically, for Russell mathematical objects, such as Dedekind's "ordinals", cannot be characterized purely by their "relational" or "structural" properties; they must have an "intrinsic" nature. More generally, Russell cannot make sense of Dedekind's appeal to "abstraction". At one point, in a seeming attempt to be

---

[6]For the rest of this section, I will draw on Reck (2013) and Reck (forthcoming a), which contain more detailed discussions of Frege's and Russell's reactions to Dedekind.

charitable, he interprets it as an unclear appeal to his own "principle of abstraction", i.e., the use of equivalence classes for introducing new mathematical objects (more on the latter below).[7]

Frege, in turn, was certainly critical of psychologistic, subjectivist views about mathematics. Yet the discussions in the sections of *Grundlagen* to which Dummett refers (a book published in 1884) and in which Frege discusses views about the nature of the natural numbers do not explicitly mention Dedekind (whose *Was sind und was sollen die Zahlen?* was published only in 1888). In his later *Grundgesetze der Arithmetik, Vol. 2* (1903), Frege does lump Dedekind with a number of writers about mathematics who talk about "creation" in connection with the real numbers; and he does bring up the psychologism charge again all of them. But even in that book, Frege is more charitable to Dedekind than Dummett. If looked at carefully, Frege's primary objection to Dedekind is that he failed to be explicit about the basic laws underlying "abstraction", not that it is problematically psychologistic.[8]

Be that as it may, the following question arises: Do we have to interpret Dedekind's appeal to "abstraction" in a psychologistic sense, whether Frege and Russell accuse him of that or not? In his 1996 paper, "Frege versus Cantor and Dedekind: On the Concept of Number", W. W. Tait takes issue with this interpretation, and especially, with Dummett's dismissal of Dedekind's position. (Similarly for Cantor's appeal to "abstraction", which Dummett dismisses as well.[9]) It is not the structuralist part of Dummett's interpretation of Dedekind that is rejected by Tait, but the psychologistic twist given to it. But what is the alternative? As Tait puts it: "[T]he abstraction in question has a strong claim to the title *logical abstraction*" (*ibid.*, p. 84). He goes on to suggest a particular way of understanding "logical abstraction", namely in terms of how the "sense" of relevant propositions is determined. However, that suggestion is not worked out in technical detail by him.[10]

Tait's suggestion to interpret Dedekind more charitably was picked up in my 2003 article, "Dedekind's Structuralism: An Interpretation and Partial Defense", including an attempt to clarify "Dedekind abstraction" further. In that article, I interpret Dedekind as a "logical structuralist"; yet my interpretation

---

[7] For an insightful, detailed discussion of Russell's reaction to Dedekind's structuralism and his talk of "abstraction", cf. Heis (forthcoming).

[8] Cf. Reck (forthcoming a); we will come back to this issue briefly in the fourth section.

[9] As Dummett writes: "It was virtually an orthodoxy, subscribed to by many philosophers and mathematicians, including Husserl and Cantor, that the mind could, by this means, create an object or system of objects lacking the features abstracted from, but not possessing any others in their place" (Dummett 1991, p. 50).

[10] For present purposes, the central remark in Tait's article is this: "[W]hat seems to me to be essential to this kind of abstraction is this: the propositions about the abstract objects translate into propositions about the things from which they are abstracted and, in particular, the truth of the former is founded upon the truth of the latter. So the abstraction in question has a strong claim to the title *logical abstraction*: the sense of a proposition about the abstract domain is given in terms of the sense of the corresponding proposition about the (relatively) concrete domain" (Tait 1996, p. 84, emphasis in the original). As we will see below, this suggestion can be elaborated in several different ways.

remained too informal and imprecise in certain respects as well. Apart from that, the suggestion to understand Dedekind in a logical rather than a psychological sense can be traced back further in time, as I found out subsequently.[11] Namely, Ernst Cassirer's 1910 book, *Substanzbegriff und Funktionsbegriff*, anticipates Tait's and my own general defense of Dedekind by almost a century. In it, one can find remarks such as the following:

> [Dedekind's form of abstraction] means logical concentration on the relational system, while rejecting all psychological accompaniments that may force themselves into the subjective stream of consciousness, which form no constitutive moment of this system (Cassirer, 1910, p. 39).

The same point is repeated in various other works by Cassirer, going back to his 1907 article, "Kant und die modern Mathematik", and forward to, e.g., his 1929 book, *The Philosophy of Symbolic Forms, Vol. III*. But while Cassirer was ahead of his time, also in other respects, his discussion of Dedekind abstraction does not include a detailed logical reconstruction of it either. After all, Cassirer was not a mathematical logician.[12]

The main goal of the present article is to supplement Tait's, Cassirer's, and my earlier readings of Dedekind, in the sense of working out a logical reconstruction of "Dedekind abstraction" in some technical detail (or indeed, several such reconstructions). But before turning to that task, let me mention another defense of Dedekind against the psychologism charge, or another way of reading Dedekind's philosophical remarks that is more charitably than Dummett's. The main suggestion in that interpretation consists of rejecting not only the supposed psychologistic side of Dedekind's position, but also the non-eliminative structuralist interpretation sketched above, i.e., the idea that "Dedekind abstraction" introduces a novel system of mathematical objects characterized by their structural properties alone. The resulting interpretation amounts, then, to a form of "eliminative structuralism".

What is the position attributed to Dedekind along such eliminative lines? Let us start again with the case of the natural numbers. (The case of the reals is parallel.) The core idea is this: After Dedekind has constructed a particular simple infinity, what his "abstraction" does is not to introduce a separate, distinguished system of objects, but simply to treat the given system in a novel way, namely by ignoring all its non-arithmetic aspects. And as should be added right away, any other simple infinity would do as well, since they are all isomorphic and the same arithmetic theorems hold of them, as Dedekind observed. What are "the natural numbers", then? Well, pick any

---

[11] In that respect, I am indebted to Michael Friedman's writings and, especially, to conversations with Pierre Keller. See also Reck (2013) and Yap (2014).

[12] Cassirer works with a broader, less formal notion of "logic" than Frege and Russell, although it is meant to encompass theirs. For more on his reception of Dedekind and the resulting "logical idealism", cf. Reck (forthcoming b) and Reck & Keller (forthcoming).

simple infinity and treat it as "the natural numbers". That is all we need to do for arithmetic purposes, i.e., all that is required for mathematical practice.

In the secondary literature the most subtle, worked-out version of this alternative, presented as a reading of Dedekind, occurs in a new paper by Wilfried Sieg & Rebecca Morris, entitled "Dedekind's Structuralism: Creating Concepts and Deriving Theorems" (this volume). The position attributed to Dedekind is still "structuralist"; but it is "eliminative", insofar as it works without introducing a *sui generis* system of objects besides the initially constructed simple infinity. With respect to the form of structuralism involved, one can say that the approach involves an "indifference to identify", in any absolute sense, the natural numbers with a particular simple infinity, since "any of them will do".[13] The same applies, *mutatis mutandis*, for the reals.

Compared to Dummett's, such a reading of Dedekind certainly has its advantages. However, one can now raise two questions: First, is the resulting position not again problematically psychologistic, since all depends on "ignoring" certain aspects of the initial simple infinity? An immediate response might be this: If we assume that the elements of the initially constructed simply infinity are not "mental objects", we do not end up with "the natural numbers" as such either. Put more positively, the "ignoring" at issue should be understood less in a psychological and more in a "pragmatic" sense. But that leads to a second question: Does this position not also have a "logical core", underneath is "pragmatic" surface; and if so, what is it? In what follows, an answer to the latter question will be suggested as well.[14]

3 Four Logical Reconstructions of Dedekind Abstraction

If one rejects uncharitable criticisms of Dedekind like Dummett's, is not fully satisfied with the "pragmatic" perspective on Dedekind in Sieg & Morris, and is intrigued by Tait's and Cassirer's alternative suggestion, one is left with a task. Namely, how can a "logical" perspective on Dedekind abstraction be spelled out in more detail (including clarifying its relationship to the pragmatic interpretation)? Actually, there is not just one option for a logical reconstruction—I will introduce four possible alternatives in this connection. These will be called the "neo-Russellian", the "neo-Hilbertian", the "neo-Fregean", and the "neo-Cantorian" reconstruction of Dedekind abstraction, respectively, for reasons that will become apparent.

Before going into specifics, let me set up a logical framework for all of these reconstructions. It will not be necessary to adopt this particular framework in the end; but starting with it will contribute to ease of formulation and clarity. (Some alternatives will be mentioned later, e.g., working within higher-order

---

[13] Compare Burgess (2015), ch. 3. In Reck & Price (2000), this position is called "relativist structuralism"; and a particular, fairly widespread version is "set-theoretic structuralism".

[14] A third question is this: Does such an "eliminative structuralist" reading do justice to Dedekind's repeated remarks about "creation"? This is doubtful, I think, although the interpretive issues are subtle. I plan to address them further in a future publication.

logic supplemented with some existential assumptions.) The kind of framework needed is some general theory of sets and functions, like in Dedekind's original approach. And the particular version of it that I will use, at least provisionally, is ZFC set theory. Actually, at certain points it will be important to talk about proper classes; and we will want to admit urelements too, collected together in a corresponding domain U, besides the pure sets in V. This background theory will allow for the construction of particular systems of objects to which Dedekind abstraction can then be applied.

Within ZFC we can, for example, introduce the finite von Neumann ordinals (starting with 1) to form a particular simple infinite system (with $\{\emptyset\}$ as the initial element, $s: x \to x \cup \{x\}$ as the successor function, and $\omega' = \omega \setminus \{\emptyset\}$ as the domain).[15] Similarly we can introduce a set-theoretic system of cuts on $\mathbb{Q}$ as a particular continuous ordered field. More generally, we can talk about "relational systems" of the form $S = \langle a_1, \ldots, a_n, f_1, \ldots, f_m, R_1, \ldots, R_l, D \rangle$, where the domain D is some set and a_i, f_i, and R_i are defined on it as usual. Each of our four reconstructions of Dedekind will now be characterized in terms of two ingredients: (i) an "abstraction operator", ab, that can be applied to such relational systems S so as to yield "abstract structures"; (ii) a related way of "analyzing" mathematical statements p, e.g., truths of arithmetic such as $2 + 3 = 5$ and $\forall n \forall m (n + m = m + n)$.

3.1 The Neo-Russellian Reconstruction

Above we saw that Russell was critical of what he took to be Dedekind's structuralist conception of mathematical objects, also that he tried to make sense of Dedekind abstraction by assimilating it to his own approach. As he wrote: "What Dedekind intended to indicate was probably a definition by means of the principle of abstraction" (Russell 1903, p. 249). Let us reconsider this suggestion. For relational systems S, let $ec(S)$ be the equivalence class of all sets isomorphic to S within V, i.e., $ec(S) = \{S' : \exists f(f : S' \cong S)\}$. Then Russell's comment leads naturally to the following definition:

(i) $ab_1(S) =_{def} ec(S)$

Here ab_1 is an "abstraction operator" that takes a relational system S as its argument and gives the equivalence class of S under isomorphism as the corresponding value.[16] We can then say that $ab_1(S)$ is "the structure" corresponding

---

[15] As Dedekind starts the natural numbers with 1 and as this will simplify some details later, I have adjusted the usual treatment in ZFC slightly.

[16] A number of variations are possible here. For example, we can consider not the whole equivalence class (a proper class), but some appropriate set-theoretic part of it; cf. the appeal to "Scott's trick" in the next subsection. We can also replace isomorphism with some weaker equivalence relation, e.g. the kind of "structure equivalence" discussed in Shapiro (1997), pp. 91–93, and then work with the resulting equivalence classes/sets instead. We will come back to some of these variants later.

to S. For instance, $\{\langle a, g, D\rangle : \exists f(f : \langle a, g, D\rangle \cong \langle \{\emptyset\}, s, \omega'\rangle)\}$ is "the structure" corresponding to the finite von Neumann ordinals.

In many mathematical cases, we work with relational systems S that are models of axiomatized theories T_S, i.e., such that $S \models T_S$. If the theory at issue is categorical—like in Dedekind's two central cases: arithmetic (with the second-order Dedekind-Peano axioms) and analysis (with the second-order axioms for complete ordered fields)—it characterizes S completely in a relevant sense. We also have: $ec(S) = \{M : M \models T_S\}$. This provides us with another way to think about "the structure" corresponding to S, namely $ab_1(S) = \{M : M \models T_S\}$; i.e., ab_1 maps S onto the model class of T_S.

As a second main ingredient of the intended position, we want to specify an analysis for every mathematical sentence p, i.e., a way of characterizing what it "really says". Let us start again with the simple case of arithmetic. Let us also assume that we are working with a formal system in which every arithmetic object, function, and relation is defined in terms of the basic constants 1, suc, and N, as usual in second-order Peano arithmetic. Given some arithmetic sentence p, we can then consider $p(1, suc, N)$, i.e., the formula in which all defined terms have been replaced by their definitions. If we let ourselves be guided by (i) above, we can re-analyze $p(1, suc, N)$ thus:

(ii) $ab_1(p) =_{def} \forall x \forall f \forall X [PA^2(x, f, X) \to p(x, f, X)]$

Here PA^2 are the usual second-order Dedekind-Peano axioms and $p(x, f, X)$ is the result of replacing 1, suc, and N by x, f, and X, respectively, wherever they occur in p (i.e., these three constants are replaced by corresponding variables of the right type over which we can then quantify). Likewise for analysis and similar theories; i.e., in those cases too we go from p to a corresponding sentence in which only the relevant basic constants, functions, and relations occur, and then we quantify these out in the universalized if-then form indicated.

This way of analyzing arithmetic sentences, or mathematical sentences more generally, was considered seriously by Russell himself, e.g., in his 1901 article "Recent Work in the Philosophy of Mathematics".[17] As he writes:

> Pure mathematics consists entirely of assertions to the effect that, if such and such a proposition is true of anything, then such and such another proposition is true of that thing (Russell 1901, p. 77).

This historical fact justifies the label "neo-Russellian reconstruction" for the current approach further. Note also that (ii) fits together naturally with (i) since we again consider all systems isomorphic to an initially given system S,

---

[17]Similar remarks can be found in Russell (1903), although in that book his views have already shifted in certain respects and he is not applying this approach to arithmetic any more; see again Heis (forthcoming) for further details.

here by quantifying over all models of T_S. (Or at least, it fits well in the case of a categorical axiom system.)[18]

Considered as a reconstruction of Dedekind, this approach has its main textual support in his remark that every theorem concerning a particular simple infinity "possesses perfectly general validity for every other simply infinite system" (Dedekind 1963, p. 96). The latter is now taken to suggest that it is really a theorem about all simple infinites at the same time, in the logical form made explicit by (ii). Similarly for the real numbers.[19]

3.2 The Neo-Hilbertian Reconstruction

In the neo-Russellian reconstruction of Dedekind abstraction, we talk, in the case of arithmetic, about all simply infinite systems together. An alternative suggestion is to assume that we talk about an arbitrary one. Similarly for other parts of mathematics, including analysis. When this is worked out more formally, it leads to our second reconstruction: the "neo-Hilbertian" one.

How could we implement this second suggestion more formally? Assuming again that we work with ZFC in the background, or with certain familiar extensions of it, one option is to make use of a global choice function for sets or classes. Assume ch is such a function, i.e., we pick one of them. Then we can give the following definition, where S is again a set-theoretic relational system:

(i) $ab_2(S) =_{def} ch(ec(S))$

In other words, as "the structure" corresponding to S we now let our choice function pick a representative from the equivalence class for it.[20] It is clear that this second form of abstraction is closely related to the first, since we have: $ab_2(S) = ch(ab_1(S))$. Note also the following: If we allow for urelements, then the systems in the equivalence class, thus also in the system picked by ch, may contain such urlements. To use the case of arithmetic again for illustration, this means that any of our urelements—say Julius Caesar or some beer mug—may now be, say, the element corresponding to 2 $(= s(\{\emptyset\})$ in $ab_2(\langle\{\emptyset\}, s, \omega'\rangle)$.

This observation reveals the first reason why this second approach deserves the name "neo-Hilbertian reconstruction" of Dedekind abstraction. As Hilbert is reported to have said in the case of an axiomatic approach to geometry: "One must be able to say at all times—instead of points, straight lines, and

---

[18] It is assumed here that we can characterize the systems S to which we apply abstraction in terms of an axiomatic theory, whether categorical or not. In the non-categorical case, (i) and (ii) can still be formulated but will correspond less closely to each other.

[19] Because of the quantification involved, the resulting position may be called "universalist structuralism"; cf. Reck & Price (2000). A well-known variant of it is Geoffrey Hellman's "modal structuralism" (Hellman 1989). Note that the "non-vacuity problem" that partly motivates Hellman's modal twist does not arise if we work with ZFC in the background.

[20] As hinted at in an earlier footnote, it is possible to avoid working with proper classes here. For example, we can use what is known as "Scott's trick", i.e., work with the set of all systems S' isomorphic to S that are of lowest rank in the ZFC hierarchy, instead of the whole equivalence class. This allows for the use of a choice function on sets alone.

planes—tables, chairs, and beer mugs." (That is to say, all our proofs must still go through if we replace geometric objects in a model of Euclidean geometry by other objects, including beer mugs etc., assuming we still deal with a relevant model.) There is also a second reason for appealing to Hilbert in this connection. Instead of using a set-theoretic choice function ch, one can work with a Hilbertian ϵ-operator for the same purpose.[21]

What about our second main ingredient in this case, i.e., a corresponding way of analyzing mathematical sentences p? This can be taken care of easily by generalizing our abstraction operator ab_2, from relational systems to their ingredients. Consider again the case of arithmetic. Assume we let $ab_2(\{\varnothing\})$ be the element corresponding to $\{\emptyset\}(=1)$ in $ab_2(\langle\{\emptyset\}, s, \omega'\rangle)$, i.e., the element onto with $\{\emptyset\}$ is mapped by ab_2; similarly for $ab_2(s)$ and $ab_2(\omega')$. Given some arithmetic sentence p, we can then stipulate:

(ii) $ab_2(p) =_{def} p(ab_2(\{\emptyset\}), ab_2(s), ab_2(\omega'))$

Here $p(ab_2(\{\emptyset\}), ab_2(s), ab_2(\omega'))$ is the result of replacing '1' by '$ab_2(\{\emptyset\})$', 'suc' by '$ab_2(s)$', and 'N' by '$ab_2(\omega')$' in $p(1, suc, N)$, parallel to above. What this means is that p is mapped onto the corresponding sentence concerning the chosen system $ab_2(\langle\{\emptyset\}, s, \omega'\rangle)$. Similarly for continuous ordered fields etc., i.e., the generalization of this approach should again be clear. (In this case there is no significant dichotomy between the categorical and the non-categorical case, since we work directly with S, or with the equivalence class it induces, not with an axiom system that characterizes S.)

Let me add a few further observations about this neo-Hilbertian reconstruction of Dedekind. As the appeal to an arbitrary representative from the relevant equivalence class indicates, this reconstruction is closely related to the pragmatic interpretation of Dedekind sketched at the end of our second section. In fact, it constitutes a way to bring out the "logical core" of that interpretation.[22] Having said that, there is also a difference. The neo-Hilbertian reconstruction of Dedekind abstraction, as a particular logical reconstruction, is tied to working within a particular formal system, such as ZFC. In contrast, the pragmatic interpretation of Dedekind, as sketched above, is naturally understood to proceed more informally.

Still, the neo-Hilbertian reconstruction clarifies the logic underlying the pragmatic approach to Dedekind, which was left implicit and somewhat ambiguous above. Note, finally, two closely related points: First, the end result is again a form of "eliminative structuralism", since no novel, *sui generis* systems are introduced by ab_2. (And the position is "semi-eliminative" in a more general

---

[21]Indeed, exactly this kind of approach has recently been explored in the literature on structuralism; cf. Schiemer & Gratzl (2016).

[22]Like the pragmatic position but more explicitly now, the neo-Hilbertian reconstruction of Dedekind amounts to a form of "relativist structuralism", and more specifically, a form of "set-theoretic structuralism"; see again Reck & Price (2000).

sense, in our setup by working with sets.) Second, the neo-Hilbertian reconstruction can point to essentially the same textual evidence as the pragmatic interpretation in terms of being a faithful interpretation of Dedekind.[23]

3.3 The Neo-Fregean Reconstruction

Given some system S, we work with all isomorphic systems in the neo-Russellian reconstruction; and we work with an arbitrary chosen representative in the neo-Hilbertian reconstruction. A variant of the latter, available in some cases, is to use a distinguished system in the equivalence class for S. Perhaps such a system can be picked for strong pragmatic reasons, e.g., when we use the finite von Neumann ordinals as "the natural numbers" since they can be generalized to the transfinite. But such pragmatic reasons are not available in general.

A third alternative is to work with a distinguished system that is new, in the sense that the "abstract" corresponding to S is introduced "purely structurally" (and is not an element of V, thus not of $ec(S)$). The following parallel may motivate such an approach. In current neo-logicism (as initiated by Crispin Wright, Bob Hale, and others), the suggestion is to use "Fregean abstraction principles" for introducing mathematical objects, e.g., "Hume's Principle" for introducing the finite cardinal numbers. Neo-logicists tend to apply these principles to concepts within the context of higher-order logic. But such an approach can be modified and generalized for our purposes, as recent work by Øystein Linnebo and Richard Pettigrew has shown.[24] This leads to our "neo-Fregean reconstruction" of Dedekind abstraction.

Adapting the approach by Linnebo & Pettigrew slightly so that it fits our basic setup, this suggests the following "structuralist abstraction principle":

(i) $ab_3(a, S) = ab_3(a', S') \leftrightarrow \exists f(f : S' \cong S \wedge f(a) = a')$

Here a is meant to be an element of S and a' an element of S'; ab_3 is meant to be a function from V to U; and all elements in U are meant to be introduced via this abstraction principle, while no "non-abstracts" exist in U. Again, the approach deserves to be called "neo-Fregean" since (i') has the form of a neo-Fregean abstraction principle modulo the differences mentioned.

What (i) does is to let an object a, considered relative to some relational system S, correspond to an element $ab_3(a, S)$ in U. As the relation on the right side of the biconditional is clearly an equivalence relation, we could also introduce an equivalence class corresponding to each pair $\langle a, S \rangle$, namely $\{\langle a', S' \rangle : \exists f(f : S' \cong S \wedge f(a) = a')\}$. This would lead us back to the vicinity of Russell. But we do not want to work with such equivalence classes here.

---

[23] Cf. Sieg & Morris for more on that evidence. A central part of it is Dedekind's remark about "leaving entirely out of consideration the special character of the elements n" in a simply infinite system; cf. his Remark 134 as quoted in Section 1.

[24] Cf. Linnebo & Pettigrew (2014).

Instead, the idea is to work with new and simple objects $ab_3(a, S)$ in U. Moreover, we want to use these new elements to construct a relational system that both "lives entirely in U" and is isomorphic to S. To achieve the latter, we need to define the domain of the resulting system and the relations and functions on it that correspond to those in S.

We can proceed as follows: Given $S = \langle a_1, \ldots, a_n, f_1, \ldots, f_m, R_1, \ldots, R_l, D \rangle$, we let $D' = \{ab_3(a, S) : a \in D\}$ be the new domain. (This modifies our definition above by extending it to $ab_3(D, S)$.) We also "lift" the structural features on S given by a_j, f_j, and R_j and transfer them to D', resulting in a'_j, f'_j, and R'_j, as follows: Let $a'_j = ab_3(a_j, S)$ ($1 \leq j \leq n$). If f_j is a k-ary function on D ($1 \leq j \leq m$), $b_1, \ldots, b_k \in D'$, and $c_1, \ldots, c_k \in D$ are such that $ab_3(c_i) = b_i$ ($1 \leq i \leq k$), we let $f'_j(b_1, \ldots, b_k) = ab_3(f_j(c_1, \ldots, c_k))$. (This extends our definition to $ab_3(f_j, S)$.) And if R_j is a k-ary relation on D ($1 \leq j \leq l$), $b_1, \ldots, b_k \in D'$, and $c_1, \ldots, c_k \in D$ are such that $ab_3(c_i) = b_i$, we let $R'_j(b_1, \ldots, b_m)$ hold if and only if $R_j(c_1, \ldots, c_m)$. (This extends the original definition further to $ab_3(R_j, S)$.) Finally, we put all of this together: $ab_3(S) =_{def} \langle a'_1, \ldots, a'_n, f'_1, \ldots, f'_m, R'_1, \ldots, R'_l, D' \rangle$. In other words:

(i') $ab_3(S) =_{def} \langle ab_3(a_1, S), \ldots, ab_3(a_n, S), ab_3(f_1, S), \ldots, ab_3(f_m, S),$
$ab_3(R_1, S), \ldots, ab_3(R_l, S), ab_3(D, S) \rangle$

I said above that our goal is for $ab_3(S)$ to "live entirely in U" and to be isomorphic to the system S from which it is derived. But there is a problem with the latter in general, as Linnebo & Pettigrew already noted. Namely, if S is non-rigid as a relational system (i.e., allows for non trivial isomorphisms), then the function ab_3 collapses distinct elements of S into the same element in $ab_3(S)$; and this prevents $ab_3(S)$ from being isomorphic to S (often already because of cardinality considerations).[25] In other words, our new abstraction principle (i'), based on (i), does not give us what we want in all cases. Then again, it works as intended for the two systems on which Dedekind focused—the natural numbers and the real numbers—since both of them are rigid.

There are several ways in which one can try to rectify the approach, at least to some degree. One can, for example, introduce additional constants for the crucial elements in S, thereby "rigidifying" the system artificially.[26] However, this has drawbacks when there are many of those crucial elements (perhaps uncountably many); and it is unsatisfactory in other ways as well. As an alternative, one can modify principle (i') by using, not abstraction on single elements a relative to S, but on corresponding sets of elements, thereby using "collective abstraction".[27] But unsatisfactory aspects remain again, which is one reason to consider the more radical forth alternative below.

---

[25] $\langle 1, +, \times, \mathbb{C} \rangle$ is often given as an example; but a simpler one is the unlabeled graph of two elements with no vertices. Cf. Linnebo & Pettigrew (2014).

[26] Examples are the introduction of the constant 'i' for the imaginary unit in the case of $\mathbb{C}$, and going from an unlabeled to a sufficiently labeled graph.

[27] Cf. Litland (unpublished).

But before moving on to that fourth approach, let me make explicit the other core aspect of the neo-Fregean reconstruction, so as to make our discussion of it parallel to those of the other reconstructions. This second aspect concerns how to analyze mathematical sentences p. Actually, after our treatment of this aspect in the neo-Hilbertian reconstruction it should be clear how to proceed in this case too. It is helpful, once again, to use the example of a sentence p for the natural numbers as a simple illustration:

(ii) $ab_3(p) =_{def} p(ab_3(\{\emptyset\}), ab_3(s), ab_3(\omega'))$

More generally, a mathematical sentence p in the language of a relational system S is analyzed as "the same" sentence for the image of S under ab_3. That is to say, we work with: $ab_3(p) =_{def} p(ab_3(a_1, S), \ldots, ab_3(a_n, S), ab_3(f_1, S), \ldots, ab_3(f_m, S), ab_3(R_1, S), \ldots, ab_3(R_l, S), ab_3(D, S))$.

As in the case of the neo-Russellian reconstruction, we can talk about $ab_3(S)$ as the "abstract structure" that corresponds to S. Note also that $ab_3(S)$ is different from all set-theoretic relational systems because, by construction, it "lives entirely in U". Indeed, since $ab_3(S)$ has been introduced as an abstract, we can say that it is "characterized by its relational or structural properties alone". In any case, ab_3 provides us with a third logical reconstruction of Dedekind's more informal talk about "abstraction". Two final comments: First, the neo-Fregean approach to Dedekind abstraction provides us not only with another reconstruction of Dedekind's remarks about "abstraction", but also of his remarks about "free creation". This corresponds to the fact that what results is a version of "non-eliminative structuralism". It also means, second, that there is additional textual evidence for our third approach to Dedekind, at least if one takes his remarks about "free creation" seriously.[28]

3.4 The Neo-Cantorian Reconstruction

The limitation we encountered for ab_3 in the case of non-rigid systems S may make one wonder if there is not another, less limited approach based on a corresponding "structural abstraction principle". In addition, there is a sense in which the neo-Fregean reconstruction works too much "from the bottom up" from a Dedekindian point of view, which is not entirely satisfactory either.[29] But how else could we proceed? To get more inspiration, let us go back to Dedekind's own remarks.

---

[28] In Sieg & Morris (2018), the claim is that Dedekind changed his mind in connection with "free creation", and thus, moved from a "non-eliminative" to an "eliminative" form of structuralism in his writings from the 1880s. But as indicated in an earlier footnote, this seems in tension with his repeated talk of "creation" in text from the 1880s.

[29] The neo-Fregean approach starts with individual elements obtained by abstraction and builds a structure out of them, as opposed to working with whole domains or systems of elements from the beginning. This is so even though individual elements a are always considered relative to a system S.

Recall that in the case of a simple infinity Dedekind proposes to "entirely neglect the special character of the elements; simply retaining their distinguishability and taking into account only the relations to one another in which they are placed by the order-setting function Φ" (Dedekind 1963, p. 68). At this point, the crucial phrase is: "simply retaining their distinguishability" when introducing new object. What that phrase suggests, on the reading to be pursued now, is to work with a set of "pure units" of the right cardinality, so as then to build "the natural numbers" out of them. Here "pure units" are meant to be mathematical objects only distinguished numerically but not in any other way, i.e., they are "qualitative indiscernibles". Is there a way to reconstruct this informal, intuitive idea more formally and logically?

In working out such an approach further, I will proceed more indirectly than in the previous three cases. Eventually, this will amount to an approach based on another "structural abstraction principle"—parallel to the neo-Fregean reconstruction but not identical with it—used to introduce a new abstraction operator ab_4. To prepare that introduction, I will first introduce two different abstraction operators, to be labeled ab_5 and ab_6, that are more closely related to the neo-Russellian and neo-Hilbertian reconstructions of Dedekind abstraction than to the neo-Fregean one. (In fact, these operators will be defined explicitly by using ab_1 and ab_2.) Eventually the approach will be given a further twist, however, which will lead to ab_4.

Suppose, once again, that we start with a set-theoretic relational system $S = \langle a_1, \ldots, a_n, f_1, \ldots, f_m, R_1, \ldots, R_l, D \rangle$, i.e., a relational system consisting of sets constructed in ZFC. In addition, suppose we have a set of urelements D' available that has the same cardinality as D, i.e., so that there exists a bijection g from D to D'. We pick such a set D' and bijection g. We now treat D' as the "abstract" that corresponds to D; i.e., we let $ab_5(D, S) = D'$. We also transfer the structural features from S to D'. This is done parallel to the neo-Fregean approach, but using g instead of ab_3. In other words: We let $a'_j = g(a_j)$, $(1 \leq j \leq n)$. (This defines $ab_5(a_j, S)$.) If f_j is a k-ary function on D $(1 \leq j \leq m)$, $b_1 \ldots, b_k \in D'$, and $c_1, \ldots, c_k \in D$ are such that $g(c_i) = b_i$ $(1 \leq i \leq k)$, we let $f'_j(b_1, \ldots, b_k) = g(f_j(c_1, \ldots, c_k))$. (This defines $ab_5(f_j, S)$.) And if R_j is a k-ary relation on D $(1 \leq j \leq l)$, $b_1, \ldots, b_k \in D'$, and $c_1, \ldots, c_k \in D$ are such that $g(c_i) = b_i$ $(1 \leq j \leq m)$, we let $R'_j(b_1, \ldots, b_m)$ hold if and only if $R_j(c_1, \ldots, c_m)$. (This defines $ab_5(R_j, S)$.) Putting all of this together, we stipulate $ab_5(S) = \langle a'_1, \ldots, a'_n, f'_1, \ldots, f'_m, R'_1, \ldots, R'_l, D' \rangle$. In other words:

(iii) $ab_5(S) =_{def} \langle ab_5(a_1, S), \ldots, ab_5(a_n, S), ab_5(f_1, S), \ldots, ab_5(f_m, S),$
$ab_5(R_1, S), \ldots, ab_5(R_l, S), ab_5(D, S) \rangle$

By proceeding thus, it is clear that $ab_5(S)$ will be isomorphic to S; and this is so even in the non-rigid case, since, as g is a bijection, it does not collapse any elements of S. Basically, we made the two sides isomorphic by construction.

While we thus avoid the main limitation of the neo-Fregean reconstruction, we end up with another problem. To see it, suppose we start with two isomorphic systems S and S'. For our purposes, or those of mathematical structuralism more generally, one would expect that $ab(S) = ab(S')$, i.e., the same "abstract" should correspond to each of them. But this is not guaranteed for ab_5 as just introduced. To be sure, we get: $ab_5(S) \cong ab_5(S')$. (This follows from three facts that are guaranteed, namely: $S \cong S'$; $S \cong ab_5(S)$; and $S' \cong ab_5(S')$.) Yet in the construction of $ab_5(S)$ and $ab_5(S')$ their respective domains may have nothing to do with each other, i.e., they may be disjoint. Having said that, it is at least consistent to assume the following: $(*)$ $ab_5(S) = ab_5(S') \leftrightarrow \exists f(f : S \cong S'))$; or so I want to argue now. I will provide three justifications for that claim, two of them informal and the third more formal. (Each will be informative in its own way.)

Assume, first and most informally, that what U contains are just "pure units", i.e., elements only distinguished numerically but not qualitatively. (Grant me for the moment that this is coherent; it will be justified further below.) Then the domains of our systems $ab_5(S)$ and $ab_5(S')$ consist of such pure units too. But then, all we could ever establish, it would seem, is that the elements of $ab_5(S)$ are distinct from each other, but not that any element of $ab_5(S)$ is distinct from any element of $ab_5(S')$. Second, put aside the idea of "pure units". Instead, assume that U contains simply ordinary urelements, but again enough of them for our purposes. In addition, assume that we do not know anything about these urelements, although they do have distinguishing properties now. Then it seems again to follow that $(*)$ can at least not be proved wrong. Both of these are relatively weak arguments and results, of course.

Next, let us proceed more formally and precisely. Assume still that we are working in ZFC with enough urelements. At this point, assume we have a way to ascertain the specific identities of these urelements, in one way or another, i.e., we know about their specific identities.[30] However, we now replace ab_5 by a closely related operator ab_6, constructed in a two=step process. Given some set-theoretic system S, we first consider $ab_2(S)$, i.e., we map S onto $ch(ec(S))$. Second, we perform the construction just described for ab_5 but now starting with $ab_2(S)$. In other words, we work with:

(iv) $ab_6(S) =_{def} ab_5(ab_2(S))$

Suppose again that we are given two isomorphic systems S and S' as arguments. Then it is clear, by construction, that ab_6 will lead to the same result in both cases, i.e., it will give the same "abstract" as their values.[31]

---

[30]For this purpose, we can use simple "duplicates" of enough elements of V. There are various ways to explicated this idea, e.g., by adding just one urelement u and then using pairs $\langle a, u \rangle$, for all a in V, to play the role of the needed new (here mixed) elements.

[31]In step one, we will be lead to the same representative $ab_2(S)$, since S and S' determine the same equivalence class, $ec(S)$, on which ch acts; and step two will coincide exactly for both. In other words, from $S \cong S'$ we get $ab_2(S) = ab_2(S')$; and hence, $ab_6(S) = ab_5(ab_2(S)) = ab_5(ab_2(S')) = ab_6(S')$.

At this point, is is tempting to work directly with ab_6, i.e., to make it our fourth main abstraction operator. But there are reasons to resist that idea. The main reason is that ab_6 depends too closely on the particular choice of the initial isomorphism g, for each relation system S, to be fully adequate.[32] Instead, this is where we switch gears and use a more "axiomatic" approach, parallel to the neo-Fregean reconstruction. Namely, we introduce an abstraction operator ab_4 that is assumed to satisfy the following abstraction principle:

(i) $[ab_4(S) = ab_4(S') \leftrightarrow \exists f(f : S \cong S')] \wedge [ab_4(S) \cong S]$

As we are now proceeding by means of an "implicit definition", as opposed to defining a_4 explicitly, the crucial question is whether introducing such an abstraction operator is consistent or not. (The same question arises for ab_3, since it was also introduced via an "implicit definition".) The answer is: It is (relatively) consistent, and the construction of ab_6 just considered can serve as a semantic consistency proof for it. (That is why we considered ab_6, which builds on ab_5, in the first place.)

Note the following right away: First, while this approach is similar to the neo-Fregean reconstruction in some respects—especially by working again with a "structuralist abstraction principle"—our new principle (i) does not have the form of a neo-Fregean abstraction principle as usually conceived (i.e., simply involving an equivalence relation on the right-hand side of a biconditional). Instead, it has a more complicated logical form. Second, our new approach provides us directly with an "abstract" for the whole system S, unlike in the neo-Fregean approach. (Note that we do not have to add a separate step (i') here, like earlier.) This corresponds to the fact that we now work "from the top down", not "from the bottom up". And third, the elements of $ab_4(S)$ will be "pure units" again, i.e., they will be "qualitative indiscernibles".

To complete the description of this fourth reconstruction of Dedekind abstraction, a few further remarks are in order. To begin with, what about a corresponding way of analyzing mathematical sentences p? Actually, after our earlier discussion it should be obvious how to come up with such an analysis, namely parallel to the neo-Hilbertian and neo-Fregean reconstructions. In the simple case of arithmetic, this means:

(ii) $ab_4(p) =_{def} p(ab_4(\{\emptyset\}), ab_4(s), ab_4(\omega'))$

Similarly for mathematical sentences p corresponding to other languages and theories, i.e., the approach generalizes just like before.

Next, what justifies calling this approach a "neo-Cantorian reconstruction" of Dedekind abstraction? The reason is this: In his well-known article "Beiträge zur Begründung der transfiniten Mengenlehre I–II" (1895–97), Cantor introduces cardinal numbers corresponding to sets M as follows:

---

[32] This becomes especially clear when we are dealing with non-rigid systems S, where non-trivial isomorphisms for both S and $ab_6(S)$ exist.

> By the "power" or "cardinal number" of M we mean the general concept, which arises [...] from the set M, in that we abstract from the nature of the particular elements of M and from the order in which they are presented. [...] Since every single element m [of M], if we abstract from its nature, becomes a 'unit', the cardinal number [...] is a definite aggregate composed of units [...] (Cantor 1932, pp. 282–283).

Later in the article, Cantor introduces order types for linearly ordered sets M in the same way:

> By this we understand the general concept which arises from M when we abstract only from the nature of the elements of M, retaining the order of precedence among them. [...] Thus, the order type [...] is itself an ordered set whose elements are pure units [...] (*ibid.*, p. 297).

If we replace the phrase "general concepts" by "relational systems" (including the case of a set with no constants, functions, and relations defined on it), these Cantorian remarks sound like a direct application, or adaptation, of Dedekind abstraction, with the language of "pure units" added.

But is it really such a good idea to appeal to "pure units" in this context? Such attempts have not been viewed positively at least since Frege's criticism of them in his *Die Grundlagen der Arithmetik* (1884). In fact, they have often been dismissed as incoherent. While this is true historically, one can respond as follows: In recent years, the pendulum has started to swing in the other direction, in the sense that several informal defenses and more formal reconstructions of "qualitative indiscernibles" have been proposed, so that they have become more respectable again.[33] My neo-Cantorian reconstruction of Dedekind abstraction constitutes a contributing to that shift. Moreover, admitting such objects seems inevitable if one want to be able to treat the case of non-rigid systems S along non-eliminative structuralist lines.[34]

4 Further Discussion and Comparisons

So far, I sketched four logical reconstructions of Dedekind abstraction, corresponding to the abstraction operators ab_1, ab_2, ab_3, and ab_4. Of these, the

---

[33] Cf. Assadian (2018) and the references in it.

[34] Besides the defense of Cantor and Dedekind in Tait (1996), cf. the more formal reconstruction of Cantor in Fine (1998), which explicitly involves "pure units". Another related approach is the form of structuralism developed in still unpublished work by Hannes Leitgeb, based on graph theory. Finally, there seems to be a connection to Univalent Foundation. Namely, the Univalence Axiom proposed in the recent literature might be seen as an analogue to, or strong generalization of, principle (i) in the neo-Cantorian approach; cf. Awodey (2014). I am planning to compare my neo-Cantorian reconstruction of Dedekind abstraction more to such approaches in the future, thereby working out its details more fully.

neo-Cantorian approach is the most original, while the others can be found, more or less explicitly, in the literature. I compared all four approaches to some degree already, but a more systematic comparison seems called for. This will concern both historical and systematic aspects.

Starting with the systematic side, let me discuss several general constraints one can adopt in connection with Dedekind's approach, or with structuralist approaches to mathematics more generally. Suppose, once more, that S and S' are relational systems constructed in ZFC, also that we are considering an abstraction operator ab on them. Now consider the following four constraints:

(1) $S \cong S' \leftrightarrow ab(S) = ab(S')$

(2) $S \cong ab(S)$

(3) $S' \neq ab(S)$ for all S' in V

(4) $ab(S)$ is characterized by its structural features alone.

How do our four reconstructions fare with respect to these constraints?

It is clear that ab_1 satisfies condition (1), since isomorphic systems determine the same equivalence class. If we work with a fixed choice function ch, as intended above, then (1) is also satisfied by ab_2, as is not hard to see. (1) is true for the operator ab_3 as well; indeed, this is ensured by construction. Concerning our fourth operator and approach, the situation was more interesting. Given the way in which we proceeded initially, via ab_5, there was no guarantee that (1) would hold. But I argued that for the modified operator ab_6 condition (1) is at least consistent. And in the end, we worked with ab_4, as introduced by the corresponding abstraction principle (i), which ensures condition (1) directly. In contrast, for the other approaches (1) is a more indirect, derived feature. That fact, in itself, reveals a main difference between ab_4 and the other three abstraction operators.

With respect to condition (2), the situation looks different in several respects. (2) is false for ab_1, since a set-theoretic system S and the corresponding equivalence class $ec(S)$ are clearly not isomorphic. On the other hand, (2) is true (in full generality) for ab_2, by construction. For ab_3, (2) holds for rigid systems S, while it fails for the non-rigid cases, as Linnebo & Pettigrew pointed out. In contrast, (2) is true for ab_4 (in full generality), once more by construction, assuming it is consistent. Note also that there is a close parallel between ab_2 and ab_4, as the argument for the (relative) consistency of (1) above indicates. On the other hand, ab_2 and ab_4 are quite different in other ways, especially insofar ab_2, like ab_1, is introduced by an explicit definition, while ab_4, like ab_3, is defined "implicitly" or "axiomatically".

Next, let us consider conditions (3) and (4) together. What (3) says is that $ab(S)$ is different from all set-theoretic relational systems. This is true for ab_1, because sets and proper classes are different.[35] In contrast, (3) is false

---

[35] If we use "Scott's trick" to replace the relevant proper classes by sets, this changes.

for ab_2, since $ab_2(S)$ is a set-theoretic relational system, even thought we don't know which one. For ab_3 and ab_4 condition (3) is true by construction, since we introduced them as functions from SV to U. Moreover, for both ab_3 and ab_4 condition (4) is true, because in each case abS is characterized completely by a structural abstraction principles (in two different ways). In contrast, for ab_1 and ab_2 condition (4) is false, since here $ab(S)$ is a set- or class-theoretic objects that has additional, non-structural characteristics. Finally, both ab_3 and ab_4 lead to "non-eliminative" structuralist positions, while ab_1 and ab_2 lead to versions of "eliminative" structuralism.

Which of our four reconstructions is grounded most firmly in Dedekind's texts, i.e., which of them is most defensible as an interpretation of Dedekind? I already indicated that each has some textual support. Thus, the neo-Russellian approach picks up on the observation, in Dedekind's Remark 134, that "every theorem in which we leave entirely out of consideration the special character of the elements n and discuss only such notions as arise from the arrangement Φ, possesses perfectly general validity for every other simply infinite system Ω". It interprets Dedekind as saying that every arithmetic theorem reveals itself, if analyzed, as a theorem about all simple infinities. Similarly, this approach can make good sense of Dedekind's suggestion to "neglect the special character of the elements" in a given simple infinity, namely by generalizing over all of them. The neo-Hilbertian reconstruction makes sense of the latter remark somewhat differently, namely by replacing an initially constructed simple infinity by an arbitrarily chosen one, where we don't know anything about the "special character of the elements", except that we are still dealing with set-theoretic objects.

Both along neo-Fregean and neo-Cantorian lines, Dedekind's remarks about "free creation", in connection with his remarks about "abstraction", are taken much more seriously than along neo-Russellian and neo-Hilbertian lines. However, doing so does not commit one to psychologism. Instead, what results in each case is a logical reconstruction, via a respective abstraction operator and corresponding structural abstraction principle. From a neo-Fregean and a neo-Cantorian perspective, what Dedekind meant to "create" was distinguished, *sui generis* "abstracts", different from the initial set-theoretic systems. For each of them, this takes the form of working with urelements, or with systems of such urelements, introduced via the structural abstraction principles. Consequently, we end up with forms of "non-eliminative structuralism", which satisfy constraints (3) and (4), in addition to (1) and (2).

Given all of this, how should Dedekind be interpreted overall? Interpretive charity would seem to require not to read him in an problematic psychologicist way, at least if there are alternatives. Whether to interpret him more along neo-Fregean or neo-Cantorian than neo-Russellin or neo-Hilbertian lines depends on how seriously we take his talk about "creation". As he only makes a few (pregnant but ambiguous) remarks in this connection, it is hard to be sure. It is also possible that Dedekind changed his relevant views over time. Whether that is the case is a subtle matter, I believe, one I do not intend to decide conclusively in the present paper.[36] Having said that, I hope that the main interpretive choices available in this context have become clearer, namely by revealing their underlying "logic".

Supposed one is inclined to interpret Dedekind as a non-eliminative structuralist. Is the neo-Fregean or the neo-Cantorian reconstruction closer to his texts? Three arguments speak in favor of the neo-Cantorian option, I believe. First, Dedekind's approach seems "top down", corresponding to the neo-Cantorian reconstruction, than "buttom up", as represented by the neo-Fregean reconstruction. Second, we already noted Dedekind's remark about "neglecting the special character of the elements" in a simple infinity while "retaining their distinguishability", which seems to point to the idea of "pure units". Third, there is the historical link between Dedekind and Cantor, including the fact that they may have influenced each other. Actually, the precise relationship between Dedekind and Cantor concerning this point is a question that seems worth more historical research; and that may lead to further insights into how to interpret either one of them.[37]

Unlike Cantor, Frege was quite skeptical about "pure units". How might Frege have reacted to a neo-Fregean reconstruction of Dedekind, though? This is a very speculative question. But there is one striking detail in Frege's *Grundgesetze der Arithmetik, Vol. II* that may be worth mentioning. After having criticized a number of views about "creation" in mathematics vehemently, in Section III of that book, Frege contrasts them with his own approach in which classes are introduced as extensions of concepts (or as value ranges of functions). Now, his crucial Basic Law V has the form of a neo-Fregean abstraction principle. Also, at this exact point in the discussion he asks the following question: "Can our procedure be called a creation?" (Frege 1903, p. 149) And quite surprisingly, he does not reject this view outright; instead he responds: "The discussion of this question can easily degenerate into a verbal quarrel. In any case, our creation, if one wishes so to call it, is not unconstrained and arbitrary, but rather the way of proceeding, and its permissibility, is settled once and for all (*ibid.*)". I take Frege's main point to be that a systematic way of introducing mathematical objects is called for, and specifically, one working with explicit basic laws. But if so, he may have found our neo-Fregean reconstruction of Dedekind not only worth investigating but even congenial.

Frege did not, of course, work within a set-theoretic framework like ZFC, but in higher-order logic supplemented by Basic Law V. This brings us back to the choice of a general framework. My discussion in this paper was framed in terms of ZFC (plus proper classes and urlements). But this was mostly for ease of presentation and because of its relative familiarity. Indeed, higher-

---

[36]In Reck (2003), I argued that Dedekind's remarks about "creation", as closely tied to "abstraction", should be taken seriously. In Sieg & Morris (2018), a careful case is made that Dedekind changed his corresponding views over time.

[37]Tait (1996) can serve as the starting point for such an investigation. A related question is whether my neo-Cantorian reconstruction of Dedekind is closest to the interpretation of both Cantor and Dedekind given in Tait (1996). I assume so, but am not altogether sure.

order logic provides an almost as familiar and general background theory for sets/classes and functions too, at least if supplemented with existence principles for them. This suggests that one could present each of my four approaches within such a framework as well, and especially, the neo-Fregean and neo-Cantorian reconstructions of Dedekind abstraction.[38] In fact, what I meant to provide in this paper was four general recipes for how to implement approaches compatible with Dedekindian remarks, given some suitable formal framework.

5 Summary and Conclusion

Let me summarize the discussion in this paper briefly. We started with Richard Dedekind's remarks about "abstraction" and "free creation" in his essay *Was sind und was sollen die Zahlen?* and related writings. While these remarks are often dismissed as a problematic form of psychologism, the paper picked up on a suggestion by W. W. Tait, earlier also by Ernst Cassirer, to interpret them in a "logical" way instead. This is not the only possible defense of Dedekind against the psychologism charge, as a brief interlude about a more pragmatic reading of Dedekind indicated. But even with respect to that reading, the question of its "logical core" arises.

As the paper then illustrated, there are four different ways to reconstruct Dedekind abstraction "logically". For reasons provided along the way, these were called the neo-Russellian, the neo-Hilbertian, the neo-Fregean, and the neo-Cantorian reconstruction, respectively. Each of them was specified in terms of two ingredients: (i) the logical form Dedekind abstraction on relational systems S takes, as spelled out in terms of a corresponding abstraction operator ab; (ii) the way in which mathematical formulas p are re-analyzed accordingly. The resulting positions were compared further, both with respect to the forms of structuralism they embody and their Dedekindian credentials.

What was my basic goal in providing all these reconstructions of Dedekind? The present paper is part of a bigger effort of providing a philosophical interpretation of Dedekind's works. There is thus a more general exegetic project in the background. Yet it seems to me that, before one can argue conclusively for a particular interpretation of any thinker, one should clarify what the main options are. It also helps to work out these options in formal detail, since certain ambiguities or fine distinctions become evident only that way. The most basic goal of this paper was, then, to do the preliminary work of exploring the space of alternatives for interpreting Dedekind abstraction.

Beyond questions of exegesis, one may wonder how Dedekind's remarks, or approaches inspired by them, fit into contemporary debates about structuralism. As we saw, several Dedekindian forms of structuralism are possible, with

---

[38] Another option would be to work within constructive type theory, as introduced by Per Martin-Löf. In that context, one may again see Univalent Foundations as a natural development of my neo-Cantorian approach to Dedekind. It might even be possible to trace a historical line from Dedekind through category theory to UF.

different strengths and weaknesses. By "Dedekindian forms of structuralism" I mean positions that work with the kind of abstraction operators our four reconstructions illustrate. Seen from that perspective, the basic outcome of the present paper is that Dedekind's talk of "abstraction" can indeed be reconstructed "logically", as W. W. Tait suggested, and that doing so reveals its continuing systematic relevance.[39]

References

Assadian, Bahram (2018): "In Defense of Utterly Indiscernible Entities", *Philosophical Studies*, online first.

Awodey, Steve (2014): "Structure, Invariance, and Univalence", *Philosophia Mathematica* 22, 1–11.

Burgess, John (2015): *Rigor and Structure*, Oxford University Press.

Cantor, Georg (1895–97): "Beiträge zur Begründung der transfiniten Mengenlehre, I, II", *Mathematische Annalen* 46, 481–512; reprinted in Cantor (1932), pp. 282–256.

——(1932): *Gesammelte Abhandlungen mathematischen und philosophischen Inhalts*, E. Zermelo, ed., Berlin: Springer.

Cassirer, Ernst (1910): *Substanzbegriff und Funktionsbegriff*, Berlin: Bruno Cassirer; English trans., *Substance and Function*, Chicago: Open Court, 1923.

Dedekind, Richard (1872): *Stetigkeit und Irrationale Zahlen*, Braunschweig: Vieweg; reprinted in Dedekind (1930–32), Vol. 3, pp. 315–334; English trans., *Continuity and Irrational Numbers*, in Beman (1963), pp. 1–27.

——(1876): "Briefe an Lipschitz", Dedekind (1930–32), Vol. 3, pp. 468–479.

——(1888a): *Was sind und was sollen die Zahlen?*, Braunschweig: Vieweg; reprinted in Dedekind (1930–32), Vol. 3, pp. 335–391; English trans., *The Nature and Meaning of Numbers*, in Beman (1963), pp. 31–115.

——(1888b): Brief an Webe, Dedekind (1930–32), Vol. 3, pp. 488–490.

---

[39]This paper has been in the works for a while. Earlier versions were presented in a number of contexts, including: the Munich Center for Mathematical Philosophy, October 2012; the Montréal Inter-University Workshop on the History and Philosophy of Mathematics, November 2012; the *Frege-Dedekind Fest*, University of California at Irvine, April 2016; and the conference *Varieties of Mathematical Abstraction*, University of Vienna, August 2018. I am grateful for the comments I received at these events, as well as for the invitations to participate in them in the first place. I am especially indebted to Øystein Linnebo and Georg Schiemer, both for their own work on this topic and for their constructive comments on my approach. The remaining problems should be attributed entirely to me, of course.

—(1930–32): *Gesammelte Mathematische Werke*, Vols. 1–3, R. Fricke, E. Noether & Ø. Ore, eds., Braunschweig: Vieweg.

—(1963): *Essays on the Theory of Numbers*, W.W. Beman, ed. and trans., Dover: New York.

Dummett, Michael (1991): *Frege: Philosophy of Mathematics*, Harvard University Press.

Fine, Kit (1998): "Cantorian Abstraction: A Defense and Reconstruction", *Journal of Philosophy* 95, 599–634.

Frege, Gottlob (1884): *Die Grundlagen der Arithmetik*, Jena: Pohle; English trans., *The Foundations of Arithmetic*, J.L. Austin, ed. and trans., Oxford: Basil Blackwell, 1950.

—(1903): *Grundgesetze der Arithmetik, Band II*, Jena, Pohle; English trans., *Basic Laws of Arithmetic*, P. Ebert & M. Rossberg, eds. and trans., Oxford University Press, 2014.

Hellman, Geoffrey (1989): *Mathematics without Numbers*, Oxford University Press.

Heis, Jeremy (forthcoming): "'If Numbers are to be Anything at All, They Must be Intrinsically Something': Bertrand Russell and Mathematical Structuralism", in *The Pre-History of Mathematical Structuralism*, E. Reck & G. Schiemer, eds., Oxford University Press.

Klev, Ansten (2018): "A Road Map to Dedekind's Theorem 66", *HOPOS: The Journal of the International Society for the History of Philosophy of Science* 8, 241–277.

Linnebo, Ø. & Pettigrew, R. (2014): "Two Types of Abstraction for Structuralism", *Philosophical Quarterly* 64, 267–283.

Litland, Jon (unpublished): "Collective Abstraction".

Parsons, Charles (1990): "The Structuralist View of Mathematical Objects", *Synthese* 84, 303–346.

Reck, Erich (2003): "Dedekind's Structuralism: An Interpretation and Partial Defense", *Synthese* 137, pp. 369–419.

—(2013): "Frege or Dedekind? Towards a Reevaluation of their Legacies", in *The Historical Turn in Analytic Philosophy*, E. Reck, ed., London: Palgrave, pp. 139–170.

—(forthcoming a): "Frege's Relation to Dedekind: Basic Laws and Beyond", in *Essays on Frege's Basic Laws of Arithmetic*, P. Ebert & M. Rossberg, eds., Oxford University Press.

——(forthcoming b): "Cassirer's Reception of Dedekind and the Structuralist Transformation of Mathematics", in *The Pre-History of Mathematical Structuralism*, E. Reck & G. Schiemer, eds., Oxford University Press.

Reck, E. & Keller, P. (forthcoming): "From Dedekind to Cassirer: Logicism and the Kantian Heritage", in *Kant's Philosophy of Mathematics, Vol. II*, O. Rechter & C. Posy, eds., Oxford University Press.

Reck, E. & Price, M. (2000): "Structures and Structuralism in Contemporary Philosophy of Mathematics", *Synthese* 125, 341–383.

Russell, Bertrand (1901): "Recent Work in the Philosophy of Mathematics", *International Monthly*, 4, 83–101; reprinted, as "Mathematics and the Metaphysicians," in *Mysticism and Logic and Other Essays*, New York: Longmans, 1918, pp. 74–96.

——(1903): *Principles of Mathematics*, Cambridge University Press.

Schiemer, G. & Gratzl, N. (2016): "The Epsilon-Reconstruction of Theories and Scientific Structuralism", *Erkenntnis* 81, 407–432.

Shapiro, Steward (1997): *Philosophy of Mathematics: Structure and Ontology*, Oxford University Press.

Sieg, W. & Morris, R. (2018): "Dedekind's Structuralism: Creating Concepts and Deriving Theorems", this volume, pp. 251–301.

Tait, W.W. (1996): "Frege versus Cantor and Dedekind: On the Concept of Number", in *Frege: Importance and Legacy*, M. Schirn, ed., Berlin: De Gruyter, pp. 70–113.

Yap, Audrey (2014): "Dedekind, Cassirer, and Mathematical Concept Formation", *Philosophica Mathematica*, online first.

Part III:
History of Logic and Philosophy of Mathematics

Carnap's Philosophy of Logic and Mathematics

Michael Friedman[*]

ABSTRACT: I argue that the application of logic and mathematics in empirical science is central to Carnap's distinctive approach to this subject. The main focus is on his monograph *Foundations of Logic and Mathematics* (1939), which is equally an important milestone in the development of Carnap's philosophy of science. Moreover, since this monograph draws essentially on his new semantical approach to metamathematics, I argue that this particular approach then allows Carnap to articulate a considerably more nuanced and sophisticated approach to the relationship between logico-mathematical foundations and empirical science than he had in *Logical Syntax of Language* (1934).

Our understanding of Carnap's philosophy of logic and mathematics has been decisively shaped by Quine's well-known criticism of the analytic/synthetic distinction. Quine's "Carnap and Logical Truth" portrays the issue as fundamentally epistemological: Carnap's problem was to explain the special kind of certainty that logic and mathematics were taken to possess, and his con-

---

[*]This paper is a companion to M. Friedman, "From Intuition to Tolerance: The Development of Carnap's Philosophy of Mathematics," in C. Posy and O. Rechter, eds., *Kant's Philosophy of Mathematics, Vol. II: Reception and Influence After Kant* (Cambridge: University Press, forthcoming). It was prompted by a question raised by W. W. Tait at the conference (in honor of Charles Parsons) at which the latter paper was originally presented in December 2013. Tait's question concerned the relationship between Carnap's earlier work in the philosophy of geometry and physics with which that paper begins and his later work in foundations of mathematics proper. The present paper attempts to address Tait's question by explaining in detail the importance of the application of mathematics in empirical science to Carnap's own conception of foundations. I am both pleased and grateful, therefore, to be able to include this paper in a volume in honor of Tait. I am also grateful for valuable comments on an earlier draft by William Demopoulos.

ception of analyticity was intended precisely to provide such an explanation. Quine begins with Kant's question how synthetic a priori judgements are possible, replaces this question (in light of the logicist reduction of mathematics to logic) with the question, "How is logical certainty possible?", and asserts that "[i]t was largely this latter question that precipitated the form of empiricism we associate with between-war Vienna—a movement which began with Wittgenstein's *Tractatus* and reached its maturity in the work of Carnap."[1] The answer it found, according to Quine, was "the linguistic doctrine of logical truth" (p. 386): "What now of the empiricist who would grant certainty to logic, and to the whole of mathematics, and yet would make a clean sweep of other non-empirical theories under the name of metaphysics? The Viennese solution of this nice problem was predicated on language. Metaphysics was meaningless through misuse of language; logic was certain through tautologous use of language."

I shall argue that Quine's portrayal is seriously misleading (compare note 1). Carnap was not motivated by the traditional concern for epistemic certainty that appears of most interest to Quine. Rather, Carnap's conception of analyticity was intended, above all, to make clear that both logic and mathematics are empty of factual content. And the point of this idea, in turn, was not to explain the special kind of epistemic security possessed by these disciplines, but rather to emphasize our complete *freedom of choice* concerning which rules of logic and mathematics to adopt. The choice of such rules, more specifically, is subject only to the purely pragmatic constraint of their utility or efficacy in formulating the genuinely contentful part of our overall system of knowledge, namely, the empirical part. This idea, as expressed in what Carnap came to call the Principle of Tolerance, is what is most important in his distinctive version of empiricism.

Carnap, Wittgenstein, and Frege

Quine portrays Carnap as combining the logicism of Frege and Russell with the basic ideas of Wittgenstein's *Tractatus*. This portrayal is not baseless, for Carnap presents the same picture in his Intellectual Autobiography.[2] He describes

---

[1] Quine's paper was written in 1954 for inclusion in the Carnap volume of the Library of Living Philosophers: see P. Schilpp, ed., *The Philosophy of Rudolf Carnap* (La Salle: Open Court, 1963), pp. 385–406. I cite the paper from this volume (here p. 385). It is worth noting that Quine appears to recognize that his construal of Carnap does not necessarily match Carnap's own view, for the paper begins with a striking disclaimer (ibid.): "My dissent from Carnap's philosophy of logical truth is hard to state and argue in Carnap's terms. This circumstance perhaps counts in favor of Carnap's position. At any rate, a practical consequence is that, though the present essay was written entirely for this occasion, the specific mentions of Carnap are few and fleeting until well past the middle. It was only by providing thus elaborately a background of my own choosing that I was able to manage the more focussed criticisms in the later pages." This disclaimer does not appear in later reprintings of the paper.

[2] See P. Schilpp, *op. cit.* (note 1), pp. 3–84; all citations are from this volume.

how a combination of Frege-Russell logicism with Wittgenstein's conception of tautology allowed the members of the Vienna Circle to arrive "at the conception that all valid statements of mathematics are analytic in the specific sense that they hold in all possible cases and therefore do not have any factual content" (p. 47), and Carnap goes on the explain how this conception resulted in a major advance over all earlier forms of empiricism:

> What was important in this conception from our point of view was the fact that it became possible for the first time to combine the basic tenet of empiricism with a satisfactory explanation of the nature of logic and mathematics. Previously, philosophers had only seen two alternative positions: either a non-empiricist conception, according to which knowledge in mathematics is based on pure intuition or pure reason, or the view held, e.g., by John Stuart Mill, that the theorems of logic and of mathematics are just as much of an empirical nature as knowledge about observed events, a view which, although it preserved empiricism, was certainly unsatisfactory. (ibid.)

This passage, at least at first sight, appears to fit Quine's portrayal rather well. The new form of empiricism developed by the Vienna Circle aimed to explain the nature of the knowledge provided by logic and mathematics in a way that preserved the traditional certainty and necessity attributed to them within the rationalist (and Kantian) tradition while also avoiding all "metaphysical" appeals to either "pure reason" or "pure intuition." Logic and mathematics consist solely of tautologies—statements that are true in all possible states of affairs and, in this sense, are therefore certainly and necessarily true. They are true, as Quine suggests, "through tautologous use of language."

Carnap explains his debt to Wittgenstein's *Tractatus* somewhat more fully earlier in his Autobiography:

> For me personally, Wittgenstein was perhaps the philosopher who, besides Russell and Frege, had the greatest influence on my own thinking. The most important insight I gained from his work was the conception that the truth of logical statements is based only on their logical structure and the meaning of the terms. Logical statements are true under all conceivable circumstances; thus their truth is independent of contingent facts of the world. On the other hand, it follows that these statements do not say anything about the world and thus have no factual content. (p. 25)

Yet here Carnap's conclusion is in tension with Quine's portrayal. For Carnap does not infer from logical statements being true "under all conceivable circumstances" that they are certainly and necessarily true. Carnap does not mention certainty at all, but instead emphasizes that logical statements are therefore in an important sense empty, lacking in "factual content," and say nothing whatsoever about the world. Indeed, Carnap emphasizes essentially the same point in the first passage quoted from his Autobiography above, where he explains that analytic statements "hold in all possible cases and *therefore* do not have any factual content" (p. 47, emphasis added).

Even more revealing, however, is the way in which Carnap describes his debt to Frege towards the beginning of the Autobiography. From Frege, Carnap says, he "gained the conviction that knowledge in mathematics is analytic in the general sense that it has essentially the same nature as knowledge in logic" (p. 12). But Carnap manifestly does not mean that we can thereby justify or explain the special epistemic status of mathematical knowledge on the basis of another type of knowledge—logical knowledge—resting on firmer or more certain grounds. Carnap instead goes on to characterize what he learned from Frege as the idea that logic and mathematics together play a distinctively formal or inferential role in framing or structuring our *empirical* knowledge:

> It is the task of logic and mathematics within the total system of knowledge to supply the forms of concepts, statements, and inferences, forms which are then applicable everywhere, hence also to non-logical knowledge. It follows from these considerations that the nature of logic and mathematics can be clearly understood only if close attention is given to their applications in non-logical fields, especially in empirical science. Although the greater part of my work belongs to the fields of pure logic and the foundations of mathematics, nevertheless great weight is given in my thinking to the application of logic to non-logical knowledge. This point of view is an important factor in the motivation for some of my philosophical positions, for example, for the choice of forms of languages, for my emphasis on the fundamental distinction between logical and non-logical knowledge. (pp. 12–13)

Thus the problem concerning "the nature of logic and mathematics" that most interests Carnap is quite different from that which most interests Quine. Understanding the nature of these fundamental disciplines, for Carnap, does not involve explaining their certainty or epistemic security, but rather involves paying "close attention [...] to their applications in non-logical fields, especially in empirical science." It is for precisely this reason, Carnap suggests, that he places particular emphasis on both the analytic/synthetic distinction ("the fundamental distinction between logical and non-logical knowledge") and the Principle of Tolerance (which involves a "choice" between "forms of languages").

Logical Syntax and the Principle of Tolerance

It is in *Logical Syntax of Language*, originally published in 1934, that Carnap first officially formulates the Principle of Tolerance.[3] It is formulated in §17 as "[*i*]*t is not our business to set up prohibitions, but to arrive at conventions*" (p. 51), and, at the end of this section, it is explained more fully (p. 52): "*In logic, there are no morals.* Everyone is at liberty to build up his own logic, i.e., his own form of language, as he wishes. All that is required of him is that, if he wishes to discuss it, he must state his methods clearly, and give syntactic rules instead of philosophical considerations." The preceding §16, "On Intuitionism," makes it clear that Carnap's primary motivation for formulating this Principle derives from the recent (and ongoing) debate about intuitionist and constructivist tendencies in the foundations of mathematics, according to which the rules of classical logic may not be applied in full generality when dealing with infinite totalities like the natural numbers. For example, establishing the falsity of 'all numbers have property *P*' by *reductio ad absurdum* and then inferring 'there exists a number with property not-*P*' via the law of excluded middle is not, according to this view, logically valid. For we are only justified in asserting the existence of a number with any property whatsoever when we have actually constructed or exhibited such a number by finitary means.

Carnap's point, against this view, is not so much that it operates with an incorrect notion of logical validity or a false picture of mathematical existence, but rather that there is no question of correct or incorrect at all. There is only the question of which rules of logic and mathematics we wish to lay down, together with the purely pragmatic question of which such choice is more expedient for this or that particular purpose. In his Autobiography, once again, Carnap provides a succinct summary of the lesson of *Logical Syntax*:

> According to my principle of tolerance, I emphasized that, whereas it is important to make distinctions between constructivist and non-constructivist definitions and proofs, it seems advisable not to prohibit certain forms of procedure but to investigate all practically useful forms. It is true that certain procedures, e.g., those admitted by constructivism or intuitionism, are safer than others. Therefore it is advisable to apply these procedures as far as possible. However, there are other forms and methods which, though less safe because we do not have a proof of their consistency, appear to be practically indispensable for physics. In such a case there seems to be no good reason for prohibiting these procedures so long as no contradictions have been found. (p. 49)

---

[3]See R. Carnap, *Logische Syntax der Sprache* (Wien: Springer, 1934); translated as *The Logical Syntax of Language* (London: Kegan Paul, 1937)—page references are to this edition.

Thus, if one is aiming above all to avoid contradictions, the weaker rules of intuitionistic logic and mathematics are safer than the stronger classical rules.[4] If, however, one is interested in a perspicuous mathematical framework for physics, one should stay with classical logic and mathematics so long as no contradictions have yet appeared. And, as his discussion makes clear, the primary purpose in accordance with which expediency is to be judged, for Carnap, is the role of logic and mathematics in formulating our total system of empirical knowledge, especially in physics. Apart from this role, in Carnap's view, there is simply no independent content conveyed by logic and mathematics at all—no independent "facts" relative to which their correctness or incorrectness may be judged.

This understanding of the Principle of Tolerance is confirmed in the later more explicitly philosophical sections of *Logical Syntax*. Section §82 concerns the "physical language," that is, the language of the total system of empirical science, especially physics. The key idea here is that we do not simply include "logical" terms within the primitive vocabulary but also "descriptive" terms, such as those for physical magnitudes like temperature, electric charge, and so on. In addition, we do not simply lay down logical and mathematical rules (L-rules) but also what Carnap calls "physical" rules (P-rules), such as, for example, Maxwell's equations. And the criteria for deciding which L-rules to lay down is basically the same as those for deciding which P-rules to accept, namely, their overall pragmatic fruitfulness and efficacy in formulating the total system of empirical science (p. 318): "No rule of the physical language is definitive; all rules are laid down with the reservation that they may be altered as soon as it seems expedient to do so. This applies not only to the P-rules but also to the L-rules, including those of mathematics. In this respect, there are only differences in degree; certain rules are more difficult to renounce than others."[5]

After an intervening §83 briefly touching on "the so-called foundations" of the various branches of empirical science (physics, biology, psychology, sociology), §84 explicitly turns to "the problem of the foundation of mathematics." Here, rather than once again discussing intuitionism, Carnap discusses logicism as founded by Frege and formalism as it has more recently been developed by Hilbert. Carnap characterizes the two apparently opposing views as follows:

> According to this [Hilbertian] view, mathematics and logic are constructed together in a common calculus; the question of freedom

---

[4]Here it is important to note that Carnap represents intuitionism and constructivism in *Logical Syntax* by his Language I: a formulation of what we now call primitive recursive arithmetic in which unbounded existential quantification over the numbers cannot even be expressed. This system is much weaker, therefore, than Heyting arithmetic.

[5]Here the question naturally arises of the precise difference between what Carnap is saying here and the holistic picture of theory testing Quine often appeals to in opposition to Carnap's analytic/synthetic distinction—most famously, of course, in "Two Dogmas of Empiricism." See note 6 below for how Carnap further distinguishes between analytic and synthetic sentences in *Logical Syntax*.

from contradiction is made the center of the investigation; the formal treatment (the so-called metamathematics) is carried out more strictly than before. As opposed to the formalist standpoint, Frege maintained that the logical foundations of mathematics has the task, not only of setting up a calculus, but also, and pre-eminently, of giving an account of the meaning of mathematical signs and sentences. He tried to perform this task by reducing the signs of mathematics to the signs of logic by means of definitions, and proving the sentences of mathematics by means of primitive sentences of logic with the help of logical fundamental laws [*Grundgesetze*]. (p. 325)

At this point, however, Carnap leaves aside the original Fregean idea of reducing mathematics to logic (p. 326): "[W]e are here not so much concerned with the question whether mathematics can be derived from logic or must be constructed simultaneously with it [à la Hilbert], as with the question whether the construction is to be of a purely formal nature, or whether the meaning of the signs must be determined."

What matters, for Carnap, is that we can now completely provide for determining the meaning of the signs by providing for the application of the mathematical calculus in empirical science. The problem with the Hilbertian view, according to Carnap, is that its "calculus does not contain all the sentences which contain mathematical signs and which are relevant for science, namely those sentences which are concerned with the *application of mathematics*, i.e., synthetic descriptive sentences with mathematical signs" (p. 326). And, when we include such sentences of applied mathematics, Carnap argues, we can simultaneously satisfy the demands of both formalism and logicism:

The [total] system must contain general rules of formation concerning the occurrence of the mathematical signs in synthetic descriptive sentences also, together with consequence-rules for such sentences. Only in this way is the application of mathematics, i.e., calculation with numbers of empirical objects and with measures of empirical magnitudes rendered possible and systematized. *A structure of this kind fulfills, simultaneously, the demands of both formalism and logicism.* For, on the one hand, the procedure is a purely formal one, and on the other, the meaning of the mathematical signs and thereby the application of mathematics in actual science is made possible, namely, by *the inclusion of the mathematical calculus in the total language*.... The *requirement of logicism is not fulfilled by a metamathematics (that is, by a syntax of mathematics) alone, but only by a syntax of the total language, which contains both logico-mathematical and synthetic sentences.* (pp. 326–27)

For Carnap, therefore, it is clear that his use of the Principle of Tolerance to settle the apparent opposition between formalism and logicism (by simultaneously satisfying the demands of both views as he understands them) essentially involves the idea that mathematics only acquires meaning or content when it is applied in empirical science—otherwise it is a mere "uninterpreted" calculus in the sense of the Hilbert school.

What kind of meaning does Carnap have in mind? What exactly is the otherwise lacking "interpretation"? Carnap is not operating here with the later idea of a semantical interpretation, for, as he suggests, the metamathematics in question is still purely syntactical. Carnap only accepts what he will then call semantical metalanguages a few years later, after learning of Tarski's new approach to semantics. I shall return to Carnap's semantical views below, but I shall first consider the kind of meaning that Carnap has in mind in *Logical Syntax*. Here it becomes clear, again in §82, that what Carnap has in mind is the *empirical meaning* provided by the inclusion in the physical language of "*protocol-sentences*, by means of which the results of observation are expressed" (p. 319). Thus, to take Carnap's own example, the descriptive sign for the electro-magnetic field is not explicitly definable from the signs that occur in protocol sentences. Nevertheless, we have ways of measuring the electro-magnetic field (using voltmeters, and the like), and these ways, for Carnap, will involve assigning (real) numbers to space-time points provisionally representing the electro-magnetic field. Although the state of the field will not deductively follow from protocol-sentences reporting the results of such measurements, these protocol-sentences will themselves deductively follow, according to Carnap, from Maxwell's equations plus other primitive sentences of classical physics (including statements of initial conditions, and so on). Statements about the electro-magnetic field can thereby be empirically tested and thus acquire empirical meaning.[6] It is precisely this circumstance that makes the sign for the electro-magnetic field in a proper formalization of Maxwell's theory a *descriptive* rather than merely *logical(-mathematical)* sign.[7]

---

[6]See §82 (p. 319): "Let protocol-sentences be the observation sentences of the usual form. The electric field vector of classical physics is not definable by means of the signs that occur in such protocol-sentences; it is introduced as a primitive sign by the Maxwell equations formulated as P-primitive sentences. There is no sentence equipollent to such an equation that contains only signs of the protocol-sentences, although, of course, sentences of the protocol form can be deduced from the Maxwell equations of classical physics; in this way the Maxwell theory is empirically tested." (According to §49, "equipollence" is defined in terms of sameness of content.) In reference to the issue raised in note 5, it is here that Carnap's conception of empirical testing differs essentially from Quine's. For, immediately after the quotation to which note 5 is appended, Carnap continues (pp. 318–19): "If, however, we assume that every new protocol-sentence which appears within a language is synthetic, there is this difference between an L-valid, and therefore analytic sentence S_1 and a P-valid sentence S_2, namely, that such a new protocol-sentence—independently of whether it is acknowledged as valid or not—can be, at most, L-incompatible with S_2 but never with S_1." This difference is related to the circumstance that logical terms, for Carnap, are *determinate* in a way that descriptive terms are not (compare note 7 below).

[7]For Carnap's general formal distinction between logical and descriptive expressions see §50. The intuitive idea is that we expect all sentences essentially containing only logical terms to be decidable in principle by the L-rules (independently of any empirical observations), but we do not expect that all sentences essentially containing descriptive terms will be similarly

Foundations of Logic and Mathematics in Semantics

In *Logical Syntax* Carnap had avoided the use of semantical concepts like truth and designation because of their apparent dependence on contingent empirical facts (§ 60b, p. 216): "[*T*]*ruth and falsity are not proper syntactical properties*; whether a sentence is true or false cannot generally be seen by its design, that is to say, by the kinds and serial order of the signs."[8] In Carnap's Autobiography § 10 ("Semantics") begins by citing "Tarski's great treatise on the concept of truth."[9] Carnap goes on to explain Tarski's impact on him in more detail (p. 60): "Even before the publication of Tarski's article I had realized, chiefly in conversations with Tarski and Gödel, that there must be a mode, different from the syntactical one, in which to speak about language. Since it is obviously admissible to speak about facts and, on the other hand, ... about expressions of a language, it cannot be inadmissible to do both in the same metalanguage." Yet the question remained how to do this within logic, without descending into empirical linguistics. Here Tarski's solution was both simple and surprising:

> When Tarski told me for the first time that he had constructed a definition of truth, I assumed that he had in mind a syntactical definition of logical truth or provability. I was surprised when he said that he meant truth in the customary sense, including contingent factual truth. Since I was thinking only in terms of a syntactical metalanguage, I wondered how it was possible to state the truth-conditions for a simple sentence like "this table is black". Tarski replied: "This is simple; the sentence 'this table is black' is true if and only if this table is black". (ibid.)

With this expedient Tarski had shown Carnap that the objection raised in *Logical Syntax* is incorrect. Although it is certainly the case that the truth of "this table is black" depends on the contingent empirical fact that the table is black, the concept of truth—which enables us to provide truth-conditions

---

decidable by the P-rules. Thus, for example, we take all sentences of arithmetic—including singular arithmetical formulas—to be decidable in principle by an adequate axiomatization of arithmetic and in this sense determinate, whereas we do not expect the same to be true for singular sentences ascribing particular values of the electro-magnetic field to particular space-time points in an adequate axiomatization of electro-magnetism. (One may well wonder, however, how the determinacy in question is consistent with Gödel's incompleteness theorem, and I shall return to this question below.)

[8]Sections 60a – e did not appear in the original German edition in 1934; their content was published in the same year as "Die Antinomien und die Unvollständigkeit der Mathematik," *Monatshefte für Mathematik und Physik* 41 (1934): 263–84, and it was then incorporated into the English translation in 1937. Here, as Carnap makes clear, he is much indebted to Gödel's recent incompleteness results, which had appeared in volume 38 of the same journal in 1931.

[9]See p. 60 of Carnap's Autobiography. The reference is of course to "Der Wahrheitsbegriff in den formalisierten Sprachen," *Studia Philosophica* I (1936): 261–405; translated as "The Concept of Truth in Formalized Languages," in A. Tarski, *Logic, Semantics, Metamathematics* (Oxford: University Press, 1956), pp. 152–278.

for any given sentence—can be explained independently of such empirical facts (independently, that is, of that sentence's actual truth-value).

All that is necessary, Carnap continues, is that the semantical metalanguage include the terms and sentences of the object-language (the language whose semantics is being explained):

> In his treatise Tarski developed a general method for constructing exact definitions of truth for deductive language systems, that is, for stating rules which determine for every sentence of such a system a necessary and sufficient condition of its truth. In order to formulate these rules it is necessary to use a metalanguage which contains the sentences of the object language or translations of them and which, therefore, may contain descriptive constants, e.g., the word "black" in the example mentioned. In this respect, the semantical metalanguage goes beyond the limits of the syntactical metalanguage. This new metalanguage evoked my strongest interest. I recognized that it provided for the first time the means for precisely explicating many concepts used in our philosophical discussions. (pp. 60–61)

And by "precise explication," it is clear, Carnap continues to mean a logical explication, one which does not invoke either the empirical facts in virtue of which particular sentences like "this table is black" is true or the empirical circumstances in virtue of which the word "black" in English comes to designate the property of being black. For empirical terms and sentences occur in such a semantical metalanguage only in Tarskian schemas like the one giving the truth-conditions for (but not the truth-value of) "this table is black."[10]

Immediately after the last quotation from his Autobiography, Carnap describes how he urged Tarski to report on the new conception of semantics at the International Congress for Scientific Philosophy to be held in Paris in September of 1935, and he promised to deliver his own paper at the congress emphasizing the importance of Tarski's work.[11] Carnap reports, to his surprise, that "there was vehement opposition even on the side of our philosophical friends" (p. 61), some of whom were aghast at the apparently "metaphysical" idea of comparing language with the world. Such objections, from Carnap's point of view, fail to appreciate the schematic and essentially disquotational character of the new semantics, in virtue of which it properly belongs to metalogic rather than metaphysics. In any case, however, what is most important for our purposes is what Carnap says next (pp. 61–62): "I began intensive work in the newly opened field. In the Encyclopedia monograph *Foundations of*

---

[10]The analogous schema for designation is "'black' designates the property of being black." I shall return to the essentially disquotational character of such Tarskian schemas below.

[11]The two papers appeared as A. Tarski, "Grundlegung der wissenschaftlichen Semantik," in *Actes du Congrès international de philosophie scientifique* (Paris: Hermann, 1936), vol. 3, pp. 1–8; and R. Carnap, "Wahrheit und Bewährung," *op. cit.*, vol. 4, pp. 18–23.

Logic and Mathematics (1939) I explained in a more elementary, non-technical way the difference between syntax and semantics and the role of semantics in the methodology of science, especially as a theory of interpretation of formal systems, e.g., axiom systems in physics."

This monograph is perhaps the single most important statement of Carnap's distinctive philosophical approach to the nature of logic and mathematics.[12] It is organized into three chapters: I. Logical Analysis of Language: Semantics and Syntax; II. Calculus and Interpretation; III. Calculi and Their Application in Empirical Science. Section 1 of Chapter I, "Theoretical Procedures in Science," emphasizes the particular importance of the application of logic and mathematics in empirical science:

> [T]he chief theoretical procedures in science—namely, testing a theory, giving an explanation for a known fact, and predicting an unknown fact—involve as an essential component deduction and calculation; in other words, the application of logic and mathematics [...]. It is one of the chief tasks of this essay to make clear the role of logic and mathematics as applied in empirical science. We shall see that they furnish instruments for deduction, that is, for the transformation of formulations of factual, contingent knowledge. However, logic and mathematics not only supply rules for transformation of factual sentences but they themselves contain sentences of a different, non-factual kind. Therefore, we shall have to deal with the question of the nature of logical and mathematical theorems. It will become clear that they do not possess any factual content. If we call them true, then another kind of truth is meant, one not dependent upon facts. A theorem of mathematics is not tested like a theorem of physics, by deriving more and more predictions with its help and then comparing them with the results of observations. But what else is the basis of their validity? We shall try to answer these questions by examining how the theorems of logic and mathematics are used in the context of empirical science. (p. 144)

The general thrust of this passage is quite similar to the perspective that we have already found in *Logical Syntax*. Yet Carnap is now prepared to speak of semantical interpretations and truth for both empirical and logico-mathematical statements. I shall argue that, although the overlap with *Logical Syntax* is considerable, Carnap's new perspective represents a significant advance.

Chapter II develops the central distinction between a *calculus* and its semantical *interpretation*. To view a particular language—the object language of

---

[12]See *Foundations of Logic and Mathematics*, in *International Encyclopedia of Unified Science*, vol. 1, part 1, no. 3 (Chicago: University Press, 1939), pp. 139–213; page references are to this volume.

our metatheory—as a mere calculus is to describe it purely formally or syntactically (p. 158): "A definition of a term in the metalanguage is called *formal* if it refers only to the expressions of the object-language (or, more exactly, to the kinds of the signs and the order in which they occur in the expressions) but not to any extralinguistic objects and especially not to the designata of the descriptive signs of the object-language." It is such a formal theory of the object-language that we refer to as its *logical syntax*.[13] Among the syntactical terms of a metalanguage are *well-formed formula* and *deduction*, both of which are defined by inductive definitions on expressions appealing to what Carnap calls *rules of transformation*, which are also purely formal in the sense indicated. Thus (p. 159): "A *syntactical system* or *calculus* (sometimes also called a formal deductive system or a formal system) is a system of formal rules which determine certain formal properties and relations of sentences, especially for the purpose of formal deduction."[14]

A *semantical system* S adds an *interpretation* to such a calculus by introducing *rules of designation* and *rules of truth* of the kind that Tarski had shown us how to construct. Briefly, we formulate rules of truth directly for the atomic sentences of the language on the basis of the rules of designation for the names and predicates occurring in such sentences, and we then proceed by induction on the logical complexity of the sentences via such rules as "'not-P' is true if and only if 'P' is not true," and so on. A Carnapian semantical system, moreover, divides all signs of the object language into *descriptive* and *logical* signs—roughly and intuitively, into "those which designate [empirical] things or properties of things" and those which "serve chiefly for connecting descriptive signs in the construction of sentences but do not themselves designate [empirical] things, properties of things, etc." (p. 149).[15] Finally, two central semantical terms which can then be defined in a semantical system are *L-true* (logically true) and *L-false* (logically false), where a sentence is "*L-true* if it is true in such a way that the semantical rules of [the system] S suffice for establishing its truth" and similarly for *L-false* (p. 155). In the case of an empirical or factual sentence like "this table is black," by contrast, the (Tarskian) rules of

---

[13]Note that Carnap's explanation of logical syntax here is completely congruent with the conception of *Logical Syntax*: compare, for example, the passage from §60b to which note 8 is appended. Of course Carnap continues to agree that "whether a sentence is true or false cannot generally be seen by [...] the kinds and serial order of the signs." What he learned from Tarski, however, is that we can *define* truth in a semantical metalanguage without determining the truth-value of the target sentence.

[14]To illustrate (ibid.): "The simplest procedure for the construction of a calculus consists in laying down some sentences as primitive sentences (sometimes called postulates or axioms) and some rules of inference. The primitive sentences and rules of inference are used for two purposes, for the construction of proofs and of derivations [i.e., deductions from sentences that are not necessarily provable from the axioms]." As we shall see, Carnap does not limit himself to this "simplest procedure" but also allows what he calls "transfinite rules" resulting in a notion of consequence—just as he did in *Logical Syntax*.

[15]I have inserted "empirical" into Carnap's text because the distinction that he draws here is clearly intended to capture the same distinction as in *Logical Syntax*. I shall return to this point below.

truth, as we have seen, do not by themselves determine either the truth or the falsity of the sentence. Thus (ibid.): "If a sentence is either L-true or L-false, it is called *L-determinate*, otherwise (L-indeterminate or) *factual*." And, Carnap continues, the property of L-determinacy is therefore centrally important in more precisely distinguishing between *logical* and *descriptive* signs (ibid.): "Every sentence which contains only logical signs is L-determinate. This is one of the chief characteristics distinguishing logical from descriptive signs."[16]

Establishing the precise relationship between syntactical and semantical terms is an important problem for Carnap. Most important, in this connection, is what Carnap calls the problem of constructing an *L-exhaustive calculus* with respect to any semantical system S—whether it is possible, that is, to find a calculus whose transformation rules allow one to establish all of the L-true sentences of S. This property is related to what we now call deductive completeness, but there is also a crucially important difference. For Carnap goes on to distinguish between two essentially different kinds of transformation rules, finite and transfinite, where the resulting "deductions" in the case of the latter rules can involve an infinite number of premises. Nevertheless, such rules still count as "syntactical" for Carnap, so long as they can be fully specified in terms of "the kinds and serial order of the signs" without mentioning the designata of these signs (compare note 13). A paradigmatic example of this situation, which had already been considered at length in *Logical Syntax*, arises in the case of the ω-*rule* for elementary arithmetic allowing us to infer a universal quantification over all numbers from the class of all instances resulting from the corresponding open formula by the substitution of *numerals* for the quantified variable. Section §14 of *Logical Syntax* refers to this kind of transformation rule as a rule of *consequence* rather than *derivation*, and it is clear that he has essentially the same distinction in mind in the 1939 monograph (p. 165): "[T]he terms 'C-implicate' and 'C-true' are applied generally with respect to both finite and to transfinite calculi. On the other hand, we shall restrict the corresponding terms 'derivable' and 'provable' to finite calculi."[17]

So Carnap's distinction between syntax and semantics does not line up with our modern distinction between proof theory and model theory. In particular, arithmetic can still be "syntactically complete," for Carnap, despite Gödel's

---

[16] For the parallel distinction in *Logical Syntax*, see note 7, together with the paragraph to which it is appended. We should note, however, that Carnap soon gives up this way of characterizing the distinction between logical and descriptive signs, beginning with *Introduction to Semantics* (Cambridge: Harvard University Press, 1942). The basic problem is that not all calculi Carnap is willing to recognize as logical are L-determinate; consider, for example, first-order logic with identity. Following Frege, however, Carnap's preferred logical framework for science throughout his career is a higher-order system of type theory with an infinity of individuals. Here we do have L-determinacy, and Carnap's characterization of the distinction between logical and descriptive signs makes sense. I shall return to this issue several times below.

[17] Note that 'C-true' is a *syntactical* term (p. 159): "We shall call the sentences to which the proofs lead *C-true* sentences (they are often called proved sentences or theorems of the calculus)." The above statement is a later refinement based on explicitly distinguishing between finite and transfinite rules of transformation.

incompleteness theorem, and this is one important reason, in particular, that Carnap can take arithmetic to be determinate in principle even after Gödel's results.[18] Similarly, it still makes sense for Carnap to appeal to the notion of L-determinacy as a distinguishing characteristic of logical as opposed to descriptive signs.[19] Indeed, Carnap extends this point of view far beyond elementary arithmetic to embrace essentially all of Zermelo-Fraenkel set theory in his discussion of Language II of *Logical Syntax*, and, on this basis, he develops a "criterion of validity" (analyticity) for all of classical mathematics.[20] The basic idea is to formulate Zermelo-Fraenkel set theory in terms of a higher-order system of types, extending into the transfinite, where the individuals are the natural numbers. We again treat the first-order arithmetical part syntactically via quantification (in the syntactical metalanguage) over *numerals*, and we extend the treatment to higher types inductively by considering classes, and classes of classes, etc., of numerals.[21] Along the way, we help ourselves to any axiom of Zermelo-Fraenkel set theory we might need (including the axiom of choice) in the (syntactical) metalanguage.[22]

It is clear that Carnap adopts the same point of view in the 1939 monograph. Chapter III begins by distinguishing between elementary logical calculi based on "the lower functional calculus" (which we now call first-order logic) and those based on "the higher functional calculus" (which we now call higher-order

---

[18] See again note 7, together with the paragraph to which it is appended. Note that the issue raised by Gödel's results is not the same as the one raised in note 16. L-determinacy is a semantical notion, and the issue raised by Gödel concerns the relationship between syntax and semantics.

[19] Although L-determinacy, defined in terms of L-truth and L-falsity, is a semantical term, the important point here is that Carnap takes such terms to be syntactically characterizable as well (pp. 158–59): "The definitions of all semantical terms refer directly or indirectly to designata. But some of these terms—e.g., 'true', 'L-true', 'L-implicate'—are attributed not to designata but only to expressions; they designate properties of, or relations between, expressions. Now our question is whether it is possible to define within syntax, i.e., in a formal way, terms which correspond more or less to those semantical terms, i.e., whose extensions coincide partly or completely with theirs. The development of syntax—chiefly in modern symbolic logic—has led to an affirmative answer to that question. Especially important is the possibility of defining in a formal way terms which completely correspond to 'L-true' and 'L-implicate' of fundamental importance. This shows that logical deduction can be completely formalized."

[20] See §§ 34a–i of *Logical Syntax*. These sections reproduce the content of "Ein Gültigkeitskriterium für die Sätze der klassischen Mathematik," *Monatshefte für Mathematik und Physik* 42 (1935): 163–90, which is a successor to the 1934 paper cited in note 8.

[21] There is an important disanalogy between first-order arithmetical quantification and higher-order quantification. In the former case quantification can be considered purely substitutionally, and so the corresponding quantification in the metalanguage is only over *expressions* of the object language (numerals). In the latter case, however, we cannot limit the metalinguistic quantification to arithmetical (or higher-level) properties definable in the object language (as in Henkin models, for example) without sacrificing the much greater strength of higher-order systems. As a result of conversations (and correspondence) with Gödel, Carnap is clear about this situation in *Logical Syntax*: see especially §34c (pp. 106–107), §34d (pp. 113–14).

[22] See, in particular, §§ 34g–h of *Logical Syntax*, and compare § 33 (pp. 97–98) for the full set of axioms of *ZFC*.

logic). The distinction between the two is that the (Tarskian) truth definition for quantification is extended to include not only quantification over individuals but also over arbitrary properties of every level (p. 175): "To the range of an individual variable belong all individuals, to the range of a predicate variable of level r belong all properties of level r." Carnap remarks "that there are some controversies and unsolved problems concerning the properties of higher levels" (ibid.), but he remains completely untroubled by such issues.[23] On the contrary, he immediately proceeds to sketch a construction of arithmetic in the style of Frege's *Grundlagen* and *Grundgesetze* within higher-order logic— one of two alternative formulations of arithmetic he considers.[24] He concludes (p. 177, emphasis added): "We see that it is possible to define within the logical calculus signs for numbers and arithmetical expression. It can further be shown that all theorems of ordinary arithmetic are provable in this calculus, *if suitable rules of transformation are established.*" Thus, although Carnap of course understands perfectly well that, since Basic Law V of the *Grundgesetze* is inconsistent, some kind of axiom of infinity is needed, he has no more qualms here concerning the properly "logical" status of such an axiom than he had in *Logical Syntax*.[25]

Carnap's seemingly blasé attitude towards the leading foundational issues of the time may strike the contemporary reader as either uninformed or irresponsible—and, in any case, of no particular interest today. As we have seen, however, Carnap was by no means uninformed or irresponsible; rather, he was deeply immersed in the cutting-edge technical work of Gödel and Tarski (and Fraenkel) on these issues and was in fact appealing to precisely this work in crafting his own characteristic response.[26] Yet Carnap's response essentially involves giving up the pursuit of any foundational program in favor of the Principle of Tolerance: *In logic there are no morals.* So it is no wonder that, closely following the discussion of his idiosyncratic conception of the relationship between syntax and semantics in Chapter II of his 1939 monograph, and immediately before his discussion of both "elementary" and "higher" logical calculi in Chapter III, Carnap inserts § 12 on the question "Is Logic a Matter of Convention?". His answer runs as follows:

---

[23] Unlike in *Logical Syntax*, however, Carnap does not attempt to construe either first-order quantification or higher-order quantification as syntactical in his sense, as involving quantification over signs or expressions rather than their designata. The "controversies and unsolved problems" to which he alludes most likely involve the issues concerning indefinable arithmetical properties and impredicativity that had arisen in his discussions with Gödel in connection with *Logical Syntax*. Compare the discussion of "indefinite" and "impredicative" terms in §§ 43–45 of *Logical Syntax*, together with the extended discussion of "incomplete and complete criteria of validity" in § 34a.

[24] Carnap compares this construction with an alternative (direct) presentation of the Peano axioms as "an elementary logical calculus" in § 17. I consider these alternatives further below.

[25] For the situation in *Logical Syntax*, see the discussion of infinity and choice in § 38a ("On Existence Assumptions in Logic"), together with the earlier discussion in § 33 (pp. 97–98).

[26] For further discussion of Carnap's interactions with Gödel, Tarski, and Fraenkel, see E. Reck, "Carnap and Modern Logic," in M. Friedman and R. Creath, eds., *The Cambridge Companion to Carnap* (Cambridge: University Press, 2007), pp. 176–199.

> It is important to be aware of the conventional components in the construction of a language system. This view leads to an unprejudiced investigation of the various forms of new logical systems which differ more or less from the customary form (e.g., the intuitionist logic of Brouwer and Heyting, the systems of logic of modalities constructed by Lewis and others, the system of plurivalued logic as constructed by Lukasiewicz and Tarski, etc.), and it encourages the construction of further new forms. The task is not to decide which of the different systems is "the right logic" but to examine their formal properties and the possibilities for their interpretation and application in science. It might turn out that a system deviating from the ordinary form will turn out to be useful as a basis for the language of science. (pp. 170–71)

That intuitionist logic leads off the list of non-customary systems, and that Carnap also emphasizes the application of logic in (empirical) science, reminds us, once again, of the point of the Principle of Tolerance in *Logical Syntax*.[27] Instead of supporting one or another foundational program, the goal of Carnap's philosophy of logic and mathematics is to clarify, on the one hand, the indispensable role of logic and mathematics in empirical science and, on the other, the fundamental difference between these disciplines and the empirical knowledge to which they contribute. It is entirely towards this end that Carnap's own technical work in both syntax and semantics is directed.

The Application of Logic and Mathematics in Empirical Science

I shall now consider Carnap's treatment of "Calculi and Their Application in Empirical Science" (Chapter III) in some detail. It is clear, from the beginning, that we are concerned with *interpreted* calculi and, for the most part, calculi considered together with "a certain interpretation or certain kind of interpretation used in the great majority of cases of its practical application[; t]his we call the *customary interpretation* (or kind of interpretation) for the calculus" (p. 171). Indeed, Carnap's overarching classification of the various calculi is guided precisely by their differing customary interpretations:

> In what follows we shall discuss some calculi and their application. We classify them according to their customary interpretation in this way: logical calculi (in the narrower sense), mathematical, geometrical, and (other) physical calculi. The customary interpretation of the logical and mathematical calculi is a logical, L-determinate

---

[27] Recall that Carnap's formulation of the Principle of Tolerance in §17 of *Logical Syntax* is a response to the discussion of intuitionism in §16—which, in turn, is preceded by the discussion of "definite" and "indefinite" terms in §15.

interpretation; that of the geometrical and physical calculi is descriptive or physical. The mathematical calculi are a special kind of logical calculi, distinguished merely by their greater complexity. The geometrical calculi are a special kind of physical calculi. (ibid.)[28]

Carnap is primarily concerned with the application of mathematical calculi (arithmetic and analysis) in physical calculi (including geometrical calculi). This is because the empirical testing of the (factual) statements of the latter calculi involve two fundamental numerical procedures, counting and measurement, where arithmetic is applied in the former and real analysis in the latter. It is here that both the subtlety and novelty of Carnap's new semantical perspective on the application of logic and mathematics becomes fully evident.

The first point worth emphasizing is that here, unlike in *Logical Syntax*, all mathematical calculi are fully interpreted. In §84 of *Logical Syntax*, as I explained above, Carnap goes beyond the purely formal Hilbertian conception (and embraces the opposed Fregean conception) by insisting on "*the inclusion of the mathematical calculus in the total language* [of science]"—otherwise mathematical calculi, considered independently of this wider context, remain mere (syntactic) calculi in the sense of Hilbert. Moreover, the kind of interpretation or meaning acquired by such inclusion within the total language is the *empirical meaning* now accruing to all sentences of the language in virtue of their relations to protocol-sentences as discussed in §82.[29] Carnap's new semantical perspective is quite different. Both arithmetic and analysis are fully interpreted, independently of empirical observations, via semantical rules, resulting, for Carnap, in L-determinate interpretations of both. Moreover, when Carnap comes to consider the empirical meaning of "abstract" theoretical concepts in physics like the electro-magnetic field, it turns out, on his preferred understanding of such concepts, that the corresponding theoretical terms are only *indirectly* interpreted via semantical rules, which are explicitly stated (as rules of designation) only for "concrete" observational terms. The theoretical terms themselves, in this sense, are strictly uninterpreted elements of a (phys-

---

[28] As I explained in note 16, Carnap gives up the generality of the claim that all logical calculi are L-determinate soon after *Foundations of Logic and Mathematics*. It would seem that Carnap, in this work, is not yet fully clear on what we now call first-order logic. For example, when Carnap describes the *lower functional calculus* in §13 (as an example of an "elementary" logical calculus), he uses three kinds of variables (sentential, individual, and predicate) but only states a semantical rule for quantification in the case of individual variables (p. 174). By contrast, Carnap's treatment of the Peano axioms as an "elementary mathematical calculus" in §17 is definitely second-order, with explicit quantification over all properties of numbers (p. 181).

[29] See the paragraph to which notes 6 and 7 are appended. However, it is now important to note that, although Carnap mentions calculation both "with numbers of empirical objects and with measures of empirical magnitudes" in the passage from §84 quoted in the preceding paragraph, the only example he presents in detail is enumerating empirical objects using Frege's definition of number. As we shall see, this case is actually quite different from the case of measuring the electro-magnetic field arising in connection with §82.

ical) calculus, which, however, still acquires an empirical (testable) meaning by means of its logical relations to sentences reporting the results of empirical measurements.[30]

Let us consider the application of arithmetic in counting and of analysis in measurement in more detail. In constructing a mathematical calculus for arithmetic, Carnap considers two alternative possibilities. We either present the Peano axioms directly, and give the customary interpretation in the domain of finite cardinal numbers, or we present Frege's construction of the finite cardinals in higher-order logic (including an axiom of infinity), and derive the Peano axioms from this construction.[31] The first case of the "application of mathematical calculi" considered in §19 is then a detailed example of calculating numbers of empirical objects using either of the two alternative formulations of arithmetic.[32] The calculation starts with empirical statements concerning the numbers of people in a room at a certain time falling under two exclusive and exhaustive (at the time) empirical predicates: there are three falling under the first and six under the second. The crucial part of the derivation is the second-order statement that, if a class F is divided (exclusively and exhaustively) into two parts G and H, the cardinal number of F is the sum of the cardinal numbers of G and H. It follows from this universal statement that there are $3 + 6$ people in the room, and thus, given the arithmetical theorem that $3 + 6 = 9$, that there are nine people in the room. So the calculation is facilitated by both a universal statement in second-order logic concerning the cardinal numbers belonging to various properties (classes) and a singular arithmetical formula.

What is important here is that Frege's characteristic conception of numbers as attributed to the extensions of arbitrary concepts is essential, and Carnap

---

[30] As Carnap developed his preferred understanding of theoretical terms further, beginning in the 1950s, they eventually became the only terms of empirical science that now relate to observational sentences as all terms whatsoever related to these same sentences (protocol sentences) in §82 of *Logical Syntax*. In this earlier discussion, as we have seen, the entire language, including its mathematical part, was treated as an uninterpreted calculus, which only acquires an empirical meaning *as a whole* by its deductive relations with protocol sentences. To be sure, Carnap there already distinguished logical from descriptive terms, logical from physical rules, so that empirical testing, properly speaking, concerned only the latter. Nevertheless, as we shall see, the account of empirical testing in the 1939 monograph is considerably more detailed and sophisticated, with the result that we not only have a new fundamental distinction between the cases of arithmetic and geometry (the latter now explicitly treated as a *physical* calculus), but also the suggestion of a more nuanced appreciation of the process of empirical testing itself—as it proceeds from concrete through ever more abstract descriptive terms: see note 40 below, together with the paragraph to which it is appended, and also, in particular, note 49 below.

[31] See note 24 for Carnap's comparison of the two alternatives, and the paragraph to which it is appended for the necessity of an axiom of infinity. In reference to the issue broached in note 28, observe that, whereas Carnap's formulation of the first alternative is second-order, the alternative formulation (as Carnap observes) is third-order.

[32] Carnap explains (§19, p. 187): "Whether we take the arithmetical calculus in the form of a part of the higher functional calculus (as in §14) or in the form of a specific calculus (as in §17) does not make any essential difference."

signals this in his derivation (§ 19, p. 187) by justifying the crucial second-order statement by reference to definitions presented within higher-order logic in §§ 13–14. These include, in particular, a recursive definition of the (finite) cardinal numbers belonging to a property (class) modeled on Frege's treatment in the *Grundlagen*.[33] And it is also worth noting that Carnap does essentially the same thing in the simpler example that he presents in § 84 of *Logical Syntax* to illustrate the advantage of a Fregean conception over Hilbertian formalism (p. 326): "For instance, the sentence 'in this room there are now two people present' cannot be derived from the sentence 'Charles and Peter are in this room now and no one else' with the help of the logico-mathematical calculus alone, as it is usually constructed by the formalists; but it can be derived with the help of the logicist system, namely on the basis of Frege's definition of '2'. A logical foundation of mathematics is only given when a system is constructed which enables derivations of this kind to be made."[34] The significance of this essentially Fregean aspect of Carnap's view will emerge below, in contrast with his treatment of empirical measurement.

Turning now to real analysis and its application in empirical measurement, I note first that Carnap does not provide a detailed formulation of the mathematical theory. Rather, he is content to present a brief sketch in § 18 ("Higher Mathematical Calculi") of how integers can be construed as pairs of natural numbers, fractions as pairs of integers, and real numbers as (infinite) sequences (or classes) of fractions in the usual way. Thus (p. 184): "On the basis of a calculus of the arithmetic of natural numbers, the whole edifice of classical mathematics can be erected [in higher-order logic as Carnap understands it] without the use of new primitive signs." Moreover, Carnap does not discuss the application of real analysis in much detail either, although he subsequently presents a derivation of a later numerical value of the length of a heated iron rod from its earlier value and its coefficient of thermal expansion in § 23 ("Physical Calculi and Their Interpretations"). As to the customary interpretation of the length of a rod at a certain time, however, Carnap merely states that it is "defined by the statement of a method of measurement" (p. 200). Nevertheless, further light is shed on what he has in mind if we return to Carnap's first mention of the "two procedures in empirical science which lead to the application of numerical expressions: counting and measurement" in § 19 (p. 186). For he

---

[33]This definition gives the two inductive clauses as follows (paraphrasing § 14, p. 176): To say that the finite cardinal number 0 belongs to F is to say that no object has property F; to say that the finite cardinal number $m+1$ belongs to F is to say that there is a property G, such that the finite cardinal number m belongs to G, and that all objects falling under G, together with one additional object (some object but no other object), fall under F. It appears from the passage quoted in note 32, therefore, that Carnap takes this definition, in effect, to belong to the formulation of the Peano axioms in § 17 as well and, more precisely, to the semantical metalanguage in which the customary interpretation in terms of finite cardinal numbers is stated (compare p. 182). It appears, in other words, that Carnap is assuming the Fregean treatment of finite cardinals in this metalanguage.

[34]This passage immediately follows the first passage (from p. 326) quoted in the paragraph preceding the one to which notes 6 and 7 are appended and immediately precedes the second passage (from pp. 326–27).

there cites a monograph in the same volume of the *International Encyclopedia, Procedures of Empirical Science*, by Victor Lenzen.[35]

Carnap refers, more specifically, to §§ 4–5 of Lenzen's monograph, which consider, respectively, counting and the measurement of length. The discussion of counting in § 4 involves both logical and empirical aspects. Lenzen states that "counting is an operation of determining similarity between collections" (p. 287), where similarity is defined in terms of one-to-one correspondence. He continues (ibid.): "Whitehead and Russell define a cardinal number as the class of all similar classes. Thus we may define the natural numbers which serve to describe collections of objects." Yet the determination of the number of objects in a given collection is a temporal empirical process, successively "performed by pointing or otherwise indicating the members of a collection to which one assigns the numbers 1, 2, etc., respectively" (p. 288). We thereby learn from experience that "[t]he number of a collection is independent of the order in which the counting occurs" (p. 287), and a similar induction from experience leads to such principles as the commutative laws of addition and multiplication. Nevertheless, while this discussion certainly appears plausible as part of an empirical genetic story, Lenzen nowhere acknowledges that a logical analysis of the empirical application of arithmetic in the style of Frege (or, for that matter, Whitehead and Russell) allows us fully to explain this application with no mention at all of such empirical procedures.[36] Indeed, it is for precisely this reason, as we shall see, that the measurement of physical magnitudes via real numbers, for Carnap, is quite different from the application of arithmetic.

Lenzen's discussion of the measurement of length in § 5 has a structure similar to § 4, but the empirical procedures in question are considerably more complex. The most basic procedure involves the determination of *coincidence*—an idealized perceptual representation of the perfectly exact contact, for example, of the two endpoints of a rigid rod with points of an object to be measured. If we begin, following Einstein, with the concept of a *practically rigid body*, then two points on one such body determine a physical interval (bounded line segment) or "stretch," which can then be determined to be *congruent* with an interval on another such body from the coincidence of the two sets of endpoints when the two intervals are adjacent to one another.[37] Moreover, if this pro-

---

[35]See *Procedures of Empirical Science*, in *International Encyclopedia of Unified Science*, vol. 1, part 1, no. 5 (Chicago: University Press, 1939), pp. 279–339; page references continue to be from this volume.

[36]Lenzen does acknowledge—and indeed emphasizes—that numbers eventually become the subject matter of a formal theory of arithmetic after their properties have been discovered empirically. But he conceives this theory as an uninterpreted theory in the style of Hilbert rather than an interpreted theory in the style of Frege (p. 288, emphasis added): "Numbers are first discovered as characters of collections; but, once principles expressing their relations are set up, the numbers acquire properties defined in terms of their relations to one another. They become the subject matter of an abstract theory which defines their properties *independently of any empirical application*."

[37]See A. Einstein, *Geometrie und Erfahrung* (Berlin: Springer, 1921). Lenzen does not cite this paper in his monograph, but he does so in his later article on "Einstein's Theory of

cedure is to result in a general concept of congruence applicable whether or not the two intervals are actually adjacent to one another, we must assume that (correcting for known distorting factors like temperature) the relation of congruence continues to hold under all possible displacements of the bodies in question.[38] Given this assumption (in effect, the existence of practically rigid bodies), we can finally define the *length* of such an interval by choosing a standard unit interval determined by a standard body (e.g., the standard meter bar in Paris) and laying it off repeatedly (end to end) along the interval of the body to be measured. The length (relative to this unit) is given by counting the number of times that the unit can be thereby laid off—where, if the two intervals are incommensurable, the result is a real number calculated as the limit of considering ever smaller sub-units of the original unit.

In my discussion of Lenzen on counting and the application of arithmetic I suggested that Carnap's Fregean conception of arithmetic involves a quite different conception of its empirical application. Carnap's treatment of the measurement of empirical magnitudes, by contrast, parallels Lenzen's rather closely. In the first place, as I have already indicated, Carnap inserts references to empirical "methods of measurement" into semantical rules giving the customary interpretation of physical calculi involving such magnitudes. In the toy theory of the dependence of length on temperature discussed in §23, for example, Carnap writes (p. 200): "The *customary interpretation*, i.e., that for whose sake the calculus is constructed, is given by the following semantical rules. '$\lg(x,t)$' designates length in centimeters of the body x at time t (defined by the statement of a method of measurement); '$te(x,t)$' designates the absolute temperature in centigrades of x at time t (likewise defined by a method of measurement);" and so on. Nothing like this occurs in Carnap's treatment of arithmetic, where the customary interpretation is either given by rules of designation for the Peano calculus in the domain of finite cardinals or by a translation of this calculus into higher-order logic together with the customary interpretation of higher-order logic.[39] In the second place, Carnap's discussion in §24 of "elementary and abstract terms" in physics also runs parallel to Lenzen's monograph. In particular, Carnap presents a series of concepts ranging from the most elementary deliverances of purely qualitative observation (red, blue, warm, cold) to increasingly abstract quantitative empirical concepts

---

Knowledge" in the Einstein Schilpp volume (1949), using the same basic notions of coincidence, practically rigid body, and "stretch" (a translation of Einstein's *Strecke*). As we shall see, Carnap explicitly draws on *Geometrie und Erfahrung* in his corresponding discussion of geometry in §§ 21–22 of his 1939 monograph.

[38] This is the principle of free mobility originally formulated by Helmholtz. For an extended discussion see my "Geometry as a Branch of Physics: Background and Context for Einstein's 'Geometry and Experience'," in D. Malament, ed., *Reading Natural Philosophy: Essays in the History and Philosophy of Science and Mathematics* (Chicago: Open Court, 2002), pp. 193–229.

[39] See again § 17. In both cases Carnap makes it clear that the customary interpretation is logical or L-true, whereas that of the toy theory of § 23, by contrast, is of course descriptive or factual.

(p. 204): "coincidence; length; length of time; mass, ...;" and so on.[40] This runs parallel to the discussion in Lenzen's monograph, which begins with qualitative observation (§ 3), and then traces the formation of increasingly abstract quantitative empirical concepts, beginning, as we have seen, with coincidence and then length (§ 5), and proceeding to time (§ 6), weight (§ 7), and so on.[41]

Lenzen treats the empirical application of arithmetic and geometry in essentially the same way. Just as, in the case, of arithmetic, he begins with empirical counting procedures and then introduces the abstract theory as a formal Hilbertian axiomatic system (see note 36), Lenzen concludes his treatment of the measurement of length in § 5 by moving from an Einsteinian description of practically rigid bodies to an Hilbertian description of the corresponding abstract formal structure.[42] The fundamental problem with this treatment, from Carnap's point of view, is that both mathematical theories are considered purely formally as Hilbertian axiomatic systems or, in Carnap's terms, as mere uninterpreted calculi. But all such theories, considered in this way, are completely on a par, whether they are logical, mathematical, or physical. Interesting differences between them can only emerge, for Carnap, when we turn to a consideration of their semantical interpretations, especially their *customary* interpretations. And here the crucial point is that, whereas the customary interpretation of the arithmetical calculus (in both versions) is logical or L-true, that of a geometrical calculus (such as Euclidean geometry) is descriptive or factual (compare note 39). It is in precisely this sense, for Carnap, that,

---

[40] On the following page Carnap makes it clear, by a reference back to p. 50 of his contribution to the symposium on *Encyclopedia and Unified Science* in the same volume, that the methods of measurement mentioned in the above semantical rules can also be formally represented within the calculus in question by a chain of *conditional* definitions or reduction sentences in which the primitive terms forming the basis of the chain are all purely qualitative. This approach to the empirical meaning of theoretical terms—which Carnap calls his "first method"—represents what I called his preferred understanding of such terms in note 30 above. On what he calls his "second method," by contrast, we take the theoretical terms as both primitive and fully interpreted (via corresponding semantical rules) and attempt rather to move down towards the more concrete terms via *explicit* definitions.

[41] In his preliminary discussion of the development of science from qualitative description to quantitative representation, Lenzen appears to have in mind Carnap's *Physikalische Begriffsbildung* (1926)—which is cited in Lenzen's bibliography. That both Carnap and Lenzen begin with coincidence and proceed to length is of course not surprising, given the centrality of Einstein's *Geometrie und Erfahrung* for both (note 37), and I shall return to this below. Meanwhile, however, I note that there is nothing corresponding to Lenzen's § 4 on counting and arithmetic in Carnap's list—because, once again, Carnap's conception of arithmetic (including applied arithmetic) is wholly logical.

[42] See § 5 (p. 295): "The propositions of geometry are to be viewed as initially generalizations from observations on the spatial properties of structures of practically rigid bodies. In Euclidean geometry the science is based on a small number of axioms from which the propositions can be deduced as theorems. In abstract geometry the original axioms have been transformed into postulates which define implicitly the fundamental concepts of geometry. From this point of view the postulates of Euclidean geometry are descriptions of the formal spatial properties of rigid bodies. That the concepts of these forms are applicable to the objects of perception is a hypothesis to be confirmed and limited by observation."

whereas arithmetic belongs among the mathematical calculi, geometry belongs among the physical.

Carnap begins his discussion in § 21 ("Geometrical Calculi and Their Interpretations") by re-emphasizing this point.[43] He then presents a fragment of a calculus for Euclidean geometry (axioms of incidence, order, parallel postulate) based on Hilbert's well-known formulation. There are many possible interpretations of this calculus. Nevertheless (§ 21, p. 194): "The *customary interpretation* is descriptive. It consists of a translation into the physical calculus (to be dealt with in the next section) together with the customary interpretation of the physical calculus." Section 22 ("The Distinction between Mathematical and Physical Geometry") then begins by explaining "a certain translation of the geometrical calculus into the mathematical calculus" (p. 195), which "leads, in combination with the customary interpretation of the mathematical calculus, to a logical interpretation of the geometrical calculus" (pp. 195–96). Here we interpret the primitive terms of the geometrical calculus (point, line, plane, and so on) in the domain of ordered triples of real numbers (three-dimensional coordinates)—where, "[o]n the basis of the customary interpretations of the mathematical calculus [real analysis], the axioms and theorems of geometry become L-true propositions" (p. 196).

Carnap continues (ibid.): "The difference between mathematical and physical geometry became clear in the historical development by the discovery of non-Euclidean geometry, i.e., of axiom systems deviating from the Euclidean form by replacing the parallel axiom [(G 4) in Carnap's fragment] by some other axiom incompatible with it." And, whereas "[m]athematicians regarded all of these systems on a par, investigating any one indifferently," this could not be satisfactory for physicists (ibid., emphasis added): "Physicists, however, are concerned with a theory of space, i.e., of *the system of possible configurations and movements of bodies*, hence with an interpretation of a geometrical calculus. When an interpretation of the specific signs is established—and, to a certain extent, this is a matter of choice—then each of the calculi yields a physical geometry as a theory with factual content [...]. The theories are factual. The truth conditions, determined by the interpretation, refer to facts. Therefore, it is the task of the physicist, and not of the mathematician, to find out whether a certain one among the theories is true, i.e., whether a certain geometrical structure is that of the space of nature."[44]

---

[43]See § 21 (p. 193): "When we referred to mathematics in the previous sections, we did not mean to include geometry but only the mathematics of numbers and numerical functions. Geometry must be dealt with separately. To be sure, geometrical calculi, aside from their interpretation, are not fundamentally different in their character and, moreover, are closely related to the mathematical calculi. That is the reason why they too have been developed by mathematicians. But the customary interpretations of geometrical calculi are descriptive, while those of the mathematical calculi are logical."

[44]Carnap will make it perfectly explicit, in the following paragraph, that he is following Einstein's *Geometrie und Erfahrung*. But there is already an implicit reference in this passage, especially in the words italicized above. Compare the corresponding passage in Einstein (pp. 5–6, emphasis added): "It is clear that the conceptual system of axiomatic geometry

The following paragraph is especially important. Carnap begins by mentioning that the conception of geometry "acknowledged by most philosophers of the past [i.e., nineteenth] century was that of Kant, saying that geometry consists of 'synthetic judgments a priori', i.e., of sentences which have factual content but which, nevertheless, are independent of experience and necessarily true" (p. 197). Although Kant held that the same was true of arithmetic (ibid.), "[m]odern logical of analysis of language, however, does not find any sentences at all of this character." Nevertheless, Kant's doctrine was "obviously meant to apply to arithmetic and geometry as theories, i.e., interpreted systems, with their customary interpretations" (p. 198):

> Then, however, the propositions of arithmetic are, to be sure, independent of experience, but only because they do not concern experience or facts at all; they are L-true (analytic), not factual (synthetic). For geometry there is also, as mentioned before, the possibility of a logico-mathematical interpretation; by it the sentences of geometry get the same character as those of mathematics. On the basis of the customary interpretation, however, the sentences of geometry, as propositions of physical geometry, are empirical. The Kantian doctrine is based on a failure to distinguish between mathematical and physical geometry. It is to this distinction that Einstein refers in his well-known dictum: "So far as the theorems of mathematics are about reality they are not certain; and so far as they are certain they are not about reality." (ibid.)

The last sentence is a direct quotation (in Carnap's translation) of the best-known statement in *Geometrie und Erfahrung*.[45] Carnap then concludes, in the

---

alone can supply no assertions about the behavior of those objects of reality that we wish to designate as practically rigid bodies. In order to be able to furnish such assertions geometry must go beyond its solely formal-logical character, so that experienceable objects of reality (experiences) are coordinated to the empty conceptual schemata of axiomatic geometry. In order to accomplish this one needs only to add the proposition: *Solid bodies relate to one another with respect to their situational possibilities [Lagerungsmöglichkeiten] as do bodies in three-dimensional Euclidean geometry*; then the propositions of Euclidean geometry contain assertions about the behavior of practically rigid bodies. The thus expanded geometry is obviously a natural science; we can actually consider it as the oldest branch of physics. Its assertions rest essentially on inductions from experience, and not only on logical inferences. We wish to call the thus expanded geometry 'practical geometry' and to distinguish it from 'pure axiomatic geometry'. The question whether the practical geometry of the world is Euclidean or not has a clear sense, and its answer can only be supplied by experience."

[45] This statement occurs early on where Einstein first begins to address the question how mathematics can be independent of experience yet still apply to reality (pp. 3–4): "In so far as the propositions of mathematics refer to reality they are not certain; and in so far as they are certain they do not refer to reality. Full clarity about this situation appears to me to have been first obtained in general by that tendency in mathematics known under the name of 'axiomatics'. The advance achieved by axiomatics consists in having cleanly separated the formal-logical element from the material or intuitive content. According to axiomatics only the formal-logical element constitutes the object of mathematics, but not the intuitive or other content connected with the formal-logical element."

final paragraph of § 22, by emphasizing, once again, the fundamental difference, on his view, between arithmetic and geometry (ibid.): that, if "we take the systems with their *customary* interpretation," we find that "the propositions of arithmetic are logical, L-true, and without factual content; those of geometry are descriptive, factual, and empirical."

There are several important points to be made here. The first concerns a significant divergence between Einstein's conception of mathematical geometry and Carnap's. When Einstein speaks of purely mathematical geometry, conceived as a "formal-logical" axiomatic system (note 44), it becomes clear very quickly that he has in mind a Hilbertian conception on which mathematical geometry is a system of "implicit definitions"—which, in the end, are merely "contentless conceptual schemata."[46] But this is manifestly not what Carnap has in mind. For Carnap, once again, all formal calculi, considered independently of semantical interpretation, are completely on a par, and the means by which the sentences of geometry acquire "the same character as those of mathematics" is a "logico-mathematical interpretation" in the domain of ordered triples of real numbers. By "mathematics" Carnap does not mean the study of arbitrary formal system, but, first and foremost, the two fundamental (interpreted) mathematical theories of arithmetic and analysis.

The second point, however, is even more important, and also all too easy to miss. In particular, when Carnap says that the propositions of arithmetic "do not concern experience or facts at all," one might easily get the impression that he is talking about pure or uninterpreted rather than applied or interpreted arithmetic. But Carnap is again talking about *interpreted* arithmetic, not some uninterpreted calculus, and, on either version of what he calls the customary interpretation of arithmetic, this theory is directly applicable empirically via Frege's conception of numbers as attributed to the extensions of arbitrary concepts. Recall the simple derivation in applied arithmetic from § 19. The crucial step in the derivation is the second-order statement that, if a class F is divided (exclusively and exhaustively) into two parts G and H, the cardinal number of F is the sum of the cardinal numbers of G and H. This statement is L-true and thus logical rather than factual. Nevertheless, it can be universally instantiated by *empirical concepts* F, G, and H—where, in the example, F is the class of all people now in this room, and G and H represent a particular empirical

---

[46]See *Geometrie und Erfahrung* (pp. 4–5). Einstein refers to Moritz Schlick's *Allgemeine Erkenntnislehre* (1915) for the notion of "implicit definition" and continues: "The conception represented by modern axiomatics purifies mathematics from all elements not belonging to it, and thus removes the mystical obscurity that previously clung to the foundations of mathematics. But such a purified mathematics makes it also evident that mathematics as such may assert nothing about either objects of intuitive representation or objects of reality. In axiomatic geometry we understand by 'point', 'line', etc. only contentless conceptual schemata. What gives them content does not belong to mathematics." See my paper cited in note 37 for further discussion of the relationship between Einstein and Schlick. Einstein's own systematic treatment of the interpretation of geometry in terms of practically rigid bodies, the coincidence of "stretches," and free mobility occurs several pages later—after he has discussed general relativity and the views of Poincaré.

division of F into two exclusive and exhaustive parts. The crucial point is then that this instantiation is also L-true and logical: even though it contains descriptive terms, it logically follows from an L-true universal statement. To be sure, the derivation begins with empirical statements concerning the number of people now in the room in the two empirically specified classes, and it ends with an empirical statement concerning the number of people now in the room. The contributions of the science of arithmetic, however, remain purely logical—perfectly illustrating, for Carnap, his claim that logico-mathematical statements are merely instruments of deduction within empirical science.[47]

It is no wonder, then, that Carnap concludes §19 by placing particular emphasis on precisely this point:

> [A] logical or mathematical theorem is, *regarded from the point of view of its application in empirical science*, a device or tool enabling us to make a very complex and long chain of applications of the rules of the calculus at one stroke, so to speak. The theorem is itself, *even when interpreted*, not a factual statement but an instrument facilitating operations with factual statement, namely, the deduction of a factual conclusion from factual premises. The service which mathematics renders to empirical science consists in furnishing these instruments; the mathematician not only produces them for any particular case of application but keeps them in store, so to speak, ready for any need that may arise. (p. 189, emphasis added)

In the arithmetical example presented in §19, as we have seen, the theorem of interest is a universally quantified second-order statement; the application is its instantiation via particular empirical concepts. It should be noted, however, that the above-quoted passage follows an example of the application of real analysis that is sketched in §19 but only explicitly presented in §23—which, as I have already suggested (in the paragraph to which note 35 is appended), concerns a derivation of a later numerical value of the length of a heated iron rod from its earlier value and its coefficient of thermal expansion. This derivation, like the one presented in §19, involves both particular statements and universal principles. But these principles, unlike in the arithmetical case, are of two essentially different kinds: proved mathematical theorems of real analysis, on the one side, universal physical laws, on the other (pp. 200–201). The derivation in §23, after all, is an example of the interpretation of a *physical* calculus—which, as such, is factual or descriptive rather than L-true or logical. Since, as the above discussion of §§21–22 has explained, the same is true of geometry (in its customary interpretation), it follows that the fundamental dif-

---

[47]This claim occurs very early in the monograph: see the passage from §1 quoted in the paragraph to which note 12 is appended.

ference between geometry and mathematics proper (arithmetic and analysis) is also of this kind.[48]

Section 19, therefore, presents Carnap's fullest explanation of the sense in which logic and mathematics function as instruments of deduction within empirical science. Carnap has thereby answered one of questions raised in §1 (see again note 47). But he also there raised an additional question concerning "the nature of logical and mathematical theorems" themselves (p. 144). It turns out, of course, that they are L-true and "do not possess any factual content," but the question remains as to "the basis of their validity" (ibid.). The answer presented in § 20 is given, once again, by the Principle of Tolerance: since there are no independent "facts" to which such L-true propositions are answerable, the only remaining basis for preferring one formulation of logic and mathematics to an alternative is its purely pragmatic fruitfulness as an instrument of deduction within empirical science. This is essentially the same point of view that was already represented in *Logical Syntax*, but Carnap's new semantical perspective has provided a much clearer and more developed account of factual content.

The title of § 20 is "The Controversies over 'Foundations' of Mathematics," which is parallel to that of § 84 of *Logical Syntax* ("The Problem of the Foundation of Mathematics"). Unlike in this parallel section of *Logical Syntax*, however, Carnap here discusses all three positions in the "foundations" of math-

---

[48] An important contribution to our understanding of this situation has recently been made in W. Demopoulos, *Logicism and its Philosophical Legacy* (Cambridge: University Press, 2013), chapter 2 ("Carnap's thesis"). Demopoulos relates Carnap to Frege using Hume's principle: that the number of Fs = the number of Gs if and only if there is a one-to-one correspondence between the Fs and the Gs. This second-order statement suffices for the derivation of the Peano axioms, and, as Demopoulos emphasizes, it also fully accounts for the application of arithmetic to empirical objects: we need merely instantiate the second-order variables in a universally quantified formula appealing only to logical notions and particular empirical concepts. As Demopoulos notes, Hume's principle supplies us with what Frege calls a *criterion of identity* for the numerical values assigned to concepts. Although applicable to empirical concepts, Hume's principle, he argues, is nevertheless *not empirically constrained*. By contrast, analogous criteria of identity appealed to by Einstein in both geometry (the assignment of real numbers to spatial lengths) and space-time physics (the assignment of real numbers to the times of occurrence of events) are manifestly empirically constrained, and this absence of empirical constraints is also the precise sense, for Carnap, that applied arithmetic—unlike applied geometry (both spatial and spatio-temporal)—is nonfactual. Demopoulos presents his analysis as a reconstruction rather than interpretation of Carnap's actual position, and it is certainly entirely novel in its use of both Hume's principle and the notion of a criterion of identity. Nevertheless, the above discussion of Carnap's treatment in his 1939 monograph shows that Demopoulos's analysis can also be regarded as an interpretation of Carnap's position. From this point of view, Hume's principle is an alternative formulation of arithmetic in second-order logic that would be perfectly congenial to Carnap: it serves as the necessary axiom of infinity and encapsulates (second-order) Peano arithmetic in a formula that controls its application. Empirical applications are comprised by direct empirical instantiations of the second-order variables in an originally logical formula, and there is again a contrast with (spatial) geometry appealing to Einstein's work. The fundamental point, on either Carnap's original formulation or Demopoulos's reconstruction, is that Carnap's characteristic conception of the non-factuality of (applied) arithmetic stands out particularly clearly when we emphasize its intellectual debts to both Frege and Einstein.

ematics: logicism, formalism, and intuitionism. The discussion of logicism and formalism is quite similar to that in *Logical Syntax*, except that Carnap here emphasizes that logicism, unlike formalism, appeals to semantical interpretations in the new sense (and, indeed, to the customary or "normal" interpretations) of the relevant mathematical calculi. And, in the case of intuitionism, Carnap here emphasizes that the latter position rejects "the interpretation of mathematics as consisting of L-true sentences without factual content" and "is rather regarded as a field of mental activities based upon 'pure intuition'" (p. 192).

The Principle of Tolerance is then applied primarily to the resulting divergence between logicism, as Carnap now understands it, and intuitionism:

> Concerning mathematics as a pure calculus there are no sharp controversies. These arise as soon as mathematics is dealt with as a system of "knowledge"; in our terminology, as an interpreted system. Now, if we regard interpreted mathematics as an instrument of deduction within the field of empirical knowledge rather than as a system of information, then many of the controversial problems are recognized as being questions not of truth but of technical expedience. The question is: Which form of the mathematical system is technically most suitable for the purpose mentioned? Which one provides the greatest safety? If we compare, e.g., the systems of classical mathematics and of intuitionistic mathematics, we find that the first is much simpler and technically more efficient, while the second is more safe from surprising occurrences, e.g., contradictions. At the present time, any estimation of the degree of safety of the system of classical mathematics, in other words, the degree of plausibility of its principles, is rather subjective. The majority of mathematicians seem to regard this degree as sufficiently high for all practical purposes and prefer the application of classical mathematics to that of intuitionistic mathematics. The latter has not, so far as I know, been seriously applied in physics by anybody. (pp. 192–93)

Leaving aside the possible applications of intuitionistic or, more generally, constructive mathematics in physics (about which significantly more is now understood than in 1939), we see that this application of the Principle of Tolerance is again quite similar to what Carnap had envisioned in *Logical Syntax* (compare note 27, together with the paragraph to which it is appended). The difference, once again, is his new semantical perspective. Since I have now explained in detail how this perspective results in a treatment of the application of logic and mathematics in empirical science (especially physics) that is considerably improved in comparison with the rather sketchy discussion in § 82 and § 84 of

Logical Syntax, it should now be clear how this same perspective thereby results in a correspondingly improved treatment of the Principle of Tolerance.[49]

References

Carnap, R. (1934a): *Logische Syntax der Sprache* (Wien: Springer); translated as *The Logical Syntax of Language* (London: Kegan Paul, 1937).

Carnap, R. (1934b): "Die Antinomien und die Unvollständigkeit der Mathematik", *Monatshefte für Mathematik und Physik* 41: 263–405.

Carnap, R. (1935): "Ein Gültigkeitskriterium für die Sätze der klassischen Mathematik", *Monatshefte für Mathematik und Physik* 42: 163–90.

Carnap, R. (1939): *Foundations of Logic and Mathematics*, in *International Encyclopedia of Unified Science*, vol. 1, part 1, no. 3 (Chicago: University Press), pp. 139–213.

Carnap, R. (1942): *Introduction to Semantics* (Cambridge: Harvard University Press).

Carnap, R. (1963a): "Intellectual Autobiography", in P. Schilpp, ed., *The Philosophy of Rudolf Carnap* (La Salle, Open Court), pp. 3–84.

Carnap, R. (1963b): "Wahrheit und Bewährung", in *Actes du Congrès international de philosophie scientifique* (Paris: Hermann), vol. 4, pp. 18–23.

Demopoulos, W. (2013): *Logicism and its Philosophical Legacy* (Cambridge: University Press).

Einstein, A. (1921): *Geometrie und Erfahrung* (Berlin: Springer).

Friedman, M. (2002): "Geometry as a Branch of Physics: Background and Context for Einstein's 'Geometry and Experience'", in D. Malament, ed., *Reading Natural Philosophy: Essays in the History and Philosophy of Science and Mathematics* (Chicago: Open Court), pp. 193–229.

Friedman, M. (forthcoming): "From Intuition to Tolerance: The Development of Carnap's Philosophy of Mathematics", in C. Posy and O. Rechter, eds., *Kant's Philosophy of Mathematics, Vol. II: Reception and Influence After Kant* (Cambridge: University Press).

---

[49]With respect to the issue broached in note 30 concerning the further development of Carnap's preferred understanding of theoretical terms in the 1950s, the crucial point is that Carnap then leaves behind his use of conditional definitions or reduction sentences (compare note 40) in favor of his mature use of Ramsey sentences to represent the empirical content of theories. Yet this latter representation, in contrast to the former, is thoroughly holistic: it thereby leaves behind the more stratified—and in this sense "more nuanced"—conception of empirical testing and measurement that Carnap had originally articulated in 1939. Further discussion of this aspect of Carnap's evolving understanding of theoretical terms, however, will have to wait for another occasion.

Lenzen, V. (1939): *Procedures of Empirical Science*, in *International Encyclopedia of Unified Science*, vol. 1, part 1, no. 5 (Chicago: University Press), pp. 279–339.

Quine, W.V.O. (1963): "Carnap and Logical Truth", in P. Schilpp, ed., *The Philosophy of Rudolf Carnap* (La Salle, Open Court), pp. 385–406.

Reck, E. (2007): "Carnap and Modern Logic", in M. Friedman and R. Creath, eds., *The Cambridge Companion to Carnap* (Cambridge University Press), pp. 176–199.

Tarski, A. (1936a): "Der Wahrheitsbegriff in den formalisierten Sprachen", in *Studia Philosophica* I: 261–405; translated as "The Concept of Truth in Formalized Languages", in A. Tarski, *Logic, Semantics, Metamathematics* (Oxford: University Press, 1956), pp. 152–278.

Tarski, A. (1936b): "Grundlegung der wissenschaftlichen Semantik", in *Actes du Congrès international de philosophie scientifique* (Paris: Hermann, 1936), vol. 3, pp. 1–8.

Wittgenstein against Logicism

WARREN GOLDFARB

ABSTRACT: In the *Tractatus*, Wittgenstein rejects the Frege-Russell logicist reduction of arithmetic, although he accepts a version of Russellian logic. His version constructs all logical structures out of truth-functions in such a way as to yield not just full first-order logic but also some infinitary functions. Those functions can capture the ancestral of a relation without use of higher-order quantifiers, and so an alternative Tractarian foundation for arithmetic can be developed using them.

I

Frege and Russell were motivated in their development of logic by the logicist thesis: that all of mathematics is reducible to logic, and hence does not require any postulation of special faculties, most pointedly no faculty of intuition. They consequently strove to formulate systems that could be claimed to articulate purely logical principles, and then to show how number theory and the theory of real numbers could be obtained in them, by means of logicist constructions. The young Wittgenstein rejected this plan: in particular, arithmetic, for Wittgenstein, was not reducible to logic, and what's usually called the Frege-Russell definition of number was, he claimed, *erroneous*. Nonetheless, Wittgenstein took the new logic of Frege and Russell, and in particular the logic of *Principia Mathematica*, very seriously, and much of the technical work of the *Tractatus* is aimed at refashioning it to show why it works. (As he wrote to Russell very early, on June 22, 1912, "Logic must turn out to be of a TOTALLY different kind than any other science.") We might characterize early logicism as having two kinds of commitment to logic: logic as giving us the general criteria for proof, justification, presupposition, contradiction, and the like, that is, as providing the standard for the inferential relationships among the claims we make in any subject area; and logic as providing us with a robust *a priori* science with its own subject matter. Indeed, it was precisely the marriage of these two strands that was emphasized by Frege and by Russell as their achievement, and it is why they avoided the label "formal logic" for the subject they were creating. Wittgenstein agrees to the first role, and much of his work leading up to the *Tractatus* is to see what kind of beast logic must be that it can play this role. But he dissents from the latter, that is, from the constructive element of logicism—the very aspect that motivated the initial project.

Wittgenstein's philosophy of logic is clear enough, at least to a first approximation. The truths of logic, he maintained, are not really truths at all: they are empty of content, mere tautologies, that do not make any claim on reality. They are artifacts of language, expressions in which the content-full claims that can be made in language cancel each other out. This is a familiar theme with respect to propositional logic, and as we shall see Wittgenstein extends this to include full first-order logic and even some higher-order logic. But logic does not provide a basis for arithmetic, that is, he rejects the logicist reduction and the notion of "logical object" that Frege introduced. What he puts in its place is hard to discern in the few cryptic remarks on arithmetic in the *Tractatus*, although it seems clear that Wittgenstein thought he had at least the basics of an account. My aim in this paper is to gain a better view of what that account might be, and to consider the philosophical grounding it might have had (on which Wittgenstein's text is pretty much silent).

Although Wittgenstein's philosophy of logic had great influence—in particular it was fully taken over by the logical positivists, and lauded as solving the ancient philosophical problem of making empiricism consonant with the existence of *a priori* science—Wittgenstein's rejection of the logicist reduction was simply ignored. The positivists maintained that arithmetic (and higher mathematics generally) did reduce to logic, and that logic was composed of tautologies. Much effort was expended, by Rudolf Carnap in particular, to give this position some plausibility, in the face of the difficulties in *Principia's* attempt to execute the reduction. Wittgenstein's own view of arithmetic (and, presumably, higher mathematics in general) was never investigated to any extent. It was was simply a road not taken.

Nonetheless, it is of interest, I think, to try to puzzle out what Wittgenstein's road was. Although by-and-large idiosyncratic, it shows a few affinities with some contemporary views. It also may evince some unexpected philosophical affinities with one of the most forceful early critics of the foundational ambitions of Frege and Russell's logic, namely, Henri Poincaré.

Poincaré's two principal objections to logicism each claim there is a *petitio principii* in it. In (1909, p. 469), after briefly explaining Russell's theory of types, Poincaré concludes, "Th[is] theory [...] is incomprehensible if one does not assume an already constituted theory of ordinals. How could one then found the theory of ordinals on the theory of types?" This can seem a reasonable objection, since one does need the integers in order to explain the hierarchy of types, and so there can seem to be some circularity here. Russell subsequently addresses Poincaré's objection.

> [His] assertion appears to me to rest upon a confusion. That the types have an order is admitted; but it is not admitted that it is necessary to study this order as an order. [...] We can make all the uses of them that are required without studying the order, just as we can distinguish a function $\varphi(x)$ from a function $\varphi(x,y)$ without knowing that the first has one argument while the second has two... So, with types, we may speak of their order in words which, strictly speaking, involve a knowledge of the ordinals, because it is obvious that we could make all the necessary uses of types without such words... Thus, although types have an order, the ordinals are not presupposed in the theory of types, and there is no logical circle in subsequently basing the theory of ordinals upon a basis which assumes the theory of types. (1910, p. 452)

Russell's reply is, perhaps, less crisp than it should be, since he does not make explicit what the "necessary uses" of types are. But the point should be clear enough to us: the theory of types is to provide proofs of the laws of arithmetic; the principles used in those proofs do not presuppose the ordinal structure of the theory of types, they presuppose only particular principles from the theory of types.

It may be objected, though, that to ascertain whether a putative proof of a law of arithmetic is in fact a correct one solely within the theory of types, we would have to be cognizant of the ordinal structure of that theory. So, the "necessary use" of types in a *particular proof* might not require the ordinal structure, but to ascertain that the proof is indeed a correct one in the theory of types would require just that knowledge.

This objection is at bottom a more general one which applies to *any* axiomatic system of logic, not particularly the theory of types. It goes thus: in order to formulate the system, we inductively define the formulas, the notion of derivation, and so on. But then, it is claimed, the very foundations of logic seem to require mathematical induction, and so it would be a *petitio* to claim that arithmetic can be grounded on such a system. (The general objection is well-formulated in Parsons 1965, pp. 167–168.)

This objection may appear to have force against logicism, but I believe the appearance is misleading. Its seeming force stems from the fact that, in contemporary logic, formal systems are automatically treated from a metatheoretical standpoint. Investigation of the status of a proposition begins with a metamathematical assertion about its provability in some system.

With respect to classical logicism, though, the formal system plays a different role. As Russell's reply indicates, we have to distinguish between what it is to employ logic (and nothing else) to justify one or another mathematical assertion, and what is involved in articulating a system of logic and verifying that we are working within it. To give the ultimate basis for a proposition is to give the actual proof inside the system, starting from first principles: it is to assert the proposition with its ground, not to assert the meta-proposition "this sentence is a theorem". The role of formality in logicism is limited: it enables us to be sure that all the principles needed for the justification of the proposition have been made explicit, but it is neither essential to the nature of the justification nor constitutive of its being a justification.

Frege explicitly mentions a similar issue:

> A delightful example of the way in which even mathematicians can confuse the grounds of proof with the mental or physical conditions to be satisfied if the proof is to be given in to be found E. Schröder. Under the heading "Special Axiom" he produces the following: The principle I have in mind might well be called the Axiom of Symbolic Stability. It guarantees us that throughout all our arguments and deductions the symbols remain constant in our memory—or preferably on paper (1884, p. viii).

Of course, if we are to be able to do mathematics, the signs will have to remain constant before our eyes; but that hardly implies that mathematics presupposes the physics of inkblots. The stability of symbols is a practical necessity, but only that. Similarly, it may be a practical necessity for us, in order to be sure that we are proceeding logically, to verify proofs by syntactic means. That may well require abilities of, e.g., counting, where this could be counting types, or counting variables, or counting parentheses; but any such practical necessity is merely that, and is irrelevant to the rational basis of the subject matter.

The distinction here undergirds the sharp divide between the logical and the psychological that Frege and Russell thought central. But this distinction goes deeper than that: for it also is needed for the distinction between the logical and the empirical more generally, and, even more, between what pertains to the justification of a claim on the one hand, and the "accidental" features of how we arrive at the claim, or the empirical necessities we have to satisfy to arrive at the claim or even the necessities—both empirical and conceptual—we have to satisfy to come to recognize the justification of a claim *as* a justification of the claim, that is, to verify it *as* a proof.

At this point, I will eventually suggest, Wittgenstein parts company with early logicism. I believe that he does not agree that there is a sharp and principled distinction between conditions that need to be satisfied if a proof is to be given or verified, and the rational basis of the proof. My ascription of this difference to Wittgenstein is highly conjectural: it is not explicit in any texts, and comes out only in a backhand way, as we shall see.

Poincaré's second objection to logicism is more mathematical, and was highly influential. In his analysis of the paradoxes, most prominently those of Richard and Burali-Forti, Poincaré descries what he calls "vicious circles", and he formulates an informal constraint to preclude them and thus block the paradoxes. He goes on to apply this analysis to criticize the logicist definition of number. In the definition of finite number as a number that belongs to every inductive set, to avoid vicious circles Poincaré claims that the inductive sets invoked cannot include those specified by reference to the set of finite numbers. But then mathematical induction for such sets cannot be derived from the definition, and so logicism crumbles. This is just Poincaré's urging that definitions must be predicative. Russell agrees, and the basic structure of *Principia* is the ramified theory of types, which prohibits impredicativity. To enable the derivation of arithmetic, Russell introduces the *ad hoc* Axiom of Reducibility. This restores the mathematical power of impredicative definitions; but has no justification other than its yielding the desired results.

II

Wittgenstein's differences with his logicist forebears come out directly in how he takes logic to be constructed, and the idiosyncratic way he has of conceiving its extent. By the extent of logic here I mean: what is the expressive power that can be captured by the logical operations as the *Tractatus* characterizes them, and what is the resulting class of validities. Wittgenstein's central thesis is that all meaningful propositions are truth-functions of elementary propositions, and the elementary propositions are logically independent of each other. So the question of the extent of logic comes down to this: what truth-functions of elementary propositions are expressible, and what can be said about the subclass which are tautologies, that is, true no matter what the truth-values of the elementary propositions of which they are composed? (I assume that there are infinitely many elementary propositions; for the question is trivial if there are only finitely many.)

At *Tractatus* 5.5, Wittgenstein tells us that every truth function comes from successive applications of the operation of joint denial to a stipulated range of propositions. Joint denial is the operation that in Russellian notation would be symbolized as a conjunction of negations. (The joint denial of p and q is $\sim p \cdot \sim q$. Wittgenstein, following Russell, is clearly impressed with Sheffer's discovery in 1912 that joint denial is expressively adequate, that is, generates all the finitary truth-functions.) But Wittgenstein allows joint denial to be applied not just to a pair of propositions, but to any range of propositions that can marked out. At 5.501, he tells us of three ways that such a range can be stipulated.

First is by direct enumeration, that is, by providing a list. This is clearly limited to the finite case; "we can simply substitute for the variable the constants that are its values". Thus one obtains all the finitary truth-functions of the elementary propositions. For example, joint denial applied first to the list p, q gives us the joint denial of p and q; applying joint denial to the list containing just that proposition yields its negation, a proposition we would express in Russellian notation as $p \vee q$.

The second way of stipulating a range of propositions is by means of a "function fx whose values for all values of x are the propositions to be described". Wittgenstein here means a propositional function, a Russellian notion that he construes linguistically. A typical such function would be obtained by starting with a proposition and turning a name into a variable, as a means to indicate the range of propositions one gets by putting all the different other names that can fit there (3.315). We might start with a proposition Fa, replace a with a variable x, obtaining a sign Fx for a function, which then can symbolize the range of all propositions obtained by putting in other names for x. Applying joint denial to this range yields the truth-function of all propositions Fb that is true if all those are false; that is, the proposition we would write as $(\forall x) \sim Fx$. From this we can obtain $(\exists x)Fx$ by negation, i.e., by denying all propositions in the range given by listing just $(\forall x) \sim Fx$ by itself. In short, using this second method of stipulating a range of propositions, Wittgenstein obtains the expressive power of the quantifiers. Let me give another example. Starting with the proposition

Jupiter is bigger than Mars

we can replace the expression "Jupiter" with a variable, obtaining

x is bigger than Mars

which is a notation, Wittgenstein thinks, for the range of propositions obtainable by replacing x with any name that yields a meaningful proposition. Applying joint denial to this range we obtain

$(\forall x) \sim (x \text{ is bigger than Mars})$

We may now take this proposition and replace "Mars" with a variable:

$(\forall x) \sim (x \text{ is bigger than } y)$

This now is a notation for a range of propositions, to which we may apply joint denial once again, obtaining the proposition that, in standard notation, reads:

$(\forall y) \sim (\forall x) \sim (x \text{ is bigger than } y)$

that is,

$(\forall y)(\exists x) \sim (x \text{ is bigger than } y)$

By this route, we obtain all of first-order logic, with arbitrarily complex nestings of quantifiers.

This is a rough-and-ready account, which has to be modified slightly to incorporate Wittgenstein's view that identity is not a relation and consequently in nested quantification the different quantifiers must take different values. How to accommodate this in the above account has been decisively described by Rogers and Wehmeier (2012). Thus, it looks like this second method of stipulation of ranges of propositions, together with Wittgenstein's conception of how variables operate, gives us first-order logic. (There was a controversy in the early 1980s between Robert Fogelin and Peter Geach about whether Wittgenstein could indeed obtain full first-order logic. This turned out to be a notational argument: in the *Tractatus* Wittgenstein does not provide a notation adequate to the distinctions needed. I believe he didn't intend to do so; he uses only enough notation to make his basic point. Rogers and Wehmeier show in rigorous detail how slight extensions of the notation Wittgenstein does use—extensions that are completely Russellian in spirit—yield an adequate language.)

The same notion of stipulation of a range by means of a propositional function would also give us some higher-order quantification. Wittgenstein does not mention this specifically at all, but it follows from his view that any expression in a meaningful proposition can be changed into a variable, yielding a propositional function. Now Wittgenstein does insist that a quantification is a truth-function of its instances; this rules out impredicative higher-order quantification. So, I take it, Wittgenstein would construct some form of ramified type theory, although it may be very restricted, that is, the higher-order variables range over smaller subclasses of Russell's orders. Ricketts (2013) gives a plausible reconstruction of such a view.

So far, despite its different point of origin, we are on common ground with Frege and Russell (and most other logicians of the early 1900s). However, Wittgenstein allows a third type of stipulation of a range of propositions to which joint denial may be applied, and this is new and otherwise unknown in that period. The method is to give "a formal law that governs the construction of the propositions, in which the bracketed expression has its members all the terms of a form-series".

Now "formal law" does not occur in any other remark in the *Tractatus*. In *Prototractatus* 5.00534 Wittgenstein elaborates "the values of the variable are all propositions that possess certain *formal* properties", and adds "This kind of generalization, which can be called the *formal* kind, was overlooked by Russell and Frege." So, it appears, the terms of anything appropriately deemed a "form-series" [*Formenreihe*] can be the range of propositions suitable for application of joint denial, that is, any succession of propositions which can be generated by formal means (in some appropriate sense of "formal"). What this third method of stipulation gives us will depend on what form-series there are. Now there is only one form-series that Wittgenstein explicitly mentions that is not truth-functional, namely (4.1253, 4.1273), the sequence

$$Rab, \quad (\exists x)(Rax \cdot Rxb), \quad (\exists x)(\exists y)(Rax \cdot Rxy \cdot Ryb), \quad \ldots$$

The joint denial of all these propositions says that b does not follow in the R-series after a. The negation of that is the infinite disjunction of those propositions; it says that b *does* follow in the R-series after a, that is, that b stands in the ancestral relation of R to a (that is, the transitive closure of R). Thus

Wittgenstein's logic yields a direct definition of the ancestral, without any use of higher order or set-theoretic quantification. Here is one support for Wittgenstein's striking assertion, "The theory of classes is completely superfluous in mathematics" (6.031).

To be more exact, the logicist reduction of arithmetic has two parts: first, the definition of numbers as classes of equinumerous classes; and second the use of the ancestral to pick out the finite numbers from among all the numbers, and to deductively yield mathematical induction. Wittgenstein rejects the first, out of his general view that there are no "logical objects"; *a priori* disciplines cannot be a matter of *entities*. (This is a view he retains throughout his later work as well.) About the second, he thinks it is not just unnecessary, for the reason I've just indicated, but it is in fact fallacious: in 4.1273 we read:

> [T]he concept 'term of the series of forms' is a *formal concept*. This is what Frege and Russell overlooked: consequently the way in which they want to express general propositions like the one above [the ancestral] is incorrect; it contains a *circulus vitiosus*.

Wittgenstein's use of the Latin phrase is, I would conjecture, a reference to Poincaré's second objection.

There is more than the ancestral that we can obtain by the third means of stipulation, still considering only first-order applications. Here we must extrapolate a bit, since Wittgenstein does not tell us much about what counts as a form-series. But it would be hard to deny that the generating relation in the series

$$(\exists_1 x)Fx, \qquad (\exists_2 x)Fx, \qquad (\exists_3 x)Fx, \qquad \ldots$$

has whatever characteristics are required. In that case, one can express "there are finitely many Fs", which is also not a first-order concept.

Thus the use of the third method of specification gives us quite a bit more power. It also renders the notion of tautology (i.e., logical truth) highly complex, computationally speaking. If we assume there are other relation signs in the language, then it is easy to find a mapping from sentences F of modern first-order arithmetic into the system of the *Tractatus* such that F is true in the standard model if and only if the image of F is a tautology. Hence the property of being a tautology is at least Δ_1^1.

Of course, given that Wittgenstein is able to capture first-order logic his notion of "being a tautology" is not decidable. Hence one has to give up the idea that when Wittgenstein says, in 5.13, "When the truth of one proposition follows from the truth of others, we perceive [*ersehen*] this from the structure of the propositions", that his "perceive" is meant to connote ascertainability. But still it could be maintained that there is a connection with ascertainability, in that the tautologies of first-order logic admit of a proof procedure in an effective way: there is an effective way of our coming to see them. (This may be in back of Pears and McGuinness's translation of *ersehen* as "can see", with the added modal.) The inclusion of the third method of stipulation, though, *completely* sunders logical truth from ascertainability. "Perceiving it in the sign", and hence Wittgenstein's general notion of what is "shown", has nothing to do with epistemology.

Of course the notion of computability was not developed until later, and I don't want to be too anachronistic here. But given that Wittgenstein knew his characterization allowed the definition of ancestral, and hence the development of arithmetic, I can't imagine that he thought it was going to follow that it would always be ascertainable that something was a tautology, in the way it is for finitary truth-functions.

I've noted that there will be a mapping from the language of (modern) first-order arithmetic to Wittgenstein's logical language such that a formula F of the former is true in the standard model iff its image is a tautology. In simple cases, this mapping can be easily framed. $n + m = p$ is mapped to $aR^n b . bR^m c \supset aR^p c$; one doesn't even need quantifiers in this other than those inside the relative products. This is in keeping with Wittgenstein's pronouncement at 6.021 that "A number is the exponent of an operation", since we are dealing with an operator that here generates the successor sequence, namely relative product. (Admittedly this rendition of the equation does not have the exact syntactic form of what he lays out in 6.02.) But I think we can now see how to extend the idea of "number as exponent of an operation" to encompass more complex number-theoretic statements (unlike the examples in 6.02). In identifying numbers with the exponent of the relation R, we are thinking of numbers as places in the R-series starting with (an arbitrary) a. So, for example, the commutative law of addition would be translated as

$$\Phi(R, A, a) \supset (\forall x)(\forall y)(\forall z)(R^* ax . R^* ay \supset (Axyz \equiv Ayxz))$$

where $\Phi(R, A, a)$ states that R and A act like successor and addition when a is taken as zero, and R^* is the ancestral of R. Here again the correctness of the arithmetical statement is claimed to be based on the tautologousness of its translation. In a way, this is a structuralist view of arithmetic, in the sense of mathematical structuralism inspired by Dedekind.

I should point out that, in my example, the commutative law of addition is not thereby reduced to a tautology. Rather, it is the tautologousness of all propositions with that form, no matter what the actual relations that appear in them, that characterize the force of the law; it is, so to speak, distributed among all its applications to the language. (So I take the "Ω" that Wittgenstein uses to indicate an arbitrary operation in 6.02 not to be a variable, but rather a schematic letter. Here I differ from the account given in Frascolla 1994.)

I do not claim that Wittgenstein worked this out. But I do think his rejection of numbers as objects, and his thinking of them as exponents of operators, does have a structuralist cast to it. And what does seem clear (as biography) is that Wittgenstein thought he had solved the problems: the ontological one by identifying numbers as exponents of operators, and the doctrinal one—that of vouchsafing mathematical induction—by the invocation of formal series.

Wittgenstein also seems to take for granted that solving the problems of philosophy of arithmetic would solve the problems of philosophy of mathematics. This is the view he grew up with, in Russell, but of course Russell could sustain it only because the theory of types (with reducibility) could interpret enough class theory to generate the real numbers from the integers. Without class theory, and without anything more than a weak form of ramified theory of types, Wittgenstein has no means to ground the classical theory of the real numbers. Nor is there any attempt to deal with the issue in the *Tractatus*.

(In later work, it should be noticed, it is when Wittgenstein talks about the real numbers that he sounds at his most revisionary. "There is no system of irrational numbers", he says in *Remarks on the Foundations of Mathematics*, II, §33.)

III

Thus Wittgenstein's logic, at the first-order level, is much stronger than standard first-order logic. The calibration of this requires tools developed only much later; and it still open exactly what its strength is. But there is also an interesting philosophical problem here in trying to make sense of his approach, namely, to see what motivated him in this direction. This is a reflection forced on us by the difference between what he was urging (that is, his third method of stipulation of a range of propositions), and the standard logic in his predecessors and in later logic as well.

Now Russell's way of describing how logic operates tends to oscillate between extensional and intensional. On my view, Wittgenstein is trying to give a consistent view based entirely intensionally, although the logic itself is entirely extensional. (For the importance of the modality in Wittgenstein's talk of "truth-possibilities", see Ricketts 2014.) Hence his requirement that to make sense of a quantified proposition, it must be given a ground-up construction, that requires an independent stipulation of values; as well as his adherence, with respect to higher order quantifiers, to a predicativity constraint.

So I take it that the implicit argument for going beyond first-order logic at the ground level, for taking the "..." of a form-series as primitive and not to be analyzed à la Frege, must work like this (this is completely speculative, as there is no trace of an argument in the *Tractatus* or in the pre-Tractarian documents): it is sufficient to understand a quantification that the range of propositions (to which the joint denial is applied) be stipulated intelligibly. But in fact such a stipulation can be done by a form-series equally intelligibly as by a propositional function. Why is this? Because our understanding of language already gives us an understanding of form-series. Hence there is no reason to limit logic to just the stipulation of ranges of propositions by explicit listings and by propositional functions. In other words: whatever is required to make sense of first-order logic will also license this more powerful type of quantifier.

That is, the argument might be framed thus: whatever is involved in the understanding (or the intelligibility) of language already gives us the understanding of (or the intelligibility of) form-series, and so gives us the more powerful quantification. Thus 6.233:

> The question whether intuition [*Anschauung*] is needed for the solution of mathematical problems must be given the answer that here language itself provides the necessary intuition.

But now we can see a contrast between Wittgenstein and the early logicists. Russell would certainly balk at any idea that *Anschauung* was at work in logical knowledge. More deeply, here we see Wittgenstein rejecting precisely the distinction that both Frege and Russell use to fend off Poincaré's first objection, that is, the distinction between what we might (as a matter of practical or even conceptual necessity) have to be equipped with to understand a proof or to verify it, and what the proof displays as the rational basis of the claim that is proved. And this distinction, I noted in the first section, was the basis for the principled distinctions Frege and Russell thought so important in the development of logic: between the logical and the psychological, between the logical and the empirical, and, in general, between the logical and anything extra-logical. So if my speculation is on target, we see here a much deeper divide between Wittgenstein and his logicist forebears than is usually noted. Let me close with three further points about this.

First: Wittgenstein may well have been aware that he was not signing on to the way of thinking that underwrote Frege's (and possibly Russell's) anti-psychologism. This comes out in 4.1121, which for many years I had found puzzling, when thinking of Wittgenstein simply as heir to Fregean and Russellian ways of thinking:

> Does not my study of sign-language correspond to the study of thought-processes, which philosophers used to consider so essential [...]? Only in most cases they got entangled in unessential psychological investigations, and with my method too there is an analogous risk.

By rejecting the Frege-Russell way of underpinning anti-psychologism, the boundary between logic and psychology is going to be less clear; yet Wittgenstein had, as he took it, an equal commitment to anti-psychologism.

The danger here shouldn't be overemphasized, though. Early logicism needs to insist on this distinction—between the rational grounds of an assertion and the conditions for coming to see that those grounds are the grounds—because for them the rational grounds of an assertion are external to it. Frege was insistent on this, and it was indeed one of his innovations: one cannot tell how the justification of an assertion will look from the content of that assertion. (See Frege 1884, §3.) This is one of his explicit divergences from Kant. In this regard, Wittgenstein is returning to an earlier tradition (most closely related to Leibniz). For him there are no external justifications; the only thing that can justify an assertion are its truth-grounds. (This relegates any further question about justification to psychological matters, where they have mostly remained.)

Second, Wittgenstein's rejection of the Fregean distinction is completely continuous with his later thought. Exactly the same point, expressed in a vein much like Poincaré, lies in his criticism in *Remarks on the Foundations of Mathematics* III, of Russell's formalization of "there are seven F's", namely, that we cannot understand the formalization without being able to count the various signs in it, and hence the formalization cannot ground our understanding of numerical discourse.

There is an important continuity between early Wittgenstein and later that this line of thought exhibits; it lies deeper yet. My speculation has the Wittgenstein of the *Tractatus* attending to what we need to be able to do, in order to operate with a language, and as such factors dictating what the logic of the language is. The priority of abilities to representations is a major theme of the *Philosophical Investigations*, expressed in the repeated third-man arguments, about color samples, signposts, mental pictures, and (most famously) rules. If I'm right, we can see some concern with this, in embryo, in the *Tractatus*. Perhaps it might be argued that in the early work he's trying, through the notion of *showing*, to put abilities into the representations, instead of recognizing the insufficiency of representations to do this or that philosophical task, because representations within language simply cannot suffice. This raises the question of what role human abilities play in the *Tractatus*. The concept is absent in any explicit way, but is suggested from time to time (e.g., in "2.1 We make pictures of facts to ourselves"). I'm not yet sure what to say about this. But I suspect that in this issue we are somewhere in the neighborhood of the deepest insights, and deepest instabilities, in the workings of the *Tractatus*.

On the technical side, Wittgenstein's idea that the "..." of an infinitary definition can be taken as primitive when (and only when) certain formal conditions are met has not been explored in later logic. Whether a viable alternative to logicist or set-theoretic foundations, at least in some areas, lies in this direction awaits further elaboration. In discussion W.W. Tait cast doubt on this, remarking with his usual acuity that mathematical induction is obtained, on such a view, simply by building it in at the start, so that no analysis of it is possible. He did not say it, but I think he would have been happy to apply Russell's famous epithet in this case: that the method has "the advantages of theft over honest toil" (Russell 1919, p. 71). But even if we remain doubtful about the usefulness of Wittgenstein's idea in providing a proper foundation for arithmetic, perhaps there are intriguing things to discover by developing it as an alternative logic.

Acknowledgments. I am grateful to many discussants at presentations I have made of this paper, who raised a large number of stimulating and helpful points.

I would like to thank most especially James Conant, Peter Hylton, Andrew Lugg, Kai Wehmeier, and, as always, Thomas Ricketts.

References

Frascolla, Pasquale (1994), *Wittgenstein's Philosophy of Mathematics*. London: Routledge.

Frege, Gottlob (1884), *Die Grundlagen der Arithmetik*, translated by J.L. Austin as *The Foundations of Arithmetic*. Oxford: Basil Blackwell, 1959.

Parsons, Charles (1965), Frege's Theory of Number. In Parsons (1983), 150–175.

Parsons, Charles (1983), *Mathematics in Philosophy*. Ithaca, NY: Cornell University Press.

Poincaré, Henri (1909), La logique de l'infini. *Revue de Métaphysique et de Morale 14*, 294–317.

Ricketts, Thomas (2013), Logical Segmentation and Generality in Wittgenstein's *Tractatus*. In P. Sullivan and M. Potter, eds., *Wittgenstein's Tractatus*, pp. 125-142. Oxford: Oxford University Press.

Ricketts, Thomas (2014), Analysis, Independence, Simplicity, and the General Sentence Form. *Philosophical Topics* 42, 263–288.

Rogers, Brian and Wehmeier, Kai (2012), Tractarian First-Order Logic: Identity and the N-Operator. *Review of Symbolic Logic 4*, 538–573.

Russell, Bertrand (1910), The Theory of Logic Types. In Russell (1973), pp. 215-52.

Russell, Bertrand (1919), *Introduction to Mathematical Philosophy*. London: George Allen and Unwin.

Russell, Bertrand (1973), *Essays in Analysis*. D. Lackey, ed. New York: Braziller.

Wittgenstein, Ludwig (1961), *Tractatus Logico-Philosophicus*. Translated by D. F. Pears and B. F. McGuinness. London: Routledge and Kegan Paul.

Wittgenstein, Ludwig (1971), *Prototractatus*. B. F. McGuinness, T. Nyberg, and G. H. von Wright, eds. London: Routledge and Kegan Paul.

Wittgenstein, Ludwig (1978), *Remarks on the Foundations of Mathematics*. Cambridge, MA: MIT Press.

Eudoxus' Theory of Proportion and His Method of Exhaustion

Stephen Menn[*]

ABSTRACT: Euclid's theory of proportion in *Elements* V is an astonishingly abstract piece of foundational reasoning. We need some historical explanation to make sense of what Euclid is doing. One clue comes from *Elements* XII's "method of exhaustion" for proving theorems about areas and volumes, which is much easier to motivate. Archimedes tell us that the results of XII go back to Eudoxus; an anonymous scholium says that Eudoxus somehow lies behind V. Did Eudoxus' versions of V and XII fit together in a single project? The methods of V and XII seem to have something in common, but it is hard to formulate precisely what; and XII makes surpris-

---

[*] I would like to thank Fabio Acerbi, Vincenzo De Risi, Henry Mendell, Howard Stein, the late Ian Mueller, and audiences at the University of Chicago (2002) and McGill University (2014), for their comments, corrections, and references to the literature. What I am presenting here comes mainly out of my reflections on an article by the late Wilbur R. Knorr, "Archimedes and the Pre-Euclidean Proportion Theory," *Archives Internationales d'Histoire des Sciences* 28 (1978), 183–244, and it is possible that Knorr had already thought, or even that he had already said, most of what I am going to say here. But if so, it was not obvious to me when I read his article, and after many rereadings I am still not sure (and neither was Mueller, and it was too late to ask Knorr). Knorr certainly said many other things that I would not want to say. Due to needs of exposition, I will have to spend much of my time going over older research, much of it antedating Knorr. This paper was originally written before the papers of Acerbi ("Drowning by Multiples: Remarks on the Fifth Book of Euclid's *Elements*," *Archive for History of Exact Sciences* 57 (2003), 175–242) and Ken Saito ("Phantom Theories of Pre-Eudoxian Proportion," *Science in Context* 16 (2003), 331–47) criticizing Knorr, and Mendell's paper developing Knorr's views ("Two Traces of Two-Step Eudoxan Proportion Theory in Aristotle," *Archive for History of Exact Sciences* 61 (2007), 3–37), were published; I cannot give full replies to these papers here, but will reply in passing to some points they raise. This paper was completed within the framework of the Topoi Excellence Cluster, research group D-3. I am pleased to dedicate the paper to Bill Tait; Bill, together with Ian Mueller and Howard Stein, first got me thinking about the history and conceptual development of mathematics, when I was a grad student at Chicago in the 1980's.

ingly little use of V. Refining and building on Wilbur Knorr's argument (using texts of Archimedes) that Eudoxus' proportion theory differed from Euclid's, I give a hypothetical reconstruction of Eudoxus' theory, show how he would be led to develop it in working on exhaustion-proofs, and show how and why Euclid would transform it into the theory of *Elements* V.

1 The problem about *Elements* V, *Elements* XII, and Eudoxus

Euclid's theory of proportions, given in *Elements* Book V, has long been an object of admiration and puzzlement. Euclid pursues his investigation at a level of abstraction, and holds himself to standards of rigor, that are in the context of ancient mathematics astonishing. Euclid apparently feels that the most elementary manipulations of proportions—the deepest result here is "alternation," that if $A : B :: C : D$, then $A : C :: B : D$—are not scientifically justified until they have been established by the difficult arguments of Book V, and in Books I-IV he reconstructs much of elementary plane geometry (including the Pythagorean theorem, the squaring of arbitrary polygons, and the construction of the regular pentagon) in such a way as to avoid any mention of proportion or similarity. Then in Book V he makes a new start, speaking not about lines or areas or any particular geometrical species of quantity, but about quantity in general. He gives abstract definitions, not precisely of ratio or proportion, but of what it is for two quantities to have a ratio, of being in the same ratio, and of being in a greater ratio (Vdef4, Vdef5, and Vdef7: in modern notation, A and B have a ratio iff $\exists p\, pA > B$ and $\exists p\, pB > A$; $A : B :: C : D$ iff $\forall p \forall q((pA > qB \leftrightarrow pC > qD)$ and $(pA = qB \leftrightarrow pC = qD)$ and $(pA < qB \leftrightarrow pC < qD))$; $A : B > C : D$ iff $\exists p \exists q(pA > qB$ and $pC \leq qD))$.[1] Then, by patient and ingenious manipulation of these definitions, he establishes alternation (V,16) and the other rules for manipulating proportions; a crucial link in the argument is the seemingly trivial V,8, if $A > B$ then $A : C > B : C$, again elaborately demonstrated from the definitions. Only after establishing these foundations does Euclid return, in Book VI, to the part of plane geometry that makes ineliminable use of similar figures, justifying his manipulations of proportions by the theorems of Book V. But the application of proportion theory to geometry cannot get off the ground unless we first show that *some* pair of geometrical quantities are proportional to some other pair (the theorems of Book V only show that *if* one proportionality holds, another holds too), and so Euclid needs to apply his definition of "being in the same ratio" to some geometrical configuration. By an amazing economy, Euclid manages to do this just once, in VI,1, "triangles or parallelograms of the same height are to each

---

[1] All lower-case variables will range over positive integers. To stay closer to Euclid, they should range over positive integers *greater than one*, which is what he means by "numbers," but I will ignore this complication.

Eudoxus' Theory of Proportion and His Method of Exhaustion

other as their bases." Euclid then builds the whole theory of similar polygons out of VI,1, using only the rules of proportion and elementary proportion-free geometry, without citing the definitions of proportion again.[2]

Given that it takes him so much work to prove the obvious-sounding rules of proportion, and given the contortions to which he must put elementary geometry to reduce its dependence on proportion theory, why does Euclid do it? There had certainly been earlier treatments of the elements based on a naive acceptance of the rules of proportion—why not continue this tradition?[3] A once popular view (the view, for instance, of T.L. Heath) held that the discovery of incommensurability (sometime around 430 BC) led to a crisis in proportion theory, which had originally been based on the assumption that all magnitudes of a given kind (e.g. all lengths or all areas) are commensurable (the definition of proportion would have been something like "$A:B::C:D$ iff $\exists p \exists q (A = pB/q$ and $C = pD/q)$"—essentially Euclid's definition of proportion among *integers*, VIIdef20).[4] This crisis would have led mathematicians to abandon the use of proportion in geometry, and to work on reconstructing as much as they could of geometry without proportion, yielding something like Euclid I-IV; the rest of geometry could not begin again until someone had worked out something like Euclid V, giving the foundations for a general theory of proportion that would apply to incommensurable as well as commensurable magnitudes.[5]

---

[2] Except in the last proposition of Book VI, VI,33, about circles rather than about rectilinear figures, whose argument closely mimics the proof of VI,1. The only other propositions of the *Elements* beyond Book V that cite the definition of proportion are XI,25 and XII,13, both of which, again, closely mimic the proof of VI,1.

[3] The squarings of lunes (figures bounded by two arcs of circles) by Hippocrates of Chios (toward the end of the fifth century BC), as described by Aristotle's student Eudemus and copied from him (with some commentary) by Simplicius *In Physica* 60,18–68,32, make free use of proportion theory. Euclid's proof of the Pythagorean theorem in I,47, a *tour de force* avoiding proportion theory, is clearly modelled on the earlier proportion-theoretic proof which Euclid gives in VI,31. (There are *other* proportion-free ways of proving I,47 which are not modelled on proportion-theoretic proofs, and may well have been older, but this is not true for the very clever and unnatural proof that Euclid actually gives.) Likewise the proportion-free squaring of the rectangle in II,14 is clearly modelled on the construction of the mean proportional given in VI,8 and VI,13 (with consequences for a square equal to a rectangle in VI,17); Aristotle testifies (*De Anima* II,2 413a13–20 and *Metaphysics* B2 996b18–22) that the problem of squaring a rectangle was seen as essentially the problem of finding a mean proportional between the two sides. Likewise the construction of the regular pentagon in *Elements* IV,10–11, using the proportion-free construction at II,11, is clearly modelled on a construction using the proportion-theoretic VI,30. Knorr, unlike everyone else, denied in *The Evolution of the Euclidean Elements* (Reidel, 1975), esp. pp. 6–9, that Euclid was transforming earlier arguments to avoid proportion; but Knorr recanted and rejoined the consensus in "What Euclid Meant" (in *Science and Philosophy in Classical Greece*, ed. Alan C. Bowen (Garland, 1991), 119–63).

[4] Heiberg prints this definition as VIIdef21, but brackets def10 as an interpolation, so it becomes def20, which is what Heath calls it; I will keep on citing it by this name.

[5] "The essence of the new theory [of proportion] was that it was applicable to incommensurable as well as commensurable quantities; and its importance cannot be overrated, for it enabled geometry to go forward again, after it had received the blow which paralysed it for the time. This was the discovery of the irrational, at a time when geometry still de-

In fact it is extremely unlikely that mathematicians ever stopped doing proportion-theoretic geometry. If there were no definition of proportion—well, Greek mathematicians had no definition of equality either. Why not just assume the rules of proportion as common notions (like "equals added to equals are equals"), and VI,1 as a postulate? It may be inelegant to have so many unproved first principles, but it is perfectly feasible, and indeed many modern textbooks of geometry, based on Euclid, skip Book V as too abstract and difficult and proceed without it. Rather than a foundational crisis caused by the discovery of incommensurability, there was something like a nagging foundational nuisance, and Aristotle gives us two bits of evidence that there were (at least two) attempts to overcome it, which I will discuss in the next section. But even if a scrupulous Greek mathematician would have found something like Euclid V logically desirable, this gives no explanation for how someone would have been able to come up with such an astonishing Dedekind-like piece of abstraction.

One clue comes in the scholia to Book V saying that "some say that this book is the discovery of Eudoxus, the teacher of Plato" (Book V, scholium 1, lines 7–8) and that "this book is said to belong to Eudoxus of Cnidus, the mathematician who lived in the time of Plato, but nonetheless it is ascribed to Euclid, and not by a false ascription: for nothing prevents it from belonging to someone else as far as discovery goes, but as far as the ordering κατὰ στοιχεῖον [= proposition-by-proposition?] of these things and the ἀκολουθία [= logical consequence?] of the things so ordered, everyone agrees that it belongs to Euclid" (scholium 3, lines 1–7).[6] Now, of itself, substituting the name "Eudoxus" for the name "Euclid" does not accomplish much. But we have other information about Eudoxus' work in geometry: the Euclid scholia also attribute to Eudoxus XII,2 and XII,10, the theorems that circles are to each other as the squares of their diameters and that a cone is equal to one-third of the cylinder of the same base and height, and Archimedes also attributes the theorem on

---

pended on the Pythagorean theory of proportion, that is, the numerical theory which was of course applicable only to commensurables. The discovery of incommensurables must have caused what Tannery calls 'un véritable scandale logique' in geometry, inasmuch as it made inconclusive all the proofs which had depended on the old theory of proportion. One effect would naturally be to make geometers avoid the use of proportions as much as possible; they would have to use other methods wherever they could" (T.L. Heath, *A History of Greek Mathematics* (2 vols., Oxford University Press, 1921), v.1 p. 326).

[6] The scholia to the *Elements* are in Euclidis *Elementa*, ed. I.L. Heiberg, vol. 5 (Teubner, 1898), which is also volume 5 of Euclidis *Opera Omnia*, edd. I.L. Heiberg and H. Menge (Teubner, beginning 1883). The scholia I have just cited, of unknown authorship, date and reliability, are the only evidence connecting *Elements* V with Eudoxus (as we will see, the case is better for *Elements* XII): the popular assumption that Euclid's theory of proportion can be straightforwardly credited to Eudoxus rests on very little. Scholium 1 to *Elements* V is a long connected text; it has often been suggested (e.g. by Heath, *Euclid's Elements* II,112) that it is an excerpt from Proclus. Scholium 3 is found only in some very late manuscripts, and it is possible that it is basing itself on no authority other than Scholium 1. It is similar to Scholium 1 to Book XIII, which likewise attempts to reconcile Euclid's authorship with reports attributing many of the theorems of the book to earlier mathematicians by saying that Euclid introduced the ordering by στοιχεῖα.

the cone (apparently also the theorem on the circle) to Eudoxus, and also says that while Democritus was the first to state the theorem on the cone, Eudoxus was the first to demonstrate it.[7] These two theorems are demonstrated by what has come to be called the "method of exhaustion," and Archimedes is clearly crediting Eudoxus with inventing this method. It is natural to suppose that the Book V theory of proportion and the Book XII method of exhaustion were somehow connected for Eudoxus. XII,2 is asserting a proportionality, and XII,10 is embedded in a group of connected results, proved by the method of exhaustion, many of which assert proportionalities. It is also easy to sense that there is a common style of foundational work in Books V and XII, and, once I have given some technical details, I will say more about what the two developments have in common.[8] And by restoring how the theory of proportion and the method of exhaustion were supposed to fit together for Eudoxus, we might gain more insight into why and how the theory of proportion was developed.

---

[7]The Euclid scholia are Book XII, scholia 12 (to XII,2) and 38 (to XII,10), each consisting of the single word "Εὐδόξου": these and the scholia cited above are the only mentions of Eudoxus in the Euclid scholia. Archimedes in the prefatory letter to *On the Sphere and the Cylinder* attributes to Eudoxus the theorems that the pyramid and the cone are one-third of the corresponding prism and cylinder; in the prefatory letter to the Method he repeats the same attributions and says that Democritus stated the theorem (about both figures?) without proof. In the prefatory letter to the *Quadrature of the Parabola* he says that "the earlier geometers" proved these theorems, and also the theorems that circles are to each other in the duplicate ratios of their diameters and that spheres are to each other in the triplicate rations of their diameters, using the "lemma" that the amount by which a greater figure exceeds a lesser figure, when added to itself sufficiently many times, will exceed any given finite figure— in other words, using the method of exhaustion. He does not name names here, but given the parallelism with the *Sphere and Cylinder* passage (to the same addressee, Dositheus), the similarity in the method of proof of the four theorems, and the fact that the circle theorem is used for the later results in *Elements* XII, Archimedes probably means to ascribe this too to Eudoxus. We also have other pieces of information about Eudoxus, and attempts have occasionally been made to connect the theory of proportions with Eudoxus' other interests, e.g. with his περὶ ταχῶν (on the speeds of the heavenly bodies) or with calendrical studies (so David Fowler, "Eudoxus: Parapegmata and Proportionality," in *Ancient and Medieval Traditions in the Exact Sciences: Essays in Memory of Wilbur Knorr*, edd. Patrick Suppes, Julius M. Moravcsik and Henry Mendell, CSLI Publications, 2000, pp. 33–48). Proclus *In primum elementorum Euclidis* 67,2–8 says that Eudoxus expanded the number of "the so-called general theorems" (possibly identified or associated with the theorems about the different kinds of means or progressions which Proclus also mentions). Proclus (or his source) might be thinking of some of the theorems about proportions which we know from *Elements* V, or of other theorems based on them; it is also possible, as Ken Saito and Fabio Acerbi have suggested, that what Proclus says here might have been the basis of a false ascription of the contents of Book V to Eudoxus. The sources on Eudoxus have been collected by François Lasserre, *Die Fragmente des Eudoxos von Knidos* (De Gruyter, 1966).

[8]So B.L. van der Waerden: "there is ... a certain unmistakable relatedness between the methods of Book V and those of Book XII. Both operate constantly with proportions, both are based on the postulate of Eudoxus cited above, and both use, in a rigorous manner, approximations and inequalities which are used ultimately to prove equalities by a *reductio ad absurdum*" (*Science Awakening*, Oxford University Press, 1961, p. 187; by "the postulate of Eudoxus" van der Waerden apparently means the Archimedean assumption that if $A > B$ then $\exists n \; n(A-B) > A$, in which case this needs some qualification). Heath, *History of Greek Mathematics*, v.1 p. 326, says something similar.

But it has proved surprisingly difficult to say precisely how Book V and Book XII fit together.

It is much easier to see where the method of exhaustion comes from than where Euclidean proportion theory comes from. "Exhaustion" is a modern and rather misleading designation, but the method must have developed out of what were more literally proofs by exhaustion. We can illustrate with XII,2, "circles are to each other as the squares of their diameters." One way to try to prove this is as follows. Take circles C and C', with diameters D and D'; we want to show that $C : C' :: D^2 : D'^2$. Inscribe in each of the circles a sequence of regular polygons with 2^n sides, the squares P_2 and P'_2, the octagons P_3 and P'_3, and so on. Since P_n is always similar to P'_n, it is easy to show that for each n, $P_n : P'_n :: D^2 : D'^2$, and indeed this is Euclid's XII,1. To get from here to XII,2, we might say that the polygons P_n *exhaust* the circle C—that is, that every part of the area of C is contained in the area of some P_n—and that the polygons P'_n likewise exhaust C'. To put it slightly differently, $C = P_2+(P_3-P_2)+(P_4-P_3)+\ldots$, and $C' = P'_2+(P'_3-P'_2)+(P'_4-P'_3)+\ldots$; since each of the successive components of C is to the corresponding component of C' as $D^2 : D'^2$, it should follow that the whole C is to the whole C' as $D^2 : D'^2$.[9] The difficulty is that ancient mathematics, unlike modern mathematics, had no rigorous theory of infinite sums. What Eudoxus' method of exhaustion does is to reconstruct the argument without infinite processes, by recasting it as a *reductio ad absurdum*.[10]

---

[9]The argument sketched would be a straightforward application of V,12, if the statement of V,12 allowed infinite sums. If, as Simplicius *In Physica* 61,5–9 seems to say, Hippocrates of Chios argued for XII,2 and did not simply hypothesize it, this might have been his argument, as suggested by Knorr, "Infinity and Continuity: The Interaction of Mathematics and Philosophy in Antiquity," in *Infinity and Continuity in Ancient and Medieval Thought*, ed. Norman Kretzmann (Cornell University Press, 1982), pp. 112–45 at 133–4. (Knorr uses very dubious methods in arguing for this attribution, but his conclusion may be right.)

[10]Of course, the modern notation of infinite sums is an abbreviation for the $\forall \varepsilon \exists N$ formulation of the convergence of a sum to a limit, and the proof of (the modern equivalent of) Euclid V,12 applied to infinite sums will still, if spelled out, turn on a *reductio ad absurdum*; but it involves quantification over infinite sequences, which the Eudoxian proofs try to avoid. I agree with Knorr, in the article cited in the previous note, that a (vaguely formulated) infinite sum argument (whether due to Hippocrates or not) must have been in the background to the Eudoxian argument given in Euclid XII,2. But Knorr also says that this reconstruction of how the method of exhaustion arose implies that Eudoxus wasn't influenced by philosophers or philosophical concerns in developing the method. As far as I can see, it implies nothing of the kind. What we can say is that Eudoxus finds the older "exhaustion" proof-technique not sufficiently demonstrative. We don't know whether reading or talking with philosophers had influenced him in this, and, even if he came to this conclusion on his own from reflecting on the mathematical situation, it is not clear what criterion we could use to conclude that his concerns were not philosophical. In general, Knorr uncritically assumes a dichotomy of mathematicians and philosophers, the former assumed to be unconcerned with "dialectical" issues, and the latter to be technically incompetent. But what is the ground for denying that e.g. Democritus or Hippias was both? Or Eudoxus himself, who took views on the Forms and on pleasure as the good? (Knorr, in *The Ancient Tradition of Geometrical Problems* (Birkhäuser, 1986), solved the Hippias problem by inventing a second Hippias, and solved the problem of Aristotle's account of the rainbow in the *Meteorology* by declaring the passage

For suppose C is *not* to C' as $D^2 : D'^2$. Then either $C : C' > D^2 : D'^2$ or $C : C' < D^2 : D'^2$, and without loss of generality we can assume the former. Book XII takes for granted the assumption that there is always a fourth proportional X, $X : C' :: D^2 : D'^2$, where (since $D^2 : D'^2 :: X : C' < C : C'$) we must have $X < C$. Eudoxus/Euclid replace the loose assertion that C is "exhausted" by the P_n by the precise assertion that, for any $X < C$, there is some $P_n > X$. (This assertion is formally justified by showing that each successive increment $P_{n+1} - P_n$ is more than half of the remainder $C - P_n$. So when we take away the successive polygons P_n from C, the remainder $C - P_n$ is diminished by more than half at each step, and so [by Euclid X,1] eventually becomes less than any given quantity. So eventually $(C - P_n) < (C - X)$, and thus $P_n > X$.) Then, taking the corresponding polygon P'_n inscribed in C', we have by XII,1 $P_n : P'_n :: D^2 : D'^2 :: X : C'$. By alternation, $P_n : X :: P'_n : C'$, which is absurd, since $P_n > X$ but $P'_n < C'$. (This argument uses several rules of proportion which Euclid justifies in Book V, but which for now I will just take as intuitively obvious.)

This proposition XII,2, and the argument for it, are trivial enough in themselves, but they are important as the easiest exercise of the method that Eudoxus developed to give rigorous proofs of the deeper results in solid geometry given later in Book XII. Indeed, this is the only justification for putting XII,1–2 in XII at all. Books XI-XIII are Euclid's treatment of solid geometry, whereas his treatment of plane geometry came in Books I-VI, and by that criterion XII,1–2, the only propositions about plane figures in a book about solids, are anomalous; we might have expected them in Book VI, and this would be logically possible, since their proofs do not rely on anything beyond Book VI.[11] But they are in XII because they use the method of exhaustion that the rest of XII will apply to solids, whereas VI consists of propositions proved by more "elementary" methods. For it is often possible to prove propositions in plane geometry by more elementary techniques than the corresponding propositions of solid geometry. That a triangle is half of a parallelogram of the same base and height can be proved by decomposing the parallelogram into two congruent triangles,[12] but that a pyramid (on a triangle or on an arbitrary polygon) is one third of a prism of the same base and height requires the method of exhaustion, which means that it cannot be proved directly but only by *reductio ad absurdum*. But while many propositions about *rectilinear* plane figures can be proved by decomposing different figures into congruent pieces, *circles* cannot be decomposed and reassembled into each other, and so propositions about areas of circles cannot be proved in this direct way, but only by a *re-*

---

an interpolation on the unusual ground that it was too good to be by Aristotle. I don't know if he ever dealt with the philosophical doctrines attributed to Eudoxus.)

[11] Except X,1; but X,1 uses nothing beyond Book V, indeed nothing beyond the *definitions* of Book V.

[12] To show that a triangle is half of *any* parallelogram of the same base and height, you also need to show that any two parallelograms of the same base and height are equal. But these two parallelograms can in turn be shown to be equal because they consist of congruent triangles, *plus* a common triangle, *minus* another common triangle (I,35).

ductio ad absurdum. The proof of XII,2 is the natural starting point for the main argument of Book XII in solid geometry: some of the proofs directly use XII,2 (notably XII,11, "cylinders and cones of the same height are to each other as their bases"), others (including the crucial XII,10, "a cone is a third of the cylinder of the same base and height") use an exhaustion of a circle by polygons closely following the model of XII,2, and other proofs use other exhaustions but still follow the logical structure seen at its simplest in the proof of XII,2.

(Let me pause to consider an objection to my way of contrasting Books VI and XII. I have said that XII,2 requires the method of exhaustion because it cannot be proved by decomposition, as the area-equalities of Books I-II can be. But statements about the *proportionality* rather than the *equality* even of rectilinear areas, such as XII,1, or its basis VI,20, "similar polygons are in the duplicate ratio of their corresponding sides," cannot be proved by decomposition either. Does it follow that these propositions can only be proved by exhaustion? If so, the distinction between them and the propositions of Book XII breaks down. My answer is that it is indeed true that propositions about the proportionality of rectilinear areas, such as most of the propositions of Book VI, cannot be proved directly but require at some stage a *reductio* argument, i.e. proving $A : B :: C : D$ by assuming $A : B > C : D$ or $A : B < C : D$ and deriving a contradiction. The ingenious logical structure of Book VI, as noted above, allows Euclid to pack the needed *reductio* into the proof of VI,1, and argue directly from there on.[13] This *reductio* does not have to be a proof by exhaustion, and Euclid's proof of VI,1 is not. But it is *possible* to prove VI,1 by exhaustion, and this may be the most straightforward way to prove it; I will come back below to describe the close logical relation between Euclid's proof of VI,1 and the exhaustion proof.)

The problem, however, is to understand the relation between the method of exhaustion and the Book V theory of proportion. Book VI depends on Book V in a clear if limited way, in that the proof of VI,1 is a direct application of Book V's definition of proportion. But there is no such clear relation between Book XII and Book V. We might expect that (say) XII,2 would prove an assertion of the form "$A : B :: C : D$" by saying "if not, then for some p and q, $pA > qB$ and $pC < qD$, or $pA < qB$ and $pC > qD$," and deriving a contradiction from either case. But in fact Book XII applies the definition of proportion only once, in the proof of XII,13, "if a cylinder is cut by a plane parallel to its base, the cut-off cylinder is to the whole cylinder as the cut-off axis is to the whole axis," which closely mimics the proof of VI,1. This makes it hard to believe what would otherwise be the attractive assumption that Eudoxus developed the proportion

---

[13]The proof of VI,1 (and of its analogues VI,33 and XI,25 and XII,13) is not stated as a *reductio ad absurdum*, but as a direct proof of the proportionality by applying Vdef5, i.e. $\forall m \forall n((mA > nB \leftrightarrow mC > nD)$ and $(mA = nB \leftrightarrow mC = nD)$ and $(mA < nB \leftrightarrow mC < nD))$. But this is logically equivalent to proving $A : B :: C : D$ by a *reductio ad absurdum* of $A : B > C : D$, implying $\exists m \exists n(mA > nB \ \& \ mC < nD)$, and of $A : B < C : D$, implying $\exists m \exists n(mA < nB \ \& \ mC > nD)$. The *reductio* formulation helps bring out the underlying logic; I will develop this point below.

theory as a foundation for his work using the method of exhaustion, presented by Euclid in Book XII. It is of course true that Book XII relies indirectly on the Book V definition of proportion, because it relies on the results of Book VI and the proof of VI,1 uses the definition of proportion, and also because Book XII uses the rules of proportion (alternation and so on) which Book V justifies on the basis of the definition. But we know that both VI,1 and the rules of proportion could be, and historically had been, justified on the basis of other definitions of proportion before the Book V definition was invented (they might also be taken as unproved principles). A mathematician might have felt dissatisfied with these earlier justifications and looked for a new one, but it is not clear why working out Book XII would lead you to do this, if you had not already been led to it in working out the much more elementary Book VI. Certainly the text of Book XII as we have it makes strikingly little direct use of Book V's definition of proportionality, citing it only in XII,13, after Eudoxus' main theorem, that the cone is one third of the cylinder, has been proved.[14] And since any proportion theory that can prove VI,1 can also prove XII,13, there is no reason why this proposition should have led Eudoxus to the Book V definition. Indeed, we do not know that Eudoxus' proof of XII,13 (if he had one) did use the Book V definition; if he proved it using an earlier proportion theory, Euclid would naturally have updated his proof using (Euclid's proof of) VI,1 as a model. Furthermore, there is evidence that the original version of Book XII did *not* presuppose Book V, since in XII,2 the step reducing to absurdity the assumptions that (in the notation above) $P_n : P'_n :: X : C', P_n > X$ and $P'_n < C'$, is proved circuitously using V,16 (alternation), when it is an immediate instance of V,14: this suggests that the proportion theory underlying the original version of Book XII had not proved V,16 via V,14—which means that it did not include the central

---

[14]In fact, it is not clear that Euclid cited the Book V definition at all in Book XII, because there is an open text-critical issue about whether XII,13 was part of Euclid's intended text of Book XII. It is present both in the "non-Theonine" Vatican manuscript P and in the "Theonine" family (the Oxford manuscript B and its relatives, here most importantly the Vienna manuscript V) but is absent in the Bologna manuscript b (which in Book XII is further from both P and the "Theonine" manuscripts than they are from each other) and in the Arabic (and Arabo-Latin) translations, which at least sometimes follow a Greek model closer (in Book XII) to b. Manuscript b contains XII,14, but the translations omit this too. The proof of XII,14 assumes the statement of XII,13; this might mean that XII,13 was in the archetype and fell out of one manuscript family by mistake, or it might mean that someone added XII,13 to fill a logical gap. In general, b (in Book XII and the last few propositions of XI) and much of the Arabic tradition (throughout) have a "thinner" text of the *Elements* (and P is in turn thinner than the "Theonine" family): it is plausible, but not certain, that at least in many cases the thinner version is closer to the original. Heiberg prints the text of b (in the section where it deviates from the main Greek tradition) in Appendix II to his *Elementa* volume IV; there is as yet no single source that collects all the information from the Arabic tradition. For an excellent survey of the problem see Bernard Vitrac, "The Euclidean Ideal of Proof in the *Elements* and philological uncertainties of Heiberg's edition of the text," in Karine Chemla, ed., *The History of Mathematical Proof in Ancient Traditions* (Cambridge University Press, 2012), pp. 69–134, building on and correcting Knorr, "The Wrong Text of Euclid," *Centaurus* 38 (1996), 208–76.

achievement of Book V. These difficulties led Oskar Becker to suggest that Eudoxus discovered (and wrote up his version of) the results of Book XII *before* he discovered his own theory of proportion. Wilbur Knorr says, instead, that only Book XII is genuinely Eudoxian, and that Book V, although starting from Eudoxus' proportion theory, contains post-Eudoxian developments which were not available for use in Book XII.[15] But while I largely agree with Knorr's reconstruction of what is Eudoxian and what is post-Eudoxian in Book V, this does not entirely solve our problem: even the genuinely Eudoxian proportion theory lying behind Book V is not being used in Book XII. To understand why this might be so, and how else (proto-versions of) Books V and XII might fit together for Eudoxus, we need to do some more thinking about the aims and origins of Eudoxus' proportion theory.

2   The prehistory of Eudoxian proportion theory

The old story about the history of proportion theory (cited from Heath above) was simple enough, if not especially explanatory. There were just two stages: an archaic proportion theory according to which $A : B :: C : D$ iff $\exists p \exists q (A = pB/q$ and $C = pD/q)$, which broke down once incommensurable magnitudes were discovered, and then the Eudoxian theory as given in Euclid Book V. On Heath's story, proportion-based mathematics stopped dead when incommensurables were discovered, and did not begin again until Eudoxus had solved the foundational problem. As I have complained above, it is very unlikely that mathematicians would have stopped doing proportion-theoretic geometry for such a reason; Heath's story also makes Eudoxus' discovery appear simply miraculous. But, beyond this, the historical record is more complicated than Heath allows. Rather than just two proportion theories, there were certainly at least three, and apparently four. Eudoxus' solution was only one in a series of solutions to the foundational problem.

One pre-Eudoxian theory of proportion has been rediscovered by several scholars, and worked out most fully by Oskar Becker, on the basis of a reference in a text of Aristotle: "In mathematics too some things are difficult to prove on account of a deficiency in definition, such as [the proposition] that the [straight line] that cuts a plane surface parallel to the side divides the line and the area similarly. Once the definition has been stated, the proposition is immediately clear: for the areas and the lines have the same mutual subtraction,

---

[15] See Becker, "Eudoxos-Studien I. Eine voreudoxische Proportionenlehre und ihre Spuren bei Aristoteles und Euklid," *Quellen und Studien zur Geschichte der Mathematik, Astronomie und Physik* 2 (1933), 311–33 (reprinted in Jean Christianidis, ed., *Classics in the History of Greek Mathematics*, Kluwer, 1994, pp. 191–209, with the original pagination in the margins), at pp. 326–9 and esp. p. 327, and Knorr, "Archimedes and the Pre-Euclidean Proportion Theory," pp. 191–200. V,14 is likewise avoided in the proofs of XII,5, XII,12, and XII,18; it is never used in Book XII. To say that *Elements* XII is "Eudoxian" is to say that it follows a work of Eudoxus which proved its central results, and does so more closely than *Elements* V follows anything in Eudoxus; this does not mean that Eudoxus wrote *Elements* XII, or that every detail of it was in a Eudoxian prototype.

and this is the definition of 'the same ratio'" (*Topics* VIII,3 158b29–35).[16] Here by "line cutting a surface" Aristotle must mean, as Alexander says in his commentary, a line cutting a parallelogram parallel to (one pair of) its sides: the claim is that this line divides the parallelogram in the same ratio that it divides its base. (Draw a diagram: this says that the parallelogram $ABCD$ is to the parallelogram $CDEF$ as the base BD is to the base DF.) This is equivalent to saying that two parallelograms of the same height, if they are in the same angle and share a side, are in the same proportion as their bases. And this in turn, given the elementary theorem that any two parallelograms of the same height and on equal bases are equal, is equivalent to Euclid VI,1, "triangles and parallelograms which are under the same height are to one another as their bases" (the triangle case immediately follows from the parallelogram case). As we have seen already, Euclid VI,1 is the crucial juncture where proportion theory needs to be applied to the geometry of rectilinear figures: once this first application of proportion theory to geometry has been made, all the other propositions of Book VI can be derived without using the definition of proportion again (using only general geometrical forms of inference and the rules of proportion), but VI,1 itself (unless we are willing to take it as an unproved assumption) can be proved only by appeal to some definition of proportion, or rather of "being in the same ratio." Euclid has one way of doing this, but Aristotle, writing perhaps 70 years earlier and using a pre-Eudoxian proportion theory, shows that people were already proposing definitions of proportion by which to prove VI,1.[17]

---

[16] ἔοικε δὲ καὶ ἐν τοῖς μαθήμασιν ἔνια δι' ὁρισμοῦ ἔλλειψιν οὐ ῥαδίως γράφεσθαι, οἷον ὅτι ἡ παρὰ τὴν πλευρὰν τέμνουσα τὸ ἐπίπεδον ὁμοίως διαιρεῖ τήν τε γραμμὴν καὶ τὴν χωρίον. τοῦ δὲ ὁρισμοῦ ῥηθέντος εὐθέως φανερὸν τὸ λεγόμενον· τὴν γὰρ αὐτὴν ἀνταναίρεσιν ἔχει τὰ χωρία καὶ αἱ γραμμαί· ἔστι δ' ὁρισμὸς τοῦ αὐτοῦ λόγου οὗτος. As Alexander of Aphrodisias says in his commentary on this passage: "[Aristotle] takes as an example of this, as something which is unclear but has been shown, [the proposition] that the [straight line] that cuts a plane surface parallel to a side divides the line and the area similarly. The meaning is that if there is a parallelogrammic surface, and there is drawn in it a parallel to one of the sides, the parallel divides the line and the whole area similarly, that is, in the same proportion. For when 'similarly' is said, this is not readily known, but when the definition of 'proportional' has been stated it becomes readily known that the line and the area are cut proportionally by the parallel that has been drawn. And the definition of proportional which the ancients used was 'magnitudes are to each other proportionally which have the same mutual subtraction [ἀνθυφαίρεσις]': for [Aristotle] here calls ἀνθυφαίρεσις 'ἀνταναίρεσις'. And things which are to each other proportionally are also said to be to each other similarly. So [Aristotle] says this 'is the definition of the same ratio' instead of 'is the definition of proportional.' When this definition has been stated it becomes clear that the parallel drawn to a side in a parallelogrammic area cuts the line and the area similarly and proportionally" (*In Topica* 545,6–21).

[17] Bernard Vitrac in Euclide, *Les Éléments*, vol.2, Livres V à IX (*Presses Universitaires de France*, 1994), pp. 518–19, suggests (at least as a possibility) that in *Topics* VIII,3 158b29–35 Aristotle does not mean to be citing an actual mathematical definition, but that it is like the causal "definition" of the squaring of a rectangle as "the finding of a mean [proportional]" (*De Anima* II,2 413a13–20 and *Metaphysics* B2 996b18–22): a mathematician would define squaring a rectangle as (roughly) finding a square equal to a given rectangle, but for Aristotle this is analogous to defining lunar eclipse as darkening of the moon at opposition, a merely nominal definition, and the deeper "scientific" definition includes the cause, "darkening of the

It may not be immediately obvious what Aristotle means by saying that the definition of "being in the same ratio" is "having the same mutual subtraction [ἀνταναίρεσις, or as Alexander paraphrases it ἀνθυφαίρεσις],"[18] but parallel passages make Aristotle's meaning clear beyond a doubt. "Mutual subtraction [ἀνθυφαίρεσις]" is the procedure described in Euclid X,2–3 whereby we take a greater quantity A and a lesser quantity B; we subtract B from A as many times as possible, leaving, if A is not an exact multiple of B, a remainder $A' < B$; we then subtract A' from B as many times as possible, leaving, if B is not an exact multiple of A', a remainder $B' < A'$; we then repeat the procedure by subtracting B' from A', and so on, until one quantity is an exact multiple of another and we subtract until nothing remains; or, if this never happens, we continue the process of mutual subtraction *ad infinitum*. As Euclid shows, if A and B are commensurable the process will eventually terminate, while if they are incommensurable it will continue *ad infinitum*. On the definition that Aristotle is reporting, $A : B :: C : D$ if A and B have the same mutual subtraction that C and D have. That is, assuming for simplicity that $A > B$ and $C > D$, we will be able to subtract B from A exactly as many times as we can subtract D from C, leaving remainders $A' < B$ and $C' < D$; we will then be able to subtract A' from B exactly as many times as we can subtract

---

moon at opposition due to the earth's blocking the sun's light from the moon" or just "the earth's blocking the sun's light from the moon," "finding a square which is equal to a given rectangle because its side is the mean proportional between the sides of the rectangle" or just "finding a line which is the mean proportional between the sides of a given rectangle." But Vitrac's analogy does not work. On Aristotle's theory of definition in *Posterior Analytics* II, a scientific definition explains why there exists something falling under the nominal definition (why the moon is sometimes darkened at opposition, why it is possible to find a square equal to a given rectangle); in our *Topics* passage, Aristotle is not explaining why there exist things in proportion, but why the rectangle is cut in the same proportion as the side. The anthyphairetic definition of being in the same ratio will be used as a premiss in proving Euclid VI,1, whereas the scientific "definition"of squaring a rectangle is not used as a premiss in showing that we can square a rectangle: the scientific definition is equivalent to the entire demonstration, "differing in arrangement from the demonstration" (*Posterior Analytics* II,10 94a12–13), rather than entering into the demonstration as a premiss. Further, and decisively against Vitrac, the *Topics* does not give *Posterior Analytics*-style causal definitions, and this is the result of a deliberate disciplinary division between non-causal dialetical reasoning in the *Topics* and scientific reasoning in the *Posterior Analytics*.

[18]Heath had dismissed this as a poetic phrase with no precise mathematical meaning (*Euclid's Elements* v.2 pp. 120–21). As Howard Stein points out to me, Heath retracted this dismissal in his posthumously published *Mathematics in Aristotle* (Oxford University Press, 1949), pp. 81–3, after seeing Becker's work. My aim is not to criticize Heath but just to point out that the text of Aristotle is not transparent, and to give due credit to Becker and H.G. Zeuthen and E.J. Dijksterhuis for seeing what it meant, and especially to Becker for reconstructing the larger theory to which this definition belonged. (Becker does not give a precise reconstruction of a whole theory, and details of the arguments he reconstructs are *exempli gratia*, but it was important to show the logical power of the theory, when some scholars thought the anthyphairetic definition was meaningless, and others thought it could not yield proofs of the central results of proportion theory.)

C' from D, and so on; and either the two process will terminate at the same time, or they will both continue *ad infinitum*.[19]

Using this definition of proportionality, it is, as Aristotle says, "immediately clear" that Euclid VI,1 holds, e.g. that the parallelogram $ABCD$ is to the parallelogram $CDEF$ as the base BD is to the base DF. For the parallelogram $ABCD$ can be subtracted from the parallelogram $CDEF$ exactly as many times as the base BD can be subtracted from the base DF; and then the remainder parallelogram $GHEF$, lying over the remainder base HF, can be subtracted from the parallelogram $ABCD$ exactly as many times as the remainder base HF can be subtracted from the base BD, and so on. Each stage of the mutual subtraction of the parallelograms will correspond exactly to a stage of the mutual subtraction of their bases. Thus the parallelograms $ABCD$ and $CDEF$ will have "the same mutual subtraction" as their bases BD and DF, and so by the anthyphairetic definition $ABCD : CDEF :: BD : DF$.[20]

But although, as our text from the *Topics* shows, the problem of proving Euclid VI,1 was a natural context for discussing how to define proportionality, proving this proposition was not the only task of a theory of proportion.[21] Eu-

---

[19] We can also state conditions for $A : B > C : D$ and for $A : B < C : D$. Write $num(A, B)$ for the number of times that B goes into A, i.e. for the greatest integer n such that $nB \leq A$, and call the remainder $rem(A, B) = A - num(A, B)B$. (This way of putting it allows $num(A, B) = 0$, which no ancient mathematician would do, but we can avoid that by restricting the rule to cases where $A > B$ and $C > D$; cases where $A < B$ and $C < D$ can be readily converted into cases where $A > B$ and $C > D$, and other cases can be decided immediately.) Then if $num(A, B) > num(C, D)$, then $A : B > C : D$; if $num(A, B) < num(C, D)$, then $A : B < C : D$; if $num(A, B) = num(C, D)$, then if $B : rem(A, B) > D : rem(C, D)$, $A : B < C : D$, and if $B : rem(A, B) < D : rem(C, D)$, $A : B > C : D$. These rules, applied finitely many times, will always yield either the result that $A : B > C : D$ or the result that $A : B < C : D$, unless it is in fact true that $A : B :: C : D$. We can also avoid talk of "num" and "rem": it is sufficient to assume that if $A \geq B$ and $C < D$, then $A : B > C : D$; if $A : B > C : D$ then $(A + B) : B > (C + D) : D$; and if $A : B > C : D$ then $B : A < D : C$. (I assume throughout that $A : B > C : D$ iff $C : D < A : B$.)

[20] Knorr, "Archimedes ...", pp. 192–3, and *Evolution of the Euclidean Elements* p. 263 and pp. 335–6, complains that this argument remains intuitive and falls short of Greek standards of demonstration. In my view it is in no worse state than any other Greek argument that depends on quantifying over sequences (of magnitudes or numbers) of arbitrary finite length, e.g. Euclid X,1 or XII,2 or even V,1.

[21] Bernard Vitrac and Ken Saito and Fabio Acerbi have objected to speaking of a "theory of proportion" before Euclid, and Saito has objected to attributing a "theory of proportion" even to Euclid. They seem to be criticizing exaggerated expectations of what such a theory can do, or of how much earlier history can be extracted from the text of Euclid. Saito says "it has been assumed that it was [Greek mathematicians'] important task to demonstrate all the propositions concerning ratio and proportion in a systematic and exhaustive manner starting from some explicit definition" ("Phantom Theories," p. 336). This assumption is, as Saito says, surely false, but I have never assumed it and I would be surprised if e.g. Becker or Knorr had assumed it either—particularly since Saito (p. 338) includes "$A : B :: C : D \leftrightarrow AD = BC$" as a fundamental theorem of proportion, and Becker and Knorr were just as aware as I am that Greek mathematics had no notion of the product of two quantities. Saito says that "The aim of [*Elements*] V and VI is ... not so much to construct a logically complete 'theory' of ratio and proportion, as to present and justify available techniques for investigation in geometry" (p. 344) and, speaking specifically of Knorr's reconstruction of a

clid shows how to derive the rest of the theory of rectilinear figures from VI,1, without another appeal to the definition of proportionality, using only the rules of proportion; but those rules must themselves be justified on the basis of the definition of proportionality. Euclid himself does this, on the basis of his own definition of proportionality, in Book V. For the anthyphairetic definition of proportionality to yield a theory capable of supporting proportion-theoretic geometry, this definition too must be shown to imply the rules of proportion. This is difficult especially in the case of alternation (Euclid V,16),[22] and Oskar Becker's main technical achievement in reconstructing the anthyphairetic

---

pre-Euclidean theory, that "these definitions were the by-product of an ingenious technique invented for justification of some particular and fundamental properties of some geometric objects concerning ratio and proportion, and not the result of a search for a fundamental definition of proportion with the ambition of demonstrating all the theorems about ratios and proportions thereof" (p. 345), all of which is true. But it is important that Greek mathematicians, rather than simply assuming *ad hoc* whatever results about proportion they needed to prove some geometrical theorem, collected these lemmas and investigated their relations of logical dependence, trying to make all assumptions explicit and to reduce them to as few as possible (where these assumptions might or might not include an explicit definition of being in the same ratio). It is also important that, by the time of Euclid and apparently by the time of Aristotle *Posterior Analytics* I,5 74a17–25 (cited below) there as an attempt to separate out universal theorems about proportion and give them proofs independent of geometrical assumptions. All this is part of the Greek inquiry into "elements," which is far older than Euclid (Aristotle cites the geometrical sense of the word at *Metaphysics* Δ3 1014a35–b2)—the fact that the Greek theory of proportions is driven backward, from the geometrical applications toward the postulates or definitions, does not mean that it is not a theory of proportions. (A similar point applies to Acerbi's remarks on Knorr's reconstruction, "Drowning by Multiples," p. 222.) Vitrac and Saito seem to assume that modern scholars' search for a pre-Euclidean theory of proportion is guided by the assumptions (i) that the Greeks were trying to deal with a "foundational crisis" about proportions, (ii) that Euclid was a mere compiler (like, say, Diodorus Siculus in history) from whose work earlier strata of the history of mathematics can be extracted by Quellenforschung, and (iii) that *Elements* VII in particular preserves an early stage of mathematics prior to the discovery of incommensurability. I believe none of those things, but *Elements* V would be inexplicable and near-miraculous without some prior history, and the available evidence (from Aristotle, Archimedes, etc.), while it calls for circumspection both about the degree of certainty of any reconstruction and about how far each "theory" had been worked out in logical detail, does allow us to reconstruct enough of the prior history to make the genesis of *Elements* V intelligible. When modern scholar decks out a reconstructed "theory" with the apparatus of definitions, postulates and common notions (as I do below), the details of the reconstruction are *exempli gratia*, but this is the way to test the power of the theory. Many of Vitrac's objections (*Éléments*, vol.2, pp. 518–22) to the idea of an anthyphairetic proportion theory just seem to be objections against the theory itself, pointing out limitations on its power. These are legitimate objections, and are why people eventually replaced it with a better method for arguing about proportions. The anthyphairetic definition was never very good at anything except proving some incommensurabilities, proving *Elements* VI,1, and proving some of the rules of proportion theory now included in *Elements* V; but that was all it was ever needed to do.

[22]The other difficult propositions are V,8 (essentially, if $A > B$ then $A : C > B : C$; this immediately entails V,9, if $A : C :: B : C$ then $A = B$, and V,10, if $A : C > B : C$ then $A > B$), and V,22, if $A : B :: D : E$ and $B : C :: E : F$ then $A : C :: D : F$. V,22 can be derived easily from V,16, but only on the assumption that A, B and C are homogeneous with D, E and F. Most of *Elements* V is constituted by the proofs of V,8, V,16, V,22 and the auxiliary lemmas they require.

theory of proportion was to show how the anthyphairetic definition could be used to derive alternation. And, on Becker's reconstruction, this is where the argument-structure of the anthyphairetic proportion theory would diverge the furthest from Euclid's theory. Euclid's theory proves alternation, a statement of universal mathematics, purely from universal mathematical considerations going back to his definition of proportionality, with no reliance on geometry, i.e. on propositions specifically about lines or areas. But it is much more difficult to prove alternation in such a universal way from the anthyphairetic definition, and Becker reconstructed a way of doing it instead by using some propositions from the theory of rectilinear figures. That is, on the anthyphairetic theory as Becker reconstructs it, some of the propositions of Book V are derived from propositions of Book VI, which are derived from VI,1, which is derived by applying the definition of proportionality. This sounds as if it might be circular, if the derivations of these later propositions of Book VI from VI,1 turned on the rules of proportion that these propositions of Book VI are being used to justify; but Becker shows how to make the arguments non-circular.

For suppose we want to show alternation for the case of straight lines: if A, B, C and D are straight lines such that $A : B :: C : D$, we claim that $A : C :: B : D$. Form a rectangle $rect(A,D)$, whose sides are equal to A and to D, and likewise form rectangles $rect(B,C)$, $rect(B,D)$ and $rect(C,D)$. Applying VI,1, we know that $rect(A,D) : rect(B,D) :: A : B$. In the same way, $rect(B,C) : rect(B,D) :: C : D$. Since the ratios $A : B$ and $C : D$ are the same, it follows that $rect(A,D) : rect(B,D) :: rect(B,C) : rect(B,D)$. We would now like to conclude that $A : B :: C : D$ implies $rect(A,D) = rect(B,C)$, which is the first half of Euclid VI,16. But we can draw this conclusion only if we know the general rule that $X : Z :: Y : Z$ implies $X = Y$ (substituting $rect(A,D) for X, rect(B,C)$ for Y, and $rect(B,D)$ for Z). This is Euclid V,9;[23] and while perhaps at an early stage in the history of proportion theory it was simply assumed, explicitly or implicitly, as a common notion, Euclid proves it from his definition of proportionality, and it can be proved from the anthyphairetic definition as well. (For suppose that $X : Z :: Y : Z$ but $X \neq Y$; without loss of generality, we can assume that $X > Y$. Then use the following lemma. Let $X : Z :: Y : W$, and suppose $X > Y$ but $Z \leq W$. Choose U such that $X > (Y+U)$. Then let X' and Z' be the next anthyphairetic differences of X and Z,[24] and let Y' and W' be the next anthyphairetic differences of Y and W. Then we continue to have the relations $X' : Z' :: Y' : W'$, $X' > (Y' + U)$,

---

[23]There is an alternative strategy for proving V,16 from VI,1 which avoids explicit reliance on V,8–9. We have seen that $A : B :: C : D$ implies $rect(A,D) : rect(B,D) :: rect(B,C) : rect(B,D)$. Trivially, $rect(B,D) : rect(C,D) :: rect(B,D) : rect(C,D)$. So by V,22 we can compound the two proportions, $rect(A,D) : rect(C,D) :: rect(B,C) : rect(C,D)$. But, by VI,1 again, $rect(A,D) : rect(C,D) :: A : C$ and $rect(B,C) : rect(C,D) :: B : D$. So $A : C :: B : D$, which is what was to be shown. Unfortunately, there seems to be no way to prove V,22 without using either V,16 or V,8–9, so there is ultimately no logical advantage to this strategy.

[24]In other words, if $X > Z$, then $X' = rem(X,Z)$ and $Z' = rem(Z,X')$, and if $X < Z$, then $Z' = rem(Z,X)$ and $X' = rem(X,Z')$.

and $Z' \leq W'$.[25] Given this lemma, we can derive an absurdity from the assumption that $X : Z :: Y : Z$ but $X > Y$. For choose U such that $X > (Y+U)$. Then, by repeated applications of the lemma, for all of the successive anthyphairetic differences X'''' from the anthyphairesis of X and Z, and Y'''' from the anthyphairesis of Y and Z, we will continue to have $X'''' > (Y'''' + U)$. But, by Euclid X,1–3, the anthyphairetic difference X'''' eventually becomes smaller than any prescribed quantity, or eventually vanishes if X and Z are commensurable. So X'''' eventually becomes smaller than U, contradicting what has just been shown.)[26] Having gone through this argument, we know that $A : B :: C : D$ implies $rect(A,D) = rect(B,C)$. This gives a geometrical criterion for when $A : B :: C : D$, and we can also reverse this criterion (as Euclid does in the second half of VI,16). For, since $rect(A,D) = rect(B,C)$, $rect(A,D) : rect(C,D) :: rect(B,C) : rect(C,D)$.[27] But, by VI,1, $rect(A,D) : rect(C,D) :: A : C$, and $rect(B,C) : rect(C,D) :: B : D$. So $A : C :: B : D$ (this is the second half of Euclid VI,16, the converse of the first half); and this is what was to be shown.

The above paragraph was a bit complicated, but all the complexity was in the proof of V,9. Given these basic rules of proportion, the proof of alternation from VI,1 by way of VI,16 is simple and straightforward (and the proof might first have been given very simply, without arguing for V,9, and then supplemented by the proof of that rule). The drawback is that it proves alternation only for straight lines A, B, C and D, whereas Euclid formulates V,16 for all (homogeneous) quantities. The practical consequences of this limitation may not be too bad. For suppose that A, B, C and D are rectilineal areas (rather than lines), such that $A : B :: C : D$. Then we can easily show that $A : C :: B : D$. For let E by any straight line, and apply A, B, C and D to E

---

[25] To prove the lemma, note that the first claim follows immediately from the anthyphairetic definition of proportionality. To verify the second and third claims, check cases. If $X > Z$, so that $Y > W$, then write $num(X,Z) = num(Y,W) = n$, $num(Z,X') = num(W,Y') = m$. Then $X' = (X - nZ) > (Y + U - nZ) = (Y' + U)$, and $Z' = (Z - mX') < (W - mY') = W'$. If $X < Z$, so that $Y < W$, then write $num(Z,X) = num(W,Y) = n$, $num(X,Z') = num(Y,W') = m$. Then $Z' = (Z - nX) < (W - nY) = W'$, and $X' = (X - mZ') > (Y + U - mW') = (Y' + U)$.

[26] Note that this reduces to absurdity not just the assumption that $X : Z :: Y : Z$ and $X > Y$, but also the more general assumption that $X : Z :: Y : W$ and $X > Y$ and $Z \leq W$, and so proves not just V,9 but also V,14. This proof (which, as I now see, is very close to that of Anders Thorup, "A Pre-Euclidean Theory of Proportions," *Archive for History of Exact Sciences* 45 (1992), 1–16, Propositions 4 and 9) is an improvement over the one given by Becker (p. 320), and helps to answer Knorr's objections ("Archimedes ...", pp. 192–3, *Evolution of the Euclidean Elements* p. 263 and pp. 338–40) to Becker's argument. It is a bit more complicated to prove V,8 on an anthyphairetic theory, because it needs the definition of being in a greater ratio, but someone might well take it as obvious that if $X > Y$ then $X : Z \geq Y : Z$, i.e. that $X > Y$ and $X : Z < Y : Z$ is absurd, but be concerned to exclude the possibility that $X > Y$ and $X : Z :: Y : Z$, i.e. to show that even a small variation in a continuous magnitude is enough to make a difference in the resulting sequence of numbers. Given this assumption, proving V,9 would also establish V,8.

[27] By Euclid V,7, "equals have the same ratio to the same." This could be taken as a common notion, like "equals added to equals are equal," or proved by considering successive anthyphairetic differences.

in the form of rectangles (as Euclid showed how to do, without using proportion theory, in I,44–5); say $A = rect(E,F)$, $B = rect(E,G)$, $C = rect(E,H)$, $D = rect(E,J)$. Then by VI,1 $A : B :: F : G$ and $C : D :: H : J$, so $F : G :: H : J$; so $F : H :: G : J$ by alternation for straight lines, so $A : C :: B : D$, which is what was to be shown.[28] But this strategy of argument will not work if A, B, C and D are non-rectilineal areas (for instance, it will not work for the instance of alternation used in the proof of XII,2, where two of the quantities are circles, although it would be possible, by making the argument more complicated, to avoid this case of alternation), or if they are three-dimensional solids.[29]

As Becker realized, this situation, where alternation is proved directly for some kinds of quantities, and then indirectly for others, rather than by a single argument from universal mathematics, is described by Aristotle in *Posterior Analytics* I,5:

> [it might wrongly be thought] that what is proportional is also so alternately [lit. "crosswise"], *qua* numbers and *qua* lines and *qua* solids and *qua* times, as used to be shown separately, although it can also be shown by a single demonstration in all cases. But because all these things, numbers and lengths and times and solids, had not been given a single name, and because they differ in species from each other, they used to be taken separately. But now it is shown

---

[28] Becker p. 323 also points out another possible argument-strategy, given by Heath (following Smith and Bryant), *Euclid's Elements*, v.2 pp. 165–6, which proves alternation directly for the case where A, B, C and D are rectilineal areas, and then (if you like) derives it for the case where they are straight lines.

[29] If we are willing to assume, as Euclid XII does, the existence of a fourth proportional in all cases—i.e., that if A and B are homogeneous and C is a third magnitude (i.e., continuous quantity) of any kind, there is a D homogeneous with C such that $A : B :: C : D$—then we can overcome this difficulty. (See below for more discussion of the postulate of the existence of a fourth proportional, in Euclid and in other theories.) For suppose that A, B, C and D are four homogeneous magnitudes, and suppose $A : B :: C : D$. Let E be any straight line, and let F, G and H be straight lines such that $A : B :: E : F$, $A : C :: E : G$, and $A : D :: E : H$. By Euclid V,22 ("*ex aequali*") we can infer $C : D :: G : H$ and thus $E : F :: G : H$, so by the straight-line case of alternation we have $E : G :: F : H$. Using V,22 again we can infer $B : D :: F : H$ and thus $A : C :: B : D$. But this argument depends on applying V,22—i.e., if $P : Q :: X : Y$ and $Q : R :: Y : Z$ then $P : R :: X : Z$—in cases where the quantities P, Q and R are not homogeneous with the quantities X, Y and Z. Euclid's proof of V,22 does work in such cases, but the anthyphairetic theory seems not to be able to do this—Becker's proof of V,22 depends on V,16, and works only where all the magnitudes are homogeneous, indeed only when they are straight lines or reducible to straight lines. It should also be possible to prove alternation for non-rectilineal areas and three-dimensional figures by exhaustion. (By exhausting a two-dimensional figure with rectilineal areas, we can reduce the general two-dimensional case to the two-dimensional rectilinear case. By exhausting a three-dimensional figure with solids all of whose segments are prisms on two-dimensional figures with height commensurable with a given height H, and then constructing, equal to these solids, a prism whose height is the greatest common measure of their heights, we can reduce the three-dimensional case to the two-dimensional case.) But these arguments will be long and tedious, and will have to be repeated separately for each instance of alternation that we want to prove.

universally: for it does not hold of them *qua* lines or *qua* numbers, but qua this [sc. quantity-in-general], which they are posited to be universally. (74a17–25)

Aristotle misleadingly makes it look as if the only difficulty in unifying the separate proofs of alternation were in coming up with a single universal name, "quantity," which would apply equally to numbers, lines and so on, so that we could formulate the universal proposition "for any four quantities A, B, C, D such that $A:B::C:D$, we have $A:C::B:D$."[30] But on the anthyphairetic theory of proportion, the difficulty is not in *formulating* this proposition, but in *proving* it.[31] Aristotle is reporting, not just the invention of a universal name, but a major technical advance in mathematics, which allowed alternation to be proved by a single proof applicable to all types of quantity. This depends on a new theory of proportion, replacing the anthyphairetic theory.[32] Presumably

---

[30] It should, of course, be "homogeneous quantities"; but even Euclid leaves this out of his statement of V,16—indeed, Vdef3 (or 3–4) is Euclid's only explicit mention of homogeneity.

[31] Note that Euclid VI,1 (and the somewhat different, earlier version of it that Aristotle states in the *Topics* passage) says that an area is to an area as a line is to a line, so it assumes that there is a single sense of ratio (expressed in a single definition of proportionality) that applies equally to ratios of lines and ratios of areas. Likewise X,5, commensurable magnitudes have the same ratio to each other that a number has to a number, assumes that $A:B::C:D$ makes sense when A and B are continuous quantities and C and D are numbers, and so assumes that there is a single sense of ratio applying both to continuous quantities and to numbers. In these cases, as with alternation, there was no difficulty in *stating* the proposition.

[32] However, a caution: it is not true, as has often been said in the wake of Becker, that it is impossible to give a universal proof of alternation using an anthyphairetic definition of being in the same ratio. Thorup, in "A Pre-Euclidean Theory of Proportions," gives such a proof. To see the basic strategy, let us use, *only for the duration of this footnote*, the notation $A/B = C/D$ for "A and B have the same ἀνθυφαίρεσις as C and D" and $A:B::C:D$ for "A and B have the same ratio as C and D *in the sense of Euclid Vdef5*". It is easy to prove by the methods of Euclid V that if $A:B::C:D$, then $A/B = C/D$ (the converse is much less obvious). We have seen in an earlier note that if $A/B = C/D$ and $A > C$, then $B > D$ (anthyphairetic V,14). It is also easy to show that $A/B = mA/mB$ (anthyphairetic V,15). Suppose $A/B = C/D$. Then $\forall m \forall n\ mA/mB = nC/nD$. So $\forall m \forall n$, if $mA > nC$ then $mB > nD$, and likewise if $mA < nC$ then $mB < nD$. So $A:C::B:D$. So $A/C = B/D$, anthyphairetic V,16. (If we apply the same argument again, it will also follow that $A:B::C:D$, so the anthyphairetic and Euclidean definitions of being in the same ratio are coextensive. It is also possible to give a more intuitive, but more notationally complicated, version of the argument using the notions of $A/B > C/D$ and $A:B > C:D$.) It is, on reflection, not surprising that it is possible using the anthyphairetic definition of proportion to show that $A/B = C/D$ implies $A/C = B/D$, since it is not surprising that it is possible using Euclid V to show that A and B have the same ἀνθυφαίρεσις as C and D iff $A:B::C:D$ (the "if" part being easy) and that $A:B::C:D$ iff $A:C::B:D$ (Euclidean V,16), and Euclid V involves no additional postulates beyond the resources of the anthyphairetic theory. (If we additionally assume the postulate of the existence of a fourth Euclidean proportional, it becomes particularly easy to show that $A/B = C/D$ entails $A:B::C:D$. For there is some E such that $A:B::C:E$. So $A/B = C/E$. So $C/D = C/E$; so $D = E$ by anthyphairetic V,9; so $C:D::C:E::A:B$.) *The question, however, is whether anything like this would occur to anyone who had not already read Euclid V*. Thorup does not claim that it would (he is only making a point about the logical power of the anthyphairetic theory), and as far as I can see, Thorup's version of the argument would not. However, later in this paper I will suggest a path by which someone might be led, starting from the anthyphairetic

the discovery is due to Eudoxus, and is either identical with, or ancestral to, the way of presenting proportion theory and of proving alternation that we have in Euclid Book V.[33] (It is odd that in the *Topics* Aristotle apparently assumes that the anthyphairetic theory is current and correct, whereas here he assumes a later development. Possibly Eudoxus made his discovery after the *Topics* passage was written and before the *Posterior Analytics* passage was written; I will suggest another possible explanation below.) Eudoxus might have wanted to give a single universal proof of alternation so as to avoid having to re-prove it in different cases, or because the cases of solids and non-rectilineal areas are technically messy on the anthyphairetic theory, or simply because he felt that a theorem about quantities as such should by proved by considerations appropriate to quantities as such, without a detour through plane geometry. This story of a succession of three proportion theories (a theory applicable only to commensurable quantities, the anthyphairetic theory, and then Eudoxus' theory) avoids positing a breakdown of mathematics after the discovery of incommensurables, it leaves a motivation for Eudoxus' discovery (in that the state of proportion theory before his discovery, while not catastrophic, was unsatisfactory), and it makes his discovery seem a bit less miraculous (in that the Euclidean definitions of sameness and inequality of ratios are not so much more abstract than the anthyphairetic definitions).[34] But it still leaves us with no real explanation of how Eudoxus discovered his definition, it does not explain the thematic connections between Euclid Books V and XII, and it also does not explain the non-connections, i.e. the fact that the "Eudoxian" definition of

---

theory and continuing to use its definition of being in the same ratio, to a universal proof of alternation (assuming the postulate of the existence of a fourth proportional) which is neither Becker's nor Euclid's (and from which a further path might lead to replacing the anthyphairetic definition of being in the same ratio with Euclid's, and to Euclid's proof of alternation). A further problem with Thorup's reconstruction is that, following Becker, he proves V,22, if $A : B :: D : E$ and $B : C :: E : F$ then $A : C :: D : F$, by three uses of V,16, so that the result follows only if all six quantities are homogeneous. It seems to be very difficult to prove V,22 directly in the framework of the anthyphairetic theory, which is very bad at compounding ratios (try multiplying two continued fractions). The theory I will give below, following Knorr and Borelli, gives a direct proof of V,22 as well as of V,16.

[33] It is unclear whether Book V as we have it is supposed to apply to numbers or not. Euclid speaks there of "magnitudes" [μεγέθη], which very often means continuous quantities as opposed to numbers, and he gives a new definition of proportionality for numbers in Book VII (as if the Book V definition did not apply to them), and re-proves there some statements that he had proved for magnitudes in Book V. On the other hand, there is no technical reason why all the definitions and all the arguments of Book V should not apply to numbers as well as to continuous quantities, except for the assumption in the proof of V,5 that a given magnitude can be divided into a given number of equal parts, and the assumption in the proof of V,18 of the existence of a fourth proportional (i.e., that given magnitudes A, B, and C, there is always an X such that $A : B :: C : X$); and these assumptions are certainly illicit on Euclid's own understanding of his project (they may possibly be vestiges of some earlier account on which they would be justifiable), and can easily be removed.

[34] It is also possible to tell plausible stories about how the anthyphairetic theory would arise in the context of the study of irrational lines: for one such account, see Knorr, *The Evolution of the Euclidean Elements*, pp. 255–73.

proportion is never used in the "Eudoxian" Book XII (except inessentially in XII,13).

3 Knorr's reconstruction of Eudoxian proportion theory and its geometrical application

Here the work of Wilbur Knorr helps us to understand more clearly how Eudoxus was able to make his great advance. Knorr pointed out that yet a *fourth* proportion theory is being applied in Archimedes' proof of the principle of the balance (weights A and B, on opposite sides of a fulcrum supporting the bar of a balance, are in equilibrium if the distance from A to the fulcrum is to the distance from B to the fulcrum as weight B is to weight A, *On the Equilibrium of Planes* I,6-7) and in Pappus' proof of a theorem (the whole circumference of a circle is to an arc of the circle as the area of the circle is to the area of the sector under the arc, *Collection* V,12 and a parallel in Pappus' commentary on Ptolemy) that is, at a minimum, immediately entailed by the argument of Archimedes' *Measurement of the Circle*.[35] Starting from these texts of Archimedes and Pappus, Knorr offered a reconstruction of the proportion

---

[35]The parallel to *Collection* V,12 is in Pappus, *Commentaria in Ptolemaei syntaxin mathematicam* 5-6, ed. Adolphe Rome, *Commentaires de Pappus et de Théon d'Alexandrie sur l'Almageste*, v.1 (Biblioteca Apostolica Vaticana, 1931), pp. 254-60. Pappus' statement is a further development of *Elements* VI,33, which says that the angles from the center in equal circles are to each other as the corresponding arcs. Theon actually added to his edition of the *Elements* VI,33 a statement and proof that sectors of equal circles are to each other as the corresponding angles, according to his own statement in his commentary on the *Almagest* (*Commentaria in Ptolemaei syntaxin mathematicam* 1-4, in Rome, op. cit., v.2, 1936), 492,6-8, and the "Theonine" manuscripts of the *Elements* duly contain at the relevant place a statement and proof that arcs of equal circles are to each other as the corresponding sectors, and draw the corollary that the sectors are to each other as the angles: Heiberg prints Theon's supplement in the Appendix to his *Elementa* v.2, pp. 233-5. (Theon's proof uses the standard Euclidean proportion theory, as is appropriate in the context, and apparently takes for granted that similar sectors of equal circles are equal.) Archimedes, unlike Euclid, is trying to determine the area of the circle or of its sectors, and to prove that the circle or the sector is equal to a triangle whose height is equal to the radius and whose base is equal to the (total or partial) circumference (for the circle see Heiberg, *Archimedes, Opera Omnia* [second edition, Teubner, 1910-1915] I,232-5; for the sector see the fragment at II,542). Pappus' statement might have been the lemma allowing Archimedes to infer from the statement about the circle to the statement about the sector. On the Archimedes and Pappus texts see Knorr, "Archimedes ...", pp. 184-91 and pp. 222-30. As Vitrac notes (p. 524), Pappus also uses the same proof-technique in proving an extension of Theodosius *Spherics* III,5 (*Collection* VI,7-9, pp. 482,23-486,24; with context, VI,5-9 pp. 480,7-486,24). Knorr pp. 186-7 and pp. 224-5 discusses a supplementation in the text of Archimedes *Plane Equilibria* I,7. In my view the text as Heiberg prints it (*Archimedes, Opera Omnia*, II,136-9), although elliptical, does not require emendation, although it is certainly possible that, as Knorr suggests, something has fallen out after ἐπὶ τῷ Δ on p. 138 line 7, probably ἀλλὰ τὸ Γ μεῖζον ἔσται ἢ ὥστε ἰσορροπεῖν τῷ Α, ὅπερ ἀδύνατον followed possibly by something like ἀφείλετο γὰρ ἀπὸ τοῦ ΑΒ ἔλασσον τᾶς ὑπεροχᾶς, ᾇ μεῖζόν ἐστι τὸ ΑΒ τοῦ Γ ἢ ὥστε ἰσορροπεῖν. The further supplement which Knorr suggests in italics (in English) on p. 225 helps to fill out the argument, but Knorr does not seriously suggest that it was part of the text of Archimedes at this point. The meaning of the text is not in doubt, and Knorr's reconstruction of Archimedes' methods for arguing

theory that they seem to presuppose, analogous to Becker's reconstruction of a theory around the anthyphairetic proof of Euclid VI,1, by showing how to prove Euclid VI,1 and the crucial propositions of Euclid V (most centrally, alternation) using the same assumptions and method that Archimedes and Pappus are using. (Knorr's reconstructions of the particular details of each proof are, of course, *exempli gratia*: as with Becker, what was important was to assess the logical power of the definition or more generally of the technique for proving results about proportions. Apparently unknown to Knorr, his reconstructed theory is extremely close to the theory of proportion given by Giovanni Alfonso Borelli in his Euclides Restitutus of 1658, Borelli's Book III and IV,1 corresponding to Euclid's Book V and VI,1—Borelli is ultimately inspired by the same texts of Archimedes and Pappus as Knorr.)[36]

---

about proportions does not depend on how, if at all, we emend the text. (Heiberg's AB for the manuscripts' A in line 8 is uncontroversial.)

[36] I discovered Borelli's *Euclides Restitutus* (Pisa, Honophri, 1658) only long after writing the main body of this paper. As Howard Stein pointed out to me in 2002, Heath *Elements* v.2 p. 193 gives a reference to the great French mathematician Adrien-Marie Legendre, who in his *Éléments de Géométrie* III,3 (pp. 63–5 in the 11th edition, Firmin Didot, 1817) proved Euclid VI,1 in the same way that Knorr reconstructed for Eudoxus; so it is very likely that Knorr was aware of Legendre at least through Heath. But, as I learned only in 2014, Legendre was drawing on a tradition going back at least to the seventeenth century of mathematicians expressing dissatisfaction with Euclid's definition of proportionality and the proofs founded on it, and trying to give alternatives. (The typical complaints are: the duplication between Vdef5 and VIIdef20, and the thought that Vdef5, if universal as it appears to be, should apply to numbers as well as to continuous quantities; somewhat strangely, an alleged duplication between Vdef3 and Vdef5; and the thought that the property cited in Vdef5, while true of all proportional magnitudes and of no non-proportional magnitudes, is too complicated and remote from common conceptions to be the essence of proportionality, but should instead be deduced from some appropriate definition of proportionality.) And these mathematicians, especially Borelli, worked out their alternatives in far greater detail than Legendre. Perhaps the only text available with a modern translation and commentary is Gerolamo Saccheri, *Euclid Vindicated from Every Blemish*, ed. and comm. Vincenzo De Risi, Latin text with English translation by G.B. Halsted and L. Allegri revised by Vincenzo De Risi (Birkhäuser, 2014). (Warning: Girolamo Saccheri, *Euclides Vindicatus*, ed. and tr. G.B. Halsted, Open Court, 1920, in fact contains only Saccheri's Book I, on parallel lines, not his Book II, on proportion theory.) Saccheri pp. 202–3 considers (and accepts with some conditions) essentially the same demonstration of Euclid VI,1 as Legendre and as Knorr's Eudoxus. But Saccheri (published originally in 1733) was responding to a sophisticated seventeenth-century discussion, which he apparently assumes will be familiar to his readers. Borelli was the high point; Borelli was mainly responding to two texts circulating only in manuscript as of 1658, Evangelista Torricelli's *De Proportionibus Liber* and the incomplete Fifth Day of Galileo's *Discorsi*, but there was an older tradition of complaints against Euclid's proportion theory and sometimes of positive reconstructions (described by Borelli pp. 117–29), and there were a number of other seventeenth-century players. De Risi has some useful remarks, and there is a full discussion of the key seventeenth-century figures in Enrico Giusti, *Euclides Reformatus: La teoria delle proporzione nella scuola galileiana* (Bollati Boringhieri, 1993). Giusti prints Galileo's text pp. 277–98, and Toricelli's pp. 299–340. (The Torricelli text is extremely interesting, although it is not the superb piece of mathematics that Borelli's is.) Borelli is currently freely downloadable from the website of the Bayerische Staatsbibliothek. Giusti is aware of the striking similarities between Galileo-Torricelli-Borelli and Knorr's Eudoxus (see his p. 114), but does not go into detail. Giusti compares Knorr's Eudoxus especially with Torricelli, but the comparison with Borelli is closer, although Torricelli gives Knorr's

Knorr also argued that it is historically more likely, not that Archimedes invented this theory (so that the theory would be post-Euclidean), but rather that Archimedes (on the occasions where he uses this theory, and not Euclid's as elsewhere) is preserving an older, pre-Euclidean proportion theory.[37] While Knorr gave different kinds of reasons here, none of them individually conclusive, his central evidence comes from similarities of Archimedes' proportion theory, not to Euclid V, but to Euclid XII, or rather to the Eudoxian original underlying Euclid XII, some details of which can be recovered (Knorr argued) from Archimedes' references to Eudoxus.[38] As we have seen, Euclid XII does not fit smoothly with Euclid V; in particular, it proves V,14 (if $A:B::C:D$ and $A > C$ then $B > D$) through V,16, alternation, rather than calling on V,14 directly, and Archimedes also does this in several passages. This suggests that both Archimedes and Euclid XII (or its Eudoxian model) are relying on a proportion theory earlier than the theory of Euclid V; presumably, this theory would have been due to Eudoxus, and would have been the theory that Euclid drew on and modified in writing *Elements* V.[39] Two points of similarity that

---

"Eudoxian" proof for Euclid VI,1 and Borelli does not. I will add some comparative footnotes, mainly to Borelli but occasionally to Torricelli and Galileo, in what follows.

[37] For some scepticism, see Vitrac, *Éléments*, vol.2, pp. 523–9; Vitrac does not quarrel with Knorr's reconstruction of the theory, but he thinks Archimedes may have invented the theory himself, beginning from the theory of *Elements* V, and that Archimedes may have had particular motivations for preferring this theory to the equimultiple-based theory of *Elements* V in some particular cases, including the law of the lever in *Plane Equilibria* I,6–7. See also Acerbi, "Drowning by Multiples," pp. 218–37.

[38] As we have seen, Archimedes describes the central results of *Elements* XII as results of Eudoxus, and he sometimes cites them in formulations differing slightly from Euclid's. Knorr argues plausibly that these formulations are older than Euclid's, and that Euclid knows them and is deliberately modifying them (Knorr, "Archimedes ...", pp. 195–7, in the context of Knorr's larger argument that the proportion theory assumed by Archimedes is pre-Euclidean, pp. 193–200); thus it is likely that Archimedes knew these results directly from a treatise of Eudoxus', the same treatise that Euclid would have taken as his model in writing *Elements* XII. There is no reason to think that Eudoxus' treatise was anything like an *Elements of Geometry* (we have references to various *Elements* before Euclid, but not to one by Eudoxus), and it is an open question how far he shared Euclid's ideal of deriving everything from a short list of explicitly stated first principles. (I will suggest below a model on which Eudoxus' treatise would have been closer in form to some extant treatises of Archimedes than to Euclid's *Elements*.) So it is an open question how far Eudoxus' text contained a *theory* of proportion—but this is partly a matter of how comprehensive we expect a "theory" to be, and for the moment I am using the word broadly.

[39] However, in the apparently later *On Spirals*, Archimedes uses the *Elements* V definition of proportion (Knorr, "Archimedes ...," p. 194), and in the apparently later *On Conoids and Spheroids* he assumes V,14 without establishing it via V,16 (ibid. pp. 197–8). (These works are later according to a chronology that Knorr had argued for independently, *before* he had the idea that Archimedes might preserve traces of a pre-Euclidean proportion theory, see his "Archimedes and the *Elements*: Proposal for a revised chronological ordering of the Archimedean corpus," *Archive for History of Exact Sciences* 19 (1978–9), 211–90.) This suggests that, at a time later than the *On the Equilibrium of Planes*, either Archimedes became aware of *Elements* V for the first time, or he decided to adopt it as his official proportion theory, presumably because he decided that it was more rigorous than whatever he had been using before; and it suggests that the proportion theory he had been using before did not prove V,16 by means of V,14. Knorr's chronology of Archimedes' works has

Knorr notes between the Archimedean proportion-theory and Euclid XII are that they both rely on what Knorr calls the "principle of bisection" (roughly Euclid X,1: if we diminish a quantity by successively subtracting at least half of what remains, it will eventually become less than any given quantity), and that they both assume the existence of a fourth proportional (for any A, B and C there is some D such that $A : B :: C : D$), whereas Euclid V makes neither of these assumptions (and indeed the existence of the fourth proportional cannot be justified on the basis of Euclid's theory). As I will show, the Archimedean proportion theory and Euclid XII are in fact related much more closely than simply by sharing these assumptions. To this extent I will be strengthening Knorr's case for attributing the Archimedean proportion theory to Eudoxus; although I will argue that the connection between Eudoxus' proportion theory and his method of exhaustion in geometry was rather different from what Knorr seems to have believed.[40] If Euclid V is not simply a presentation of Eudoxus' theory of proportion, we can see why Euclid XII does not depend more deeply on Euclid V; at the same time, if Eudoxus' theory of proportion was connected with his method of exhaustion in the way I will suggest, we can see how Eudoxus was led to discover his theory of proportion, and why and in what sense the methods of Books V and XII are akin. (I will also, below, sketch how and why Euclid would have turned Eudoxus' exposition of proportion theory into Book V as we now have it.)

To see how this proportion theory works, we can look at Pappus' proof that the whole circumference of a circle is to an arc of the circle as the area of the circle is to the area of the sector under the arc. First, suppose that arc AB is commensurable with the whole circumference ABC; thus suppose that the circumference can be divided into p equal arcs, AD etc., and that the arc AB is the sum of q of those equal arcs. Draw lines from the center O to each of the points D etc., dividing the circle into p equal sectors OAD etc.; the sector OAB will be the sum of q of these equal sectors. Therefore the circumference of the circle ABC is to the arc AB as the area of the circle ABC is to the area of the sector OAB. (To justify this inference, we can supply the principle, "magnitudes are proportional when the first is the same multiple, or the same part, or the same parts, of the second that the third is of the fourth"— Euclid's definition of proportionality for numbers, VIIdef20,

---

been contested in an article I have not seen, Bernard Vitrac, "À propos de la chronologie des œuvres d'Archimède," in *Mathématiques dans l'Antiquité*, ed. Jean-Yves Guillaumin (Publications de l'Université de Saint-Etienne, 1992), pp. 59–93.

[40]Knorr also gave a further argument that Archimedes' theory of proportion is pre-Euclidean, namely that Aristotle is referring to this theory in several passages including *De Caelo* I,6 (the argument that an infinite body cannot have a finite weight) and *Physics* VI,2 (the account of "faster"). If Knorr is right about the Aristotle texts, this would be very strong evidence that Archimedes' proportion theory comes from Eudoxus, but while the Aristotle passages are suggestive, I think it is underdetermined what proportion theory they are using. Knorr's argument here has been further developed by Henry Mendell, "Two Traces of Eudoxan Two-Step Proportion Theory in Aristotle," *Archive for History of Exact Sciences* 61 (2007), 3–37; but I still think the evidence is not decisive.

with the word "magnitude" substituted for "number."[41] To paraphrase this in more precise terms, $X : Y :: Z : W$ if, for some numbers m and n, $X = nY$ and $Z = nW$, or $X = Y/n$ and $Z = W/n$, or $X = mY/n$ and $Z = mW/n$ [since from a modern point of view we allow our integer values to be 1 we can just say that $X : Y :: Z : W$ if $X = mY/n$ and $Z = mW/n$].)

Now suppose that the arc AB is not commensurable with the whole circumference ABC, and suppose that the area of the circle ABC is not to the area of the sector OAB as the circumference of the circle ABC is to the arc AB, but, rather, as the circumference ABC is to some other arc AE, which must be either greater or less than the arc AB. (To justify the existence of this arc AE, we need to assume the postulate of the existence of the fourth proportional; note that this step is quite non-constructive.) Suppose arc AE is less than arc AB. Then there is some arc AF, greater than AE and less than AB, such that arc AF is commensurable with the circumference ABC. (Pappus justifies the existence of this arc AF by explicitly referring to "a lemma of the *Spherics*," i.e. the proposition supplied as a lemma to Theodosius' *Spherics* III,9 and now preserved as a scholium on that text, which says that for any three homogeneous magnitudes X, Y and Z, if $X > Y$ there is some W such that $X > W > Y$ and W is commensurable with Z.) Thus, by the commensurable case which we have already proved, the area of the circle ABC is to the area of the sector OAF as the circumference ABC is to the arc AF. But the ratio of the circumference ABC to arc AF is less than the ratio of the circumference ABC to the shorter arc AE, which is (by the definition of AE) the ratio of the area of the circle ABC to sector OAB. So the ratio of the area of the circle ABC to sector AOF is less than the ratio of the area of the circle ABC to the larger sector OAB, which is absurd. (Pappus then considers the other case, where arc AE is greater than arc AB, and derives a contradiction in an exactly similar way.)

From this argument, and from the kindred argument in Archimedes *On the Equilibrium of Planes* I,6–7, we can extract the principles that are being assumed in this Archimedean proportion theory. We could axiomatize the theory in a variety of slightly differing ways, without there being sufficient ground to attribute any one particular version to Eudoxus (or even to Archimedes), but it is important to give *some* axiomatization, in order to see how much the theory can prove and what kinds of assumptions it must make. The crucial lesson, which will hold good on any version, is that the theory can prove everything that Euclid can prove with the theory of *Elements* V, and is not simply an *ad hoc* invention for proving *Equilibrium of Planes* I,7 and the supplement to *Elements* VI,33; but also that it must make some assumptions (most importantly the existence of a fourth proportional, also the "lemma of the *Spherics*"

---

[41] However, Euclid may intend "when" to be glossed as "iff," whereas here it must be glossed as "if"—if we interpret "when" as "iff," then the principle will imply that $X : Y :: Z : W$ *never* holds when X and Y are incommensurable. It gives a necessary and sufficient condition for proportionality assuming either that X and Y are commensurable or that Z and W are commensurable.

or something like it) which Euclid is able to avoid in *Elements* V, but which he draws on in later books of the *Elements*. My presentation is broadly similar to Knorr's, with some significant differences,[42] and attempts a full axiomatization where Knorr does not. I am closer to Borelli, who does give a full axiomatization, with the difference that I do not *define*, but only *axiomatize*, what it is for A to be in the same, or a greater or lesser, ratio to B than C is to D: why this difference is significant, and why I prefer my formulation, will become clear further on.

The theory must contain the principle: if A and B are commensurable, say $A = pB/q$,[43] then $A : B :: C : D$ if $C = pD/q$ (a modification of Euclid VIIdef20 for commensurable magnitudes instead of numbers). It would be natural to strengthen this to the biconditional: if A and B are commensurable, say $A = pB/q$, then $A : B :: C : D$ iff $C = pD/q$. Thus if $C <> pD/q$, $A : B <> C : D$. It would be natural to include the strengthening that if $C < pD/q$, then $A : B < C : D$, and if $C > pD/q$, then $A : B > C : D$. This adds up to *Postulate I: if A and B are commensurable, say $A = pB/q$, then $A : B :: C : D$ iff $C = pD/q$, $A : B < C : D$ iff $C > pD/q$, and $A : B > C : D$ iff $C < pD/q$.*[44]

The theory needs the principle: if $A > B$, then there is some D such that $A > D > B$ and D is commensurable with C (the "lemma of the *Spherics*", Lemma II'). Rather than taking this as an unproved postulate, it is better to derive it from *Postulate II*, the "bisection principle": *if we diminish a quantity by successively subtracting at least half of what remains, it will eventually become less than any given quantity* (roughly Euclid X,1). To prove Lemma II' from Postulate II: either $C < A$ or not. If not, divide C in half, and repeat this division until we reach some $C' < A$. Then subtract C' from A as many times as it can be subtracted; clearly this will be subtracting at least half of A, perhaps leaving some remainder A'. Repeat the process by dividing C' in half until we reach some $C'' < A'$, and subtracting C'' from A' as many times as it can be subtracted; clearly this will be subtracting at least half of A', perhaps

---

[42]Knorr's reconstruction is given at "Archimedes ...", pp. 230–5. Knorr does not have anything much like my Postulate I, and, inspired by *Elements* X,5–6, makes liberal use of proportions among numbers and of *ex aequali* in his proofs.

[43]By pB/q I will always mean $p(B/q)$, not $(pB)/q$. I am including the cases where $p = 1$ or $q = 1$ (or both). From the ancient point of view, this subsumes four cases: pB/q is either equal to B, or a multiple of B, or a part of B, or parts of B.

[44]We could get the same results by assuming two postulates: *if A and B are commensurable, say $A = pB/q$, then $A : B :: C : D$ if $C = pD/q$, and if $A > B$, then $A : C > B : C$* (Euclid V,8). Borelli gives first a definition of when $A : B :: C : D$ if A where A is commensurable with B and C is commensurable with D (his IIIdef8); then a definition of when $A : B > C : D$ or $A : B < C : D$ where C and D are commensurable (IIIdef9—these two together are essentially equivalent to my Postulate I); then definitions of when $A : B > C : D$ or $A : B < C : D$ where A and B are incommensurable and C and D are incommensurable (IIIdefs10-11—the inequalities hold iff a commensurable ratio can be interpolated between the two incommensurable ratios); then a definition of when $A : B :: C : D$ where A and B are incommensurable and C and D are incommensurable (IIIdef12—$A : B :: C : D$ iff neither $A : B > C : D$ nor $A : B < C : D$). Borelli intends this compound definition to hold for all kinds of quantities, and in particular for numbers (which are, of course, all commensuable).

leaving some further remainder A''. Repeat the process by dividing C''' in half until we reach some $C'''' < A''$, and so on. Since we are diminishing the quantity A by successively subtracting at least half of what remains, it will eventually become less than $A - B$, so the total amount D that we have subtracted will be greater than B. But D is a multiple of C' plus a multiple of C''' and so on, and C', C''' etc. are all submultiples of C; so D is commensurable with C, and $A > D > B$, as was to be shown. (Or, equivalently: divide C in half, and repeat this division until we reach some C' such that $C' < (A - B)$. Then let D be the first multiple of C' that is greater than B. We must have $D < A$, since if $D \geq A$ we would have $(D - C') > (A - (A - B)) = B$, and so D would not be the first multiple of C' that is greater than B. Thus $A > D > B$ and D is commensurable with C, as was to be shown.)

The theory also needs the postulate of the existence of the fourth proportional, *Postulate III: for any three magnitudes B, C and D, there is some X such that $X : B :: C : D$.*[45]

Finally, the theory will also need either to assume or to demonstrate a number of obvious-sounding assumptions about the ordering of ratios, which at an early stage might not have been explicitly formulated as assumptions. I will describe all these as *Common Notions*. They will include: given magnitudes A, B, C, D, etc., *exactly one of $A : B :: C : D$, $A : B > C : D$, and $A : B < C : D$ is true*; $A : B > C : D$ iff $C : D < A : B$; if $A : B :: C : D$ then $C : D :: A : B$; if $A : B :: C : D$ then $B : A :: D : C$; if $A = B$ then $A : C :: B : C$ and $C : A :: C : B$; if $A : B :: C : D$, then $A : B :: E : F$ if $C : D :: E : F$, $A : B > E : F$ if $C : D > E : F$, and $A : B < E : F$ if $C : D < E : F$; if $A : B > C : D$ and $C : D > E : F$ then $A : B > E : F$. There would be ways to eliminate some of these assumptions, but I am not going to worry about this.

Notice that this theory does not include a definition of "the ratio $A : B$". Neither does Euclid's, but Euclid does give definitions of "A is in the same ratio to B as C is to D" and "A is in a greater ratio to B than C is to D"; and so did the anthyphairetic theory as Becker reconstructs it,[46] and so presumably did

---

[45]Obviously I am assuming that C and D are homogeneous, so that they have a ratio. Howard Stein points out that B must be assumed to be a continuous rather than a discrete quantity, and that without some further restrictions on what kind of quantity B is, the assumption may still be implausibly strong. As Stein says, the crucial logical work can be done if we restrict the postulate to the case where B is a length. However, Euclid (who never explicitly states the postulate, and so never makes clear how broad a form of it he is assuming) uses it for (not necessarily rectilinear) areas in XII,2 and solids in XII,5, XII,11 and XII,12; in XII,18 he uses the extraordinarily strong assumption that if B is a sphere and C and D are lengths there is some *sphere* X such that $X : B :: C : D$, which could be justified only by a strong principle of continuity. Also the unnecessary use of the postulate in Euclid V,18 (which I will discuss toward the end of this paper) applies it to magnitudes of any kind without restriction. So I will assume that early geometers who used the postulate simply had not reflected on whether it could or should be restricted to particular kinds of magnitudes, beyond the obvious restriction to continuous rather than discrete quantities.

[46]This needs a caveat, depending on how we think $A : B > C : D$ was "defined": perhaps its "definition" is just an algorithm for testing whether $A : B > C : D$ or $A : B < C : D$ (such that, if the procedure continues for ever, then A and B have the same ἀνθυφαίρεσις as

the original proportion theory that all these theories are responding to (namely $A : B :: C : D$ iff $\exists p \exists q$ such that $A = pB/q$ and $C = pD/q$; perhaps supplemented by $A : B > C : D$ iff $\exists p \exists p' \exists q$ such that $A = pB/q$ and $C = p'D/q$ and $p > p'$ or the like). It would be possible to say that the "Eudoxian" theory includes an "implicit" or "operational" definition of being in the same (or in a greater) ratio, but I would prefer to say that this is not a full-scale "theory" of proportion, comparable to the original proportion theory or the anthyphairetic theory or the Euclidean theory, but rather a technique for proving results about proportion, compatible with several different definitions of proportion. If, as I agree with Knorr in believing, Eudoxus did put forward roughly this "theory," he may well have accepted the anthyphairetic definition of proportion, from which many of his assumptions can be derived. But the anthyphairetic theory of proportion is a very awkward instrument for proving almost anything other than Euclid VI,1 and a few incommensurability claims, and Eudoxus might well have been motivated, without necessarily rejecting this theory, to come up with a more efficient proof-technique, for proving alternation or any other results that Eudoxus may have been interested in. (This would solve the problem, mentioned above, of why Aristotle seems to think in the *Topics* that the anthyphairetic definition of being in the same ratio is current, and in the *Posterior Analytics* that Eudoxus' proof of alternation has displaced the old proof, without having to date Eudoxus' discovery in the presumably short interval between the composition of the two works; Eudoxus might simply have accepted the anthyphairetic definition of being in the same ratio, and although he would have given an alternative to the anthyphairetic proof of Euclid VI,1, he need not have rejected the anthyphairetic proof.)

We can illustrate how this theory would work by showing how to prove Euclid VI,1 with it. This can be done by mimicking Pappus' proof that the area of a circle is to the area of a sector as the circumference of the circle is to the arc bounding the sector.

For simplicity, we will prove only the version for rectangles: *a rectangle of height H and base A is to a rectangle of height H and base B as the base A is to the base B*, or $rect(H, A) : rect(H, B) :: A : B$. But the argument would work in very much the same way for parallelograms or triangles.

First suppose that the bases A and B are commensurable, say $A = pD$, $B = qD$. Then, dividing the bases A and B into segments of length D, draw lines from the endpoints of each of these segments parallel to the sides of the rectangle; we can thus divide the rectangle with height H and base A into p rectangles with height H and base D, and divide the rectangle with height H and base B into q rectangles with height H and base D. So $rect(H, A) = p \ rect(H, D)$ and $rect(H, B) = q \ rect(H, D)$. So, by Postulate I, we have $rect(H, A) : rect(H, B) :: A : B$, as was to be shown.

C and D, and $A : B :: C : D$). But at least being in the same ratio has a definition, namely having the same ἀνθυφαίρεσις.

Now suppose that the bases A and B are not commensurable, and suppose that $rect(H, A)$ is not to $rect(H, B)$ as the base A is to the base B. Then, by Postulate III, there is some other area X such that $X : rect(H, B) :: A : B$, where X must be either greater or less than $rect(H, A)$.[47] Suppose that $X < rect(H, A)$. Then, by Lemma II$'$, there is some area Y such that $X < Y < rect(H, A)$ and Y is commensurable with $rect(H, B)$. Let the line Z be such that $Z : B :: Y : rect(H, B)$ (because Y is commensurable with $rect(H, B)$, this is constructive; if we write $Y = p\, rect(H, B)/q$, then $Z = pB/q$). Mark out a line of length Z on the base A (produced if necessary), and draw a perpendicular from its endpoint to yield a rectangle of height H and base Z. By the commensurable case of Euclid VI,1, which we have already proved, we know that $rect(H, Z) : rect(H, B) :: Z : B$, and thus (by the "only if" side of Postulate I) $rect(H, Z) = p(rect(H, B)/q) = Y$. Since $Y < rect(H, A)$, clearly the rectangle $rect(H, Z)$ must be contained within the rectangle $rect(H, A)$ rather than vice versa, and so $Z < A$. But since $Y > X$, we have $p(rect(H, B)/q) : rect(H, B) :: Y : rect(H, B) > X : rect(H, B) :: A : B$. So, by Postulate I, we have $A < (pB/q) = Z$, a contradiction.

So consider the remaining case, $X > rect(H, A)$. We could give a similar argument for this case, finding an area Y such that $X > Y > rect(H, A)$ and Y is commensurable with $rect(H, B)$, and deriving a contradiction in a similar way. Alternatively, if we have some of the results of Euclid V at our disposal, we could simply reduce this case to the previous case. For, by Postulate III, there must be some area Z such that $X : rect(H, B) :: rect(H, A) : Z$. Since $X > rect(H, A)$, we must have $rect(H, B) > Z$ by Euclid V,14 (which we can prove, if we like, from alternation). So $Z : rect(H, A) :: rect(H, B) : X :: B : A$, and $Z < rect(H, B)$; and this is the same as the assumption that we reduced to absurdity in the previous case.

This is only a minor deviation from Pappus' proof about the arcs and sectors of a circle, and is clearly an illustration of the same method that Pappus and Archimedes are using. And it gives a simple and straightforward proof of Euclid VI,1.[48] But what is striking is that the proof, for the incommensurable case, is closely parallel to Euclid's proof of XII,2. It is, in fact, a classic proof by exhaustion. In Euclid XII,2, we exhaust the circle C by polygons, which have been shown by Euclid XII,1 to bear the desired proportions to polygons

---

[47] I am here deviating from Pappus' proof (if I were following Pappus I would say $rect(H, A) : rect(H, B) :: A : C$ for some C greater or less than B). It would also work to follow Pappus more closely; my reason for preferring this way of doing it will appear shortly. My reconstruction is essentially the same as Knorr's, "Archimedes ...", p. 231.

[48] A bit surprisingly Borelli, who is usually the closest to Knorr's reconstruction of Eudoxus, does not give this proof for his IV,1 = Euclid VI,1; instead, he proves (in my terms) that $rect(H, A) > rect(H, pB/q) = p\, rect(H, B)/q$ iff $A > pB/q$ (and likewise for = and <). But Torricelli, *De Proportionibus Liber* propositions 2–3 (in Giusti op. cit. pp. 312–14), gives essentially the proof I have given. Saccheri, *Euclid Vindicated*, ed. De Risi, pp. 202–3, also proposes this as one possible proof for VI,1, but complains (pp. 204–5) about its dependence on the postulate of the existence of a fourth proportional (and compare Saccheri's discussion from p. 198 on). De Risi's notes discuss earlier geometers who had given proofs in this family.

inscribed in the circle C'.[49] In the present Archimedean-Eudoxian proof of the incommensurable case of Euclid VI,1, we exhaust the rectangle $rect(H, A)$ by rectangles $rect(H, pB/q)$, which have been shown by the commensurable case of Euclid VI,1 to bear the desired proportions to the rectangle $rect(H, B)$. The function of Lemma II' is to show (in this case) that $rect(H, A)$ can be exhausted by rectangles with height H and base commensurable with B, just as the circle can be exhausted by inscribed regular polygons with 2^n sides. In both cases, we prove this from the bisection principle, successively removing from $rect(H, A)$ rectangles of the form $rect(H, B/2^n)$, and successively removing from the circle the triangles based on the sides of a 2^n-gon. Both proofs proceed by a *reductio ad absurdum*, supposing that the desired proportionality does not hold, and using the postulate of the existence of the fourth proportional to find an area $X < rect(H, A)$ such that $X : rect(H, B) :: A : B$, or an area $X < C$ such that $X : C' :: D^2 : D'^2$. We then interpolate a Y between X and $rect(H, A)$ which is the area of a rectangle with height H and base commensurable with B, or a Y between X and C which is the area of a regular polygon with 2^n sides inscribed in C, and then derive a contradiction by using the case we have already proved, that rectangles with the same height and commensurable bases are to each other as their bases, or that similar polygons inscribed in circles are to each other as the squares of their diameters. And these easier propositions were proved by breaking down the two rectangles into simpler units, rectangles with the same height and base, or by decomposing the two polygons into simpler units, similar triangles. In both cases we are proving a proportionality of areas by exhausting a figure (the rectangle or the circle) by figures that can be decomposed into a finite number of figures for which the desired proportionality can be proven directly.

In other words, the Eudoxian proof for Euclid VI,1, as Knorr and I reconstruct it, is a straightforward application of the method of exhaustion. And this leads to what may be my main disagreement with Knorr. When he is trying to determine the historical relations between the two kindred theories, Archimedes' proportion theory and the Euclid XII theory of measurement by exhaustion (Knorr's own view being that they were both invented by Eudoxus), Knorr says that "in devising a theory of proportion, one would not look to the rather different field of geometric measurement for technical models."[50] So Knorr says that common features of the two theories, such as the nonconstructive (and ultimately unnecessary) assumption of the existence of the fourth proportional, and the proof of Euclid V,14 via alternation rather than vice versa, would have begun in the proportion theory, and thence been taken up into exhaustion proofs of (Eudoxus' version of) Euclid XII, since many of the statements of Euclid XII express proportionalities and so depend on a prior proportion theory: "it was the prior studies in proportion theory which influenced the proof techniques in the 'exhaustion' method" (ibid.). At the same

---

[49] I borrow the symbols C, C', D and D' from my formulation of XII,2 in Section I above.
[50] This and the two following quotes from Knorr are from "Archimedes ...", p. 198.

time, Knorr notes that the "principle of bisection" (Euclid X,1, my Postulate II) would more naturally occur to someone in the context of the method of exhaustion (most obviously in the exhaustion of a circle by inscribed 2^n-gons) rather than in the theory of proportion, so that in this respect the proportion theory seems to have been inspired by an idea in the exhaustion-geometry. Knorr solves the problem of the apparent mutual dependence of the two theories by saying that they were both invented by Eudoxus, and that "this form of proportion theory [was] designed with regard for the specific needs of the 'exhaustion' theory" (ibid.). In other words, while the proportion theory is logically prior to the exhaustion-geometry and is presupposed by it, Eudoxus designed the proportion theory specifically to provide the foundations for the method of exhaustion, and in this sense the proportion theory and the exhaustion-geometry are a package deal.

I think this way of thinking about the relation between the proportion theory and the exhaustion-geometry is wrong, and is not what emerges from Knorr's own reconstruction of Eudoxus' proportion theory. While Euclid XII does not use the proportion theory of Euclid V (except inessentially in XII,13, where the Eudoxian precursor of Euclid XII may well have used some other proportion theory), there is no sign that it used Knorr's reconstructed Eudoxian proportion theory either. It logically presupposes *some* proportion theory, but it makes little difference *what* proportion theory: it might originally have been the anthyphairetic theory.[51] Eudoxus' proportion theory is not the *foundation* of the method of exhaustion, but a straightforward *application* of the method of exhaustion to help resolve the longstanding foundational nuisance better than the anthyphairetic theory did (although, as I noted above, Eudoxus has no need to *reject* the anthyphairetic theory). Knorr says that "in devising a theory of proportion, one would not look to the rather different field of geometric measurement for technical models," but Knorr's own reconstruction of the Eudoxian proof of Euclid VI,1 shows that this is exactly what Eudoxus did.[52]

---

[51]Except that, as noted above, the difficulty in proving alternation for non-rectilineal figures or for solids on the anthyphairetic theory might have motivated Eudoxus to look for a proportion theory that would give a more general proof of alternation. It remains that *Elements* XII shows no signs of any such concern, or (apart from XII,13) of the use of Eudoxus' proportion theory as a foundation; hence Becker's conclusion that *Elements* XII reflects an early work of Eudoxus, and *Elements* V a later work of Eudoxus. (And *Elements* XII makes no more use of Knorr's Eudoxian theory of proportion than of *Elements* V.)

[52]Fabio Acerbi says, "The method of equimultiples took shape as an evolution of proof techniques within a relatively well established field, viz. that of quadratures of non-rectilineal figures. Transfer and adaptation of the method to the developing field of general theorems on proportions was immediate, but perhaps it is better to say that the techniques really created the new field. The idea of replacing successive division with multiples led only as a final consequence to the elaboration of a theory. It is in fact plausible that defs. 4 and 5 of book V have not been the starting point of the whole reworking enterprise but that they have resulted as a crystallization of a proof technique, thereby creating the mathematical object 'proportion'" ("Drowning by Multiples," p. 222). I am in agreement with the spirit of this: the definitions of being in the same or a greater ratio, or postulates for determining whether something is in the same or a greater ratio, are a byproduct of a proof-technique, namely the

As the passage from Aristotle's *Topics* shows, a standard question in the foundations of geometry was to ask what was the right way of proving Euclid VI,1. Eudoxus, having invented a new technique for proving geometrical proportionalities and equalities that cannot be proved by decompositions, the method of exhaustion (or rather, having refined a crude method of "passing to the limit" and made it rigorous), points out that his technique can also be used to give a new solution to the old worry about Euclid VI,1. And we can also see how, starting from his proof of Euclid VI,1, Eudoxus might also be led to give new proofs for the rules of proportion, proofs that are directly ancestral to the proofs now given in Euclid V.

Since, according to Becker's reconstruction of the anthyphairetic proportion theory, alternation used to be proved by Euclid VI,16, which was in turn derived from VI,1, it would be natural for Eudoxus to investigate what implications his new method for proving VI,1 would have for proving alternation.

We can start by seeing how to prove VI,16 (for any four straight lines A, B, C, D, if $A:B :: C:D$ then $rect(A,D) = rect(B,C)$, and conversely) using the same techniques we have just used for VI,1.

Suppose $A:B :: C:D$, and suppose A and B are commensurable. First suppose A is a multiple of B, say $A = pB$. Then $C = pD$ and $rect(A,D) = p\,rect(B,D) = rect(B,C)$, as was to be shown. A similar argument works if B is a multiple of A. Now suppose that neither A nor B is a multiple of the other, but that they have a common measure E, say $A = pE$ and $B = qE$. By Postulate I, there is a magnitude F such that $C = pF$ and $D = qF$. Then $rect(E,D) = q\,rect(E,F) = rect(B,F)$. So $rect(A,D) = p\,rect(E,D) = p\,rect(B,F) = rect(B,C)$, as was to be shown.

Now consider the general case, and suppose $A:B :: C:D$ but $rect(A,D) \neq rect(B,C)$; without loss of generality, we can assume $rect(A,D) > rect(B,C)$. Application of Postulate II shows that $rect(A,D)$ is exhausted by rectangles of height D and base commensurable with B, so that for some E commensurable with B we have $E < A$ and $rect(E,D) > rect(B,C)$. By the existence of a fourth proportional, there is an F such that $E:B :: F:D$ (this is constructive: writing $E = pB/q$, we have $F = pD/q$). So, by the commensurable case which we have proved, we have $rect(B,F) = rect(E,D) > rect(B,C)$; thus $F > C$. So (using Euclid V,8 twice) we have $A:B > E:B :: F:D > C:D$, contradicting the assumption that $A:B :: C:D$.[53]

"method of exhaustion," for determining areas or volumes. But this process did not have to wait for the reformulation in terms of equimultiples; it can perfectly well have begun with the technique of exhaustion by commensurable magnitudes.

[53]With a little more difficulty, we can get the same result by exhausting $rect(A,D)$ by rectangles of height D and base commensurable with C (rather than B). For some E commensurable with C we have $E < A$ and $rect(E,D) > rect(B,C)$. It is now again, with a little more difficulty, constructive that there is an F such that $E:B :: C:F$ (writing $C = rE/s$, we have $F = rB/s$; this depends on two applications of the rule $X:Y :: nX:nY$, Euclid V,15, whose status I will discuss below). It is then automatic (without appeal to alternation) that $E:C :: B:F$. Then, applying again the commensurable case which we have proved, we

The converse can be shown immediately using the existence of the fourth proportional. For suppose $rect(A, D) = rect(B, C)$, and suppose $A : B \neq C : D$. Then there is some $E \neq A$ such that $E : B :: C : D$. Then $rect(E, D) = rect(B, C) = rect(A, D)$; so $E = A$, contradicting the assumption.

Note that, in proving the incommensurable case of VI,16, I allowed myself two uses of Euclid V,8, inferring from $A > E$ to $A : B > E : B$ and from $F > C$ to $F : D > C : D$.[54] Both are instances of the "commensurable case" of V,8, i.e. the case inferring from $X > Y$ to $X : Z > Y : Z$ where at least one of the magnitudes X and Y is commensurable with Z. As noted above, it would be possible to take Euclid V,8 as an unproved postulate, and Eudoxus may have done so (or, if he accepted the anthyphairetic definition of proportion, he may have accepted an anthyphairetic proof such as was sketched above). But all we need for VI,16 is the commensurable case of V,8, and this is simply a subcase of our Postulate I. For suppose that X and Z are commensurable, say $X = pZ/q$. Then since $X > Y$, we have $Y < pZ/q$, and thus $X : Z > Y : Z$. A similar argument works if it is Y that is commensurable with Z.[55]

Now, as we have seen above, a proof of VI,16 immediately yields a proof of alternation for straight lines. For let A, B, C and D be four straight lines such that $A : B :: C : D$. By VI,16, $rect(A, D) = rect(B, C)$; so $rect(A, D) = rect(C, B)$; so by VI,16 in the converse direction, $A : C :: B : D$. In that sense, we already have an "exhaustion" proof of alternation for straight lines. But if we fill in all the steps, this is a quite complicated proof, using two exhaustions. It is natural to try to simplify the argument.

---

have $rect(E, F) = rect(B, C) < rect(E, D)$, so $F < D$. So $A : B > E : B :: C : F > C : D$, contradicting the assumption that $A : B :: C : D$.

[54]The use of V,8 or V,9 or some kindred proposition is unavoidable in proving V,16.

[55]It is easy to prove the general case of V,8 given the commensurable case. For if $X > Y$, by Lemma II' there is some W commensurable with Z such that $X > W > Y$. So, by the commensurable case of V,8, $X : Z > W : Z > Y : Z$, as was to be shown. Acerbi, "Drowning by Multiples," pp. 221–2, suggests that there is a circle on any Knorr-like reconstruction of Eudoxus, that the proof of V,8 will depend on the "rational interpolation lemma," $A : B > C : D$ iff $\exists m \exists n\ A : B > m : n > C : D$, and that the proof of this will in turn depend on V,8. It would be perfectly possible simply to *define* $A : B > C : D$, in the case where A and B are incommensurable and C and D are incommensurable, as $\exists m \exists n\ A : B > m : n > C : D$, or $\exists m \exists n\ (A > mB/n$ and $C < mD/n)$, and this is what Borelli does. But it is also straightforward to prove the "rational interpolation lemma" from Postulates I-III. (The "if" direction requires only Postulate I and the Common Notions. For the "only if" direction, by Postulate III there is an X such that $A : B :: C : X$ and if $X \neq D$, by Lemma II' there will be some Y between X and D commensurable with C, and the result will follow from two applications of Postulate I.) And V,8, if $A > B$ then $A : C > B : C$, can be proved by Lemma II' to find D commensurable with C such that $A > D > B$, and then using Postulate I twice (and the Common Notion of the transitivity of "greater ratio") to prove $A : C > D : C > B : C$. There is no circularity. Postulate I is (rather, the inequality cases of Postulate I are) a special case of V,8, but there is nothing objectionable in proving a general proposition from a special case of it, as the "Eudoxian" theory routinely does in inferring from the commensurable to the incommensurable case of any proposition. We could even weaken Postulate I to "if $A = pB/q$, then: if $C = pD/q$, then $A : B :: C : D$; if $C > pD/q$, then $A : B \leq C : D$; if $C \geq (p+1)D/q$, then $A : B < C : D$; if $C < pD/q$, then $A : B \geq C : D$; and if $C < (p-1)D/q$, then $A : B > C : D$" and get the same results.

The general strategy in arguing from the assumption $A : B :: C : D$ to the conclusion $A : C :: B : D$ is to assume the contradictory $A : c \neq B : D$, infer that, for some $E \neq A$, $E : C :: B : D$, and then try to prove that $E = A$ and thus derive a contradiction. We can do this by proving that $rect(E, D) = rect(A, D)$. We did this above by showing that each term is equal to $rect(B, C)$, by supposing that $rect(E, D)$ or $rect(A, D)$ is greater or less than $rect(B, C)$, interpolating a rectangle of the form $rect(F, D)$, for F commensurable with B or C, between $rect(E, D)$ or $rect(A, D)$ and $rect(B, C)$, and deriving a contradiction. But it is easier to do just one interpolation rather than two. So suppose that (say) $rect(E, D) > rect(A, D)$, and interpolate between them a rectangle $rect(F, D)$ where F is commensurable with B or C; we can then derive a contradiction by showing that $rect(F, D)$ is both greater and less than $rect(B, C)$.

There are several ways to do this, by varying the arguments for VI,16. One way is as follows. Let F such that $E > F > A$ be commensurable with C. Then there is a G such that $F : C :: B : G$; since F and C are commensurable, this is constructive (if we write $C = pF/q$, then $G = pB/q$). Then $B : G ::$ $F : C < E : C :: B : D$, so $G > D$. So $rect(F, D) < rect(F, G)$, and by the trivial commensurable case of VI,16, $rect(F, G) = rect(B, C)$, so $rect(F, D) < rect(B, C)$. On the other hand, there is also constructively an H such that $F : H :: C : D$ (as we argued in proving VI,16, if we write $F = qC/p$, then $H = qD/p$; as before, this depends on Euclid V,15). Since $F : H :: C : D :: A :$ $B < F : B$, we have $H > B$, and thus $rect(C, H) > rect(C, B)$. Since (as in the proof of VI,16) it is automatic (without alternation) that $F : C :: H : D$, by the trivial commensurable case of VI,16 we have $rect(F, D) = rect(C, H)$, it follows that $rect(F, D) > rect(C, B)$. So $rect(F, D)$ is both greater and less than $rect(B, C)$, a contradiction.[56]

This last argument has an interesting feature. Since F is throughout commensurable with C, for $rect(F, D)$ to be greater or less than $rect(B, C)$ is just (by the trivial commensurable case of VI,16) for the ratio $F : C$ to be greater or less than the ratio $B : D$. What the argument is in effect doing is interpolating F such that $E > F > A$ and F is commensurable with C, and then showing that $F : C$ is both greater and less than $B : D$. The reference to rectangles is incidental and can be eliminated. Deleting the rectangles, we can rewrite the argument as follows. Suppose that $A : B :: C : D$ but $A : C \neq B : D$, so that, for some $E \neq A$, $E : C :: B : D$; assume without loss of generality that $E > A$. Then let F such that $E > F > A$ be commensurable with C. Then $F : C < E : C :: B : D$. On the other hand, there is constructively an H such that $F : H :: C : D$ (if we write $F = qC/p$, then $H = qD/p$; this

---

[56] Another strategy would be to show first that if F such that $E > F > A$ is commensurable with C, then $rect(F, D) < rect(B, C)$, and then that if F' such that $E > F' > A$ is commensurable with B, then $rect(F', D) > rect(B, C)$. We can then get a contradiction by choosing F such that $E > F > A$ and F is commensurable with C, and then F' such that $F > F' > A$ and F' is commensurable with B. This is logically simpler in that it does not depend on *Elements* V,15, but it involves two interpolations rather than one.

depends on Euclid V,15). Since $F : H :: C : D :: A : B < F : B$, we have $H > B$. Since it is automatic (without alternation) that $F : C :: H : D$, we have $F : C :: H : D > B : D$. So the ratio $F : C$ is both greater and less than $B : D$, a contradiction.[57]

This proof of alternation, in dispensing with rectangles, becomes independent of plane geometry, and in particular of VI,1 and VI,16. The proof belongs purely to universal mathematics, and consequently it establishes alternation for any four homogeneous quantities A, B, C and D, whether they are straight lines or not. The proof could arise naturally by reflection on the exhaustion proofs of VI,1 and VI,16 (and of alternation by way of VI,16), and it can itself be described as an exhaustion proof. The steps of taking the fourth proportional $E \neq A$ and assuming $E > A$, interpolating F such that $E > F > A$ and F is commensurable with C, and deriving a contradiction, are analogous to the steps in Euclid's proof of XII,2 (in my notation in Section I above) of taking the fourth proportional X, assuming $X < C$, interpolating P_n such that $C > P_n > X$ and P_n is inscribed in the circle C, and deriving a contradiction. The difference is that we are no longer exhausting a curvilinear area by rectilinear areas, or even exhausting a rectangle by rectangles whose bases are commensurable with a given base, but exhausting a line (or an arbitrary quantity) by lines (or quantities of whatever kind) commensurable with a given line (or quantity). But this is a natural generalization of the method of exhaustion, and a mathematician might be led to it particularly naturally by reflection on the procedure of exhausting a rectangle by rectangles whose bases are commensurable with a given base (as carried out in the proof of VI,1), since after all this is mathematically equivalent to exhausting the base of the rectangle by lines commensurable with a given line.

One structural feature that this proof of alternation shares with the "Eudoxian" proofs of VI,1 and VI,16, and with Pappus' proof about the arcs and sectors of the circle (and also with Archimedes' proof of the principle of the balance in *On the Equilibrium of Planes* I,6–7, where a *weight* is exhausted by weights commensurable with a given weight) is that the theorem is proved first

---

[57] We can do something similar with the alternative argument from the previous footnote. Suppose that $A : B :: C : D$ but $A : C \neq B : D$, so that, for some $E \neq A$, $E : C :: B : D$; assume without loss of generality that $E > A$. Let F such that $E > F > A$ be commensurable with C. Then $F : C < E : C :: B : D$. Now let F' such that $F > F' > A$ be commensurable with B. Then $F' : B > A : B :: C : D$. In the usual constructive way, there is a G such that $F' : B :: C : G$; since $C : G :: F' : B > A : B :: C : D$, we have $G < D$. We can then infer that $F' : C :: B : G > B : D$. Since $F > F'$, we have $F : C > F' : C > B : D$, so the ratio $F : C$ is both greater and less than $B : D$, a contradiction. The inference from $F' : B :: C : G$ to $F' : C :: B : G$ (the commensurable case of alternation) works because if $F' = pB/q$, so that $C = pG/q$, we have $pB/q : pG/q :: B : G$. (Once again, this depends on two applications of Euclid V,15, $X : Y :: nX : nY$. There seems to be no way to avoid using this rule, or the commensurable case of alternation, in proving alternation.) On the other hand, there is constructively an H such that $F : H :: C : D$ (if we write $F = qC/p$, then $H = qD/p$; this depends on Euclid V,15). Since $F : H :: C : D :: A : B < F : B$, we have $H > B$. Since it is automatic (without using alternation) that $F : C :: H : D$, we have $F : C :: H : D > B : D$. So the ratio $F : C$ is both greater and less than $B : D$, a contradiction.

Eudoxus' Theory of Proportion and His Method of Exhaustion 219

for a commensurable case and then for the general or incommensurable case. This may not have been clear above, since the commensurable case of alternation was disguised as an appeal to the rule $X : Y :: nX : nY$, Euclid V,15. But we can see that this is equivalent to the commensurable case of the proposition to be proved; and in any case the rule must be proved. A further complication will be that the commensurable case will itself have a commensurable and an incommensurable subcase.[58] To see all this, let us set out the argument in full.

Suppose that $A : B :: C : D$; we must show that $A : C :: B : D$.

First suppose that A and B are commensurable. Then we can write $A = pE$, $B = qE$; we can therefore also write $C = pF$, $D = qF$. So we must show that $pE : pF :: qE : qF$. It is sufficient to show that $pE : pF :: E : F$, Euclid V,15.[59]

This claim has a commensurable and an incommensurable subcase. First suppose that E and F are commensurable, $E = rG$, $F = sG$. Then $pE = p(rG)$, and $p(rG) = r(pG)$; this is a special case of Euclid V,1, $p(\sum G_i) = \sum(pG_i)$, taking all the G_i to be equal. Likewise $pF = p(sG) = s(pG)$. So $pE : pF :: p(rG) : p(sG) :: r(pG) : s(pG) :: rG : sG :: E : F$.[60]

Now suppose E and F are incommensurable. We must show that $pE : pF :: E : F$. Suppose not. Without loss of generality, suppose that $pE : pF < E : F$. Then, by the existence of the fourth proportional, $pE : pF :: G : F$ for some $G < E$. Take H commensurable with F such that $G < H < E$. Then $pE : pF :: G : F < H : F :: pH : pF$, by the commensurable subcase which we have proved. So $pE < pH$, contradicting the assumption that $H < E$.

This completes the proof that $A : C :: B : D$ in the case where A and B are commensurable. So now suppose that A and B are incommensurable, and suppose that $A : C \neq B : D$. Without loss of generality, suppose that $A : C < B : D$. By the existence of a fourth proportional, there is an E such that $E : C :: B : D$; so $E > A$. Then let F such that $E > F > A$ be commensurable with C. Then $F : C < E : C :: B : D$. Now there is (constructively) an H such that $F : C :: H : D$. Since $H : D :: F : C < B : D$,

---

[58]My treatment of the commensurable case and its two subcases is quite different from Knorr's ("Archimedes ...", p. 232), but is, as it turns out, almost the same as Borelli's (*Euclides Restitutus* III,10–12). Mine uses (what Knorr and I think to be) the typical Eudoxian two-step method, and Knorr's does not; mine does not involve *ex aequali* or proportions between numbers; and mine gives a much easier transition to the Euclidean proof. The advantage of Knorr's proof is that it stays closer to the pattern of Euclid's arguments in X,5–6. (One way to explain the breakdown of the commensurable case into commensurable and incommensurable subcases is that alternation is the abstract analogue, not of VI,16, but of VI,16 applied twice, one in the forward and once in the converse direction. The commensurable case of alternation corresponds to the commensurable case of the *first* application of VI,16; the commensurable and incommensurable subcases correspond to the commensurable and incommensurable cases of the *second* application of VI,16.)

[59]The more historically authentic formulation would be first to prove the theorem where A is a multiple of B, then when B is a multiple of A, then when A and B are both multiples of a common measure E. I am simplifying in the hope of keeping things surveyable.

[60]All sums are of course finite. This proof can be generalized to give a "Eudoxian" proof of Euclid V,12 (if, for each i, $G_i : H_i :: G : H$, then $(\sum G_i) : (\sum H_i) :: G : H$), building on V,1 as Euclid's proof of V,12 does. Euclid proves V,12 from V,1, and derives V,15 as the special case of V,12 where all the G_i are equal and all the H_i are equal.

we have $H < B$. But since F and C are commensurable, by the commensurable case of alternation which we have proved we have $F : H :: C : D$. Since $H < B$, we have $C : D :: F : H > F : B > A : B$, contradicting the assumption that $A : B :: C : D$.[61] So the assumption that $A : B :: C : D$ implies that $A : C :: B : D$.[62]

4 From Eudoxus to Euclid and Archimedes

This proof of alternation is similar to Euclid's in rigor and abstraction, and in giving a plausible account of how Eudoxus might have arrived at (something like) this proof, I hope to have helped to solve the problem of how Greek mathematicians could have come up with something so abstract as Euclidean proportion theory. Nonetheless, this proof is importantly different from Euclid's. In particular, it does not turn on Euclid's definition of proportion (or rather, of being in the same or a greater or lesser ratio), or indeed on any other definition. But it is not hard to see how a mathematician would get

---

[61]Or "Now there is (constructively) an H such that $F : C :: H : D$. Since F and C are commensurable, by the commensurable case of alternation which we have proved we have $F : H :: C : D :: A : B$. Since $F > A$, $H > B$. So $F : C :: H : D > B : D$, contradicting what we have already shown, that $F : C < B : D$."

[62]This is essentially Borelli's proof: the commensurable subcase of the commensurable case is his *Euclides Restitutus* III,10, the incommensurable subcase of the commensurable case is 11, and the incommensurable case is 12. As far as I know, this is original to Borelli. Borelli is, as noted above, responding to a long tradition of complaints about Euclid's definition(s) of proportionality. Both Galileo and Torricelli try to show how to do without Euclid Vdef5. Galileo tries to base himself on Vdef3, "a ratio [λόγος] is a certain relation according to quantity between two magnitudes of the same kind," with its supplement (probably spurious but found in the text or margin of many manuscripts and in Campanus' Latin translation) "proportion [ἀναλογία] is sameness [or similarity] of ratios [λόγοι]." Since this is much too vague to deduce anything from it, Galileo winds up basing himself on some supplemental axioms. In effect, he assumes V,4 (if $A : B :: C : D$ then $mA : nB :: mC : nD$) and V,8 or V,14 as axioms and deduces Vdef5 and Vdef7; but then (in the unfinished text) he does not seem to worry about the rest of *Elements* V, perhaps assuming that once we have proved Euclid's definitions we can accept the rest of his proofs. Torricelli, however, goes further in reconstructing Euclid, proving VI,1 in the "Eudoxian" style by *reductio ad absurdum* starting from the commensurable case. He then, with Becker's anthyphairetic theory rather than with Knorr's Eudoxian theory, proves V,16 for straight lines from VI,1 (not in the obvious way through VI,16, but in a very clever way from Euclid VI,2-3, see Torricelli *De Proportionibus Liber* propositions 4-6, pp. 315-17 in Giusti—but ultimately he is drawing on the power of VI,1 in the same way that Becker's anthyphairetic theory does). Borelli complains that Torricelli's argument will work only for lines, when we are trying to prove a universal proposition about all kinds of quantity (*Euclides Restitutus* p. 123; it is not clear to me whether Borelli is thinking of Aristotle *Posterior Analytics* I,5 74a17-25). Borelli, while granting some of the complaints against Vdef5, sees the uselessness of Vdef3 and suggests very reasonably that Euclid intended Vdef5 to fill in what *kind* of relation between magnitudes a ratio is, or what *kind* of similarity constitutes proportionality. Rather than axiomatizing like Galileo, he *defines* being in the same or a greater or lesser ratio, starting from commensurable cases and proceeding to incommensurable cases (see a note above: his definitions, taken together, have essentially the same content as my Postulate I), and proves V,16 universally as we have seen, starting from commensurable and proceeding to incommensurable cases.

from something like this proof of alternation to the Euclidean proof. Since it is easy to give similar "exhaustion"-style proofs of V,22 (and its "perturbed" version V,23), and of the connected group of theorems V,17–19 and V,24 on adding and subtracting magnitudes in proportion, and then to transform these into (something close to) the Euclidean proofs, and since *Elements* V just is the proofs of these theorems (together with the necessary lemma V,8 and its immediate consequences V,9–10, and the concluding flourish V,25, on which more below), we can reconstruct a plausible genesis for *Elements* V as a whole.

To turn a "Eudoxian" proof like the one I have given for alternation into a Euclidean proof, we must do the following. Take the general case of the proof (there is no need to call it the "incommensurable" case, since it will work whether the magnitudes are commensurable or not). First, we must rewrite the argument to avoid the assumption of the existence of a fourth proportional. Second, instead of speaking of commensurable magnitudes, write one magnitude as a fractional multiple of another. Third, eliminate the fractional multiples by multiplying through (in inequalities, proportions etc.) by the denominator. Fourth, the argument can be converted from a *reductio ad absurdum* to a direct proof. The commensurable case will be treated in essentially the same way (except that proportions can often be eliminated in favor of explicit equalities between multiples), and will become a lemma. I will show how to go through all these steps in the case of alternation.

To begin with, the assumption of the existence of a fourth proportional can be eliminated. In proving the general case of alternation, I wrote "by the existence of a fourth proportional, there is an E such that $E : C :: B : D$; so [since $B : D > A : C$] $E > A$. Then let F such that $E > F > A$ be commensurable with C. Then $F : C < E : C :: B : D$." But the rest of the proof does not need E, only F, and I can get F without using E if I simply write "since $B : D > A : C$, there is an F commensurable with C such that $B : D > F : C > A : C$ [or, equivalently, $B : D > F : C$ and $F > A$]." It is not immediately obvious how we *justify* this assumption without relying on the existence of a fourth proportional, and I will return to this question shortly, but for now we can simply note that this is an assumption weaker than the assumption of the existence of a fourth proportional. (The *second* time I took a fourth proportional, taking H such that $F : C :: H : D$, does not involve any special assumption, since F is commensurable with C; writing $F = pC/q$, we have $H = pD/q$.)

Writing explicitly $F = pC/q$, the assumption becomes "since $B : D > A : C$, there are p, q such that $B : D > (pC/q) : C > A : C$." The rest of the proof then follows. By V,8–10, "$(pC/q) : C > A : C$" is equivalent to "$(pC/q) > A$" ["$F > A$" in the formulation above]. We can also manipulate the inequality "$B : D > (pC/q) : C$"; since, by Postulate I, we have $(pC/q) : C :: (pD/q) : D$, it follows that $B : D > (pD/q) : D$, thus $B > (pD/q)$ ["$B > H$" in the formulation above]. Then, by the commensurable case of alternation, we have $C : D :: (pC/q) : (pD/q)$. So, using the inequalities we have established,

$C : D :: (pC/q) : (pD/q) > (pC/q) : B > A : B$, contradicting the assumption that $A : B :: C : D$.

We can then eliminate the fractions. By V,15 (which is a special case of the commensurable case of alternation), $X : Y :: qX : qY$, so for instance $(pC/q) : C :: pC : qC$. So, since there are p, q such that $B : D > (pC/q) : C > A : C$, there are p, q such that $B : D :: qB : qD > pD : qD :: pC : qC > qA : qC :: A : C$. So $qB > pD$ and $pC > qA$, and thus $C : D :: pC : pD > pC : qB > qA : qB :: A : B$, contradicting the assumption that $A : B :: C : D$.

This remains very close to the "Eudoxian" argument given above. We prove the proportion $A : C :: B : D$ by assuming a disproportion $A : C < B : D$ and interpolating between the two given ratios a ratio of commensurable magnitudes $(pC/q) : C$ (equivalently, $pC : qC$, or $pD : qD$), and then deriving a contradiction. And a similar strategy will be followed in proving all the other theorems of pure proportion theory. Now in the "Eudoxian" theory the interpolation of the ratio of commensurable magnitudes was justified by the assumption of a fourth proportional, followed by an application of Lemma II'. But we can dispense with this dubious machinery. We can look back at our various proofs in proportion theory, and note that in disproving disproportions, and thus proving proportions, we are habitually using as the criterion or *de facto* definition of "$B : D > A : C$" "there are p, q such that $B : D > (pC/q) : C > A : C$." Now in defining the notion of being in a greater ratio, we should not use the notion of being in a greater ratio. But this is eliminable, since as we saw just above, "$B : D > (pC/q) : C$" is equivalent to "$qB > pD$," and "$(pC/q) : C > A : C$" is equivalent to "$pC > qA$." So the definition of "$B : D > A : C$" becomes "there are p, q such that $qB > pD$ and $pC > qA$"; which is essentially Euclid Vdef7.[63] The "Eudoxian" theory that I have given does not have any definition of being in a greater ratio; as I have said, Eudoxus himself may well have accepted the anthyphairetic definition of being in a greater ratio, but his practice of proof in proportion theory (as Knorr and I reconstruct it) does not use this definition, and reflection on his practice of proof shows that what it is implicitly presupposing is just Euclid Vdef7. Once we make this definition explicit, we no longer need to invoke the existence of a fourth proportional. And we can also now turn the *reductio ad absurdum* into a direct proof. The way we have habitually been proving the proportion $A : C :: B : D$ is to refute the disproportions $A : C < B : D$ and $A : C > B : D$, in other words to show "there are no p, q such that $qB > pD$ and $pC > qA$, or such that $qB < pD$ and $pC < qA$." So we can give a direct proof if we take this criterion as the definition of the proportion $A : C :: B : D$: and this definition is essentially Euclid Vdef5.[64] We can restate the above reasoning as a direct

---

[63] By "essentially" I mean "ignoring cases of equality."

[64] This is precisely how Borelli reaches his definition of being in the same ratio for incommensurable magnitudes, as the joint negation of being in a greater or lesser ratio (IIIdef12). His definition of $A : B > C : D$ where C and D are commensurable amounts to $\exists p \exists q$ ($A > pB/q$ and $C = pD/q$), and his definition of $A : B > C : D$ where everything is incommensurable amounts to $\exists p \exists q$ ($A > pB/q$ and $C < pD/q$). So, although Borelli uses

proof as follows. Suppose $A : B :: C : D$, and let p and q be any numbers. Then, by two uses of V,15, we have $pC : pD :: qA : qB$. So if $qA < pC$, we have (by V,8) $pC : qB > qA : qB :: pC : pD$, and thus (by V,10) $qB < pD$; and a similar argument shows that if $qA > pC$, then $qB > pD$. And this is essentially Euclid's proof of V,16.[65]

To complete the argument, we can also see how to transform the "Eudoxian" proof of V,15 into a Euclidean proof. The claim is that for any E and F, and any p, $pE : pF :: E : F$. The "Eudoxian" argument, for the general case, went: "suppose not. Without loss of generality, suppose that $pE : pF < E : F$. Then, by the existence of the fourth proportional, $pE : pF :: G : F$ for some $G < E$. Take H commensurable with F such that $G < H < E$. Then $pE : pF :: G : F < H : F :: pH : pF$, by the commensurable subcase. So $pE < pH$, contradicting the assumption that $H < E$." To eliminate the assumption of the fourth proportional, "by the existence of the fourth proportional, $pE : pF :: G : F$ for some $G < E$. Take H commensurable with F such that $G < H < E$" can be rewritten as "for some q, r, $pE : pF < (qF/r) : F < E : F$"; or, eliminating fractions and using Vdef7, "for some q, r, $r(pE) < q(pF)$ and $rE > qF$". Now, instead of the commensurable subcase $(qF/r) : F :: p(qF/r) : pF$, we need only the non-proportion-theoretic fact that $r(pE) = p(rE)$ (and likewise $q(pF) = p(qF)$), a special case of Euclid V,1, which we used above in proving the commensurable subcase. From this it follows that since $rE > qF$, $r(pE) = p(rE) > p(qF) = q(pF)$, contradicting the assumption. Rewriting this as a direct proof, what we get is essentially Euclid's proof of V,15, except that Euclid, instead of proving V,15 directly from the special case of V,1, instead proves V,12 in an exactly parallel way from the more general V,1, and then specializes from V,12 to V,15.

---

submultiples (though not when everything is commensurable, thus not for numbers) and Euclid does not, he first gives his equivalent of Euclid Vdef7, and then derives his equivalent of Vdef5, defining $A : B :: C : D$ as the joint negation of $A : B > C : D$ and $A : B < C : D$. And presumably Borelli crystallizes these definitions out of the argument-strategy that Torricelli and others had used to prove or disprove $A : B > C : D$ in particular contexts, and to prove $A : B :: C : D$ by disproving $A : B > C : D$ and $A : B < C : D$. Borelli isolates the assumptions that Torricelli in fact uses (and shows how to do more with them than Torricelli had, notably in proving alternation). Despite his criticisms of Euclid Vdef5, Borelli winds up recapitulating the logic that had, in all likelihood, led Euclid to this definition, and his own definition winds up being mostly notationally different in the incommensurable case. (Borelli gets simpler definitions than Euclid where some or all of the quantities are commensurable, and so avoids the need for a separate VIIdef20, and avoids the logical gap in Euclid X,5 and following, where Euclid lumps both magnitudes and numbers under the same notion of proportion. The price is that Borelli's definition has multiple cases, depending on which quantities are commensurable. That is fine on an axiomatic approach such as I have adopted here, but is awkward in a definition, and means that, to be logically correct, even a proposition like "if $A : B > C : D$ and $C : D > E : F$, then $A : B > E : F$" would have to be proved by examining multiple cases—here, five non-trivial cases.)

[65] The differences are: Euclid adds the case where $qA = pC$; he condenses the uses of V,8 and V,10 into the single proposition V,14 (V,14, as we saw above, seems to have been lacking in Eudoxus, since in Euclid XII it is rederived from V,16); and he proves the transitivity relations V,11 and V,13, which I have implicitly used as common notions.

It is worth noting that the key lemma V,8 (if $A > B$ then $A : C > B : C$), with its immediate corollaries V,9–10, has a somewhat different status on Euclid's theory and on my reconstruction of the Eudoxian theory. On the "Eudoxian" theory, the "commensurable case" (i.e. the case where either A or B is commensurable with C) follows immediately from Postulate I; if neither A nor B is commensurable with C, we use Lemma II$'$ to interpolate D such that $A > D > B$ and D is commensurable with C, and then the result follows from the commensurable case. Euclid cannot use exactly this argument, because he does not have Postulate I.[66] On the other hand, if we take Lemma II$'$, writing $D = pC/q$, and multiply the inequalities through by q, we have that for some p, q, $qA > pC > qB$; and by Vdef7, this is exactly to say that $A : C > B : C$. So for Euclid, V,8 comes in much the same logical place that Lemma II$'$ does on the "Eudoxian" theory. Euclid does not state Lemma II$'$ in Book V: indeed, Book V does not use the notions of commensurable and incommensurable, which are not introduced until Book X. Indeed, more generally, Euclid's Book V contrasts with the "Eudoxian" theory in that it does not break the given magnitudes down into parts or submultiples (e.g. by saying "D is commensurable with C," i.e. "D and C have some common measure or submultiple"), but rather considers only multiples of the given magnitudes. For the same reason, Euclid does not introduce our Postulate II, the "principle of bisection," in Book V, and so cannot use it in proving V,8 as we used it in proving Lemma II$'$. Rather, when Euclid finally does introduce the principle of bisection in X,1, he *proves* it from the assumption that if $X > Y$ then $\exists p\ pY > X$, that is, from an assumption about multiplying quantities, not about breaking them into smaller pieces:[67] this must be a deliberate decision on Euclid's part, since the principle of bisection was certainly the older and the more natural assumption, having presumably first arisen in the context of exhaustion proofs. When we proved Lemma II$'$ above, we successively bisected C until it was less than A, less than the remainder, and so on, or alternatively until it is less than $A - B$. Euclid, instead, multiplies $A - B$ until it is greater than C, implicitly using Archimedes' postulate that this can always be done.[68] Then if $p(A - B) > C$, and qC is the first multiple of C that is greater than or

---

[66]Euclid does in fact assume Postulate I, in the case of proportionality, in X,5–6. (He assumes that $(pA/q) : A :: p : q$, or rather he assumes that $pA : A :: p : 1$ and then deduces this by *ex aequali*; in X,6 he uses this together with V,9 to infer that if $A : B :: p : q$ then $B = qA/p$.) But he does not have a right to assume it: its logical place has been taken by Vdef5 and Vdef7, and it would be ugly and unnecessary to add it as an extra unproved assumption. He does not use it in Book V.

[67]Presumably Euclid thinks this follows from Vdef4, which says that this holds if X has a ratio to Y. Euclid cannot simply suppress the principle of bisection, since he needs it for proving incommensurablity by ἀνθυφαίρεσις in X,2, as well as for the exhaustion proofs of Book XII.

[68]Again Euclid presumably thinks that this follows from Vdef4—wrongly, since A and B might each have a ratio to C without $A - B$ having a ratio to C. (For instance, let B and C be any rectilinear angles, and let A be B plus a "horn angle," i.e. the angle between a circle and a line tangent to it.)

equal to pB, we will have $(qC - pB) < C < p(A - B) = (pA - pB)$, and thus $pA > qC \geq pB$, which is what Euclid needs for V,8.[69]

We can give a similar treatment, constructing first a "Eudoxian" proof and then a Euclidean version, for all the other propositions of pure proportion theory. Thus the other key result of Euclid Book V, V,22 (if $A : B :: D : E$ and $B : C :: E : F$, then $A : C :: D : F$—compounding of ratios is well-defined) can be given a "Eudoxian" proof similar in structure to the "Eudoxian" proof of alternation, and then this proof can be converted into a Euclidean proof in a similar way; the commensurable case in the "Eudoxian" version becomes Euclid V,4 (if $A : B :: D : E$ then $pA : qB :: pD : qE$), and the commensurable subcase of the commensurable case becomes the non-proportion-theoretic Euclid V,3, which is a special case of Euclid V,2 in the same way that, in the proof of alternation, $r(pE) = p(rE)$ was a special case of Euclid V,1.[70] We can also do the same with the "perturbed" version Euclid V,23 (if $A : B :: E : F$ and $B : C :: D : E$, then $A : C :: D : F$—compounding of ratios is commutative); here the commensurable case can be replaced by V,4 together with an application of V,15, and the proof that results is not actually Euclid's proof of V,23 (i.e. the proof transmitted in all the manuscripts), which takes a "short cut" through V,16 and thus makes unnecessary homogeneity assumptions, but rather the alternative proof added in most of the "Theonine" manuscripts.[71]

---

[69]The $\geq$ in $pA > qC \geq pB$ is enough by Euclid's formulation of Vdef7 to imply $A : C > B : C$; but it would be easy enough to eliminate the possibility of equality if we wanted. I am here simplifying Euclid's unnecessarily long proof of V,8.

[70]Borelli gives a "Eudoxian" proof of V,22, *Euclides Restitutus* III,17 (commensurable subcase of the commensurable case), III,18 (incommensurable subcase of the commensurable case), and III,19 (incommensurable case). To prove V,22, suppose $A : B :: D : E$ and $B : C :: E : F$ but (say) $A : C > D : F$. The "commensurable case" is when B and C (thus also E and F) are commensurable. To prove the general case, for some $X < A$ we have $X : C :: D : F$; choosing some Y such that $A > Y > X$ and Y is commensurable with C, we have $Y : C > D : F$; there is (constructively) a $Z > D$ such that $Y : C :: Z : F$ and Z is commensurable with F. Since $B : C :: E : F$, $C : Y :: F : Z$, and C and Y are commensurable, by the commensurable case we have $B : Y :: E : Z$, so $Y : B :: Z : E$, $A : B > Y : B :: Z : E > D : E$, contradicting the assumption that $A : B :: D : E$. The commensurable case is a form of V,4—writing $C = pB/q$ and $F = pE/q$, the claim is that $A : B :: D : E$ implies $A : (pB/q) :: D : (pE/q)$, which holds if $A : B :: D : E$ implies $A : pB :: D : pE$ and conversely. This in turn has a commensurable case equivalent to V,3, saying that $p(qA)$ is the same multiple of A that $p(qB)$ is of B.

[71]Borelli gives a "Eudoxian" proof of V,23, *Euclides Restitutus* III,20. Here we can take as the commensurable case that $A : B :: E : F$ implies $A : (pB/q) :: (qE/p) : F$, which follows from $A : (pB/q) :: E : (pF/q)$ (the version of V,4 used as the commensurable case for V,22) and $E : (pF/q) :: (qE/p) : F$, which follows from V,15, $A : B :: pA : pB$. For the alternative proof in the manuscripts of Euclid, see Heiberg's apparatus to *Elementa* v.2 p. 34 line 17; the passage Heiberg prints there in the apparatus gives a (much shorter and simpler as well as logically preferable) alternative to the segment of the proof from the first comma on p. 34 line 17 to the full stop on p. 35 line 8. See Heath *Elements* v.2 pp. 182–3 for discussion; but Heath misleadingly suggests that the "Theonine" manuscripts transmit *only* the alternative proof ("Simson's proof"). Rather, according to Heiberg's apparatus, they transmit both (or rather the Oxford manuscript and a later manuscript do, the Vienna manuscript V adds the alternative proof in the margins, and the "Theonine" Florence manuscript F, like the "non-Theonine" Vatican manuscript P, gives only the standard proof). This makes it far

We can also give the same treatment to the propositions about the addition and subtraction of magnitudes in proportion, V,17–19 and V,24 (V,25 is a corollary of V,19, and would be proved the same way on any theory). And since these are the only remaining propositions of Book V, apart from the non-proportion-theoretic lemmas V,1–6 and the lemmas V,7, V,11 and V,13 (which were all treated as primitive common notions in my reconstruction of the Eudoxian theory, but are immediately provable once we have Euclid's Vdef5 and Vdef7), we will in fact have sketched a plausible genesis for Book V as a whole. V,17–19 and V,24 are less exciting than V,16 and V,22 (Knorr does not discuss them at all), but a brief discussion will bring out some interesting logical points, and may shed some new light on the structure of Book V.

V,17 says that if $(A+B) : B :: (C+D) : D$, then $A : B :: C : D$, and V,18 says the converse, that if $A : B :: C : D$, then $(A+B) : B :: (C+D) : D$. V,24, building on V,18, says that if $A : B :: C : D$ and $A' : B :: C' : D$, then $(A + A') : B :: (C + C') : D$.[72] To give the "Eudoxian" proof of V,17, suppose that $(A + B) : B :: (C + D) : D$ but $A : B \neq C : D$; without loss of generality, suppose $A : B < C : D$. Then by Postulate III there is some $E > A$ such that $E : B :: C : D$, and by Lemma II' there is some F commensurable with B such that $E > F > A$. Thus $F : B < E : B :: C : D$; there is (constructively) some $G < C$ such that $F : B :: G : D$. Then, by the commensurable case of V,18 (which follows immediately from V,2 and Postulate I), we have $(F + B) : B :: (G + D) : D$. But since $F > A$ and $G < C$ it follows that $(A + B) : B < (F + B) : B :: (G + D) : D < (C + D) : D$, contradicting the assumption. This is all straightforward enough, and we can transform it in the usual way, eliminating fractional multiples and Postulate III, into something like what Euclid actually does.[73] What is worth noting is that the lemma for commensurable quantities that is needed for proving V,17 is not the commensurable case of V,17, but rather the commensurable case of V,18; likewise, the lemma needed for V,18 will be the commensurable case of V,17. This is important, because Knorr speaks as if the "Eudoxian" method of proving a proportion-theoretic proposition were first to prove its commensurable case, and then to derive the incommensurable case. This is indeed very often how it works, but in some cases the commensurable lemma

---

more likely that the alternative proof was a later scholar's supplement (written as a marginal scholium in some manuscripts, and creeping into the main text in others) than that it was Euclid's original proof.

[72]Note that V,17–18 are automatic on an anthyphairetic theory, but take some work on the "Eudoxian" and Euclidean theories. V,17 could equally well be symbolized by saying that if $A : B :: C : D$, then $(A - B) : B :: (C - D) : D$. Euclid derives V,19, that if $A : B :: C : D$, then $(A - B) : (C - D) :: A : B$, from V,17 by two applications of alternation. V,19 could also be derived from V,12 using the postulate of the existence of the fourth proportional, in the same way that Euclid derives V,18 from V,17 (see below).

[73]Euclid uses not only V,2 (crudely, $pA+qA = (p+q)A$), but also V,1, $p(A+B) = pA+pB$, because he needs to take multiples of $A + B$ in order to apply Vdef5.

that we need for the *reductio ad absurdum* proof of the general case will not be a case of the proposition to be proved, but rather of some converse.[74]

Interestingly, the "Euclidean" proof of V,18 that we would generate this way (using V,5–6, both propositions that are in fact unused in Book V) is not the proof that Euclid actually gives:[75] instead, where the "Eudoxian" proof would prove V,18 by *reductio ad absurdum*, using Postulate III, from the commensurable case of V,17, Euclid proves V,18 by *reductio ad absurdum*, using Postulate III, from the *general* case of V,17. This is remarkable, since Euclid has no right to Postulate III, has succeeded in avoiding it everywhere else in Book V (although he does use it in Book XII), and could perfectly well have avoided it here too. But given that Euclid has already proved V,17, rather than prove V,18 by an elaborate argument turning on the commensurable case of V,17 (or something equivalent to that case), it may have seemed easier just to use V,17 itself, ignoring the fact that Postulate III is not justified on Euclid's theory. So Euclid's proof of V,18 may be a trace of the Eudoxian proof, even though the most distinctively "Eudoxian" feature, the distinction of commensurable and incommensurable cases, has been eliminated.[76]

---

[74] There is actually another way to prove V,17–18 which does meet Knorr's description. In the case of V,17, proceed as above up to the construction of F. Then $(F+B):B > (A+B):B :: (C+D):D$. Since $F+B$ is commensurable with B, there is (constructively) an H commensurable with D such that $(F+B):B :: H:D$. Clearly $H > D$, so we can write $H = J + D$, and J too will be commensurable with D. Since $(J+D):D :: (F+B):B > (C+D):D$, we have $J + D > C + D$ and thus $J > C$. Then, by the commensurable case of V,17, $(F+B):B :: (J+D):D$ entails $F:B :: J:D$. Thus $E:B > F:B :: J:D > C:D$, contradicting the assumption that $E:B :: C:D$. But this way of proceeding seems less natural than the one I suggest in the text. In any case, what Euclid actually does corresponds to what I suggest in the text, not to this alternative.

[75] For the "Eudoxian" proof of V,18, suppose that $A:B :: C:D$ but $(A+B):B > (C+D):D$. Then by Postulate III there is some $E < A+B$ such that $E:B :: (C+D):D$, and by Lemma II' there is some F commensurable with B such that $E < F < A+B$. Thus $F:B > E:B :: (C+D):D$, and there is (constructively) some $G > C+D$ such that $F:B :: G:D$. By the commensurable case of V,17, $(F-B):B :: (G-D):D$; so $A:B > (F-B):B :: (G-D):D > C:D$, contradicting the assumption that $A:B :: C:D$. Converting this in the standard way into a "Euclidean" proof: suppose that $A:B :: C:D$ but $(A+B):B > (C+D):D$. For some p and q, $p(A+B) > qB$ but $p(C+D) \leq qD$. Because $pD < p(C+D) \leq qD$, we must have $p < q$. Then subtract pB from both sides of the inequality $p(A+B) > qB$, and subtract pD from both sides of the inequality $p(C+D) \leq qD$. By V,5, $p(A+B) - pB = pA$ and $p(C+D) - pD = pC$. By V,6, $qB - pB$ and $qD - pD$ are equimultiples of B and D, say $qB - pB = rB$ and $qD - pD = rD$ (obviously, $r = q - p$, although Euclid would have difficulty expressing this). So $pA > rB$ and $pC \leq rD$, contradicting the assumption that $A:B :: C:D$. The suggestion that Euclid should have proved (and, on the assumption that Euclid was perfect, originally did prove) V,18 using the otherwise unused V,5 and V,6, and not using the postulate of the existence of a fourth proportional, goes back at least to Robert Simson, *The Elements of Euclid* (sixth edition, Edinburgh, Nourse and Balfour, 1781), p. 322.

[76] This is similar to the way that Euclid proves V,5, $pA - pB = p(A - B)$, by a *reductio ad absurdum* from V,1, $p(A+B) = pA + pB$. Letting $pA - pB = pC$, we have $pA = pB + pC = p(B+C)$ by V,1, and thus $A = B + C$ (it can be neither greater nor less), so $C = A - B$ and $pA - pB = p(A - B)$ as was to be proved. Euclid has, of course, no right to divide by p, i.e. to assume that $pA - pB$ is p times anything, although such a step would be normal on a

There may be a similar "Eudoxian" relationship between V,18 and V,24, with V,18 establishing a commensurable case of V,24. V,24 says that if $A : B :: C : D$ and $A' : B :: C' : D$, then $(A + A') : B :: (C + C') : D$, and Euclid gives a straightforward proof using V,18 and V,22: under the given assumptions, we have by V,22 $A : A' :: C : C'$, and therefore by V,18 $(A+A') : A' :: (C+C') : C'$; but since $A' : B :: C' : D$, by V,22 again we have $(A + A') : B :: (C + C') : D$, as was to be shown. This is a perfectly reasonable way to prove V,24, and it would work just as well on the "Eudoxian" theory as on Euclid's. But it is also possible that, in a "Eudoxian" precursor of Book V, V,18 was used to prove a commensurable case of V,24. For, under the assumptions of V,24, suppose that A' is commensurable with B, say $A' = pE$, $B = qE$; thus likewise $C' = pF, D = qF$. We must then prove that $(A + pE) : qE :: (C + pF) : qF$. But since $A : qE :: C : qF$, something like V,4 shows that $A : E :: C : F$ (on Euclid's version, use V,4 to show $qA : qE :: qC : qF$, and then apply V,15). Then V,18, repeated p times, shows that $(A + pE) : E :: (C + pF) : F$, and then V,4 again shows that $(A + pE) : qE :: (C + pF) : qF$, as was to be shown; and then a typical *reductio ad absurdum* allows us to reduce the general case of V,24 to this commensurable case.[77] If this methodical but cumbersome argument was Eudoxus' proof of V,24, perhaps Euclid streamlined it, keeping the lemma V,18 but then using V,22 to jump immediately to the conclusion. This would be similar to Euclid's use of alternation in his proofs of V,19 (which Eudoxus may have proved from V,12 and Postulate III) and of V,23, and is logically harmless (like the use of alternation in V,19, but unlike its notoriously unfortunate use in V,23).[78] If this is right, V,24 would originally have been independent of V,22 (although using its "commensurable case" V,4), and would be the culmination of the group V,17–18. For Euclid, instead, after V,17–18 (and V,19, proved from V,17 and V,16), we need another key theorem, V,22 (proved with its perturbed version V,23 and their lemmas V,20–21). Then, as V,17 combines with V,16 to yield V,19, so V,18 combines with V,22 to yield V,24, which emerges as the final step of Book V—except for V,25, dependent only on V,19, but saved for the end as a final flourish.[79]

"Eudoxian" approach. This is a special case of Postulate III: Euclid is assuming that there is a C such that $pA : A :: (pA - pB) : C$.

[77]For suppose $A : B :: C : D$ and $A' : B :: C' : D$ but $(A + A') : B > (C + C') : D$. By Postulate III there is some $E < (A + A')$ such that $E : B :: (C + C') : D$ Since $E : B :: (C + C') : D > C : D :: A : B$, we have $E > A$, so we can write $E = (A + F)$, where $F < A'$, and thus $(A + F) : B :: (C + C') : D$. By Lemma II', take G such that $F < G < A'$ and G is commensurable with B, and take (constructively) H such that $G : B :: H : D$; since $C' : D :: A' : B > G : B :: H : D$, we have $C' > H$. Then, by the commensurable case of V,24, $(A+G) : B :: (C+H) : D$. So $(A+F) : B < (A+G) : B :: (C+H) : D < (C+C') : D$, contradicting the assumption that $(A + F) : B :: (C + C') : D$.

[78]V,24 could also be proved by V,12 and alternation, which is a more obvious procedure than Euclid's, but Euclid avoids it, presumably in order to avoid unnecessary homogeneity assumptions.

[79]V,25 says that if $A : B :: C : D$, with $A > B, A > C$ (and thus $B > D, C > D$), then $(A + D) > (B + C)$. This follows since, by V,19, $(A - B) : (C - D) :: A : B$, and thus $(A - B) > (C - D)$; adding $(B + D)$ to both sides, we have $(A + D) > (B + C)$,

We have seen that the hypothesis of a historically intermediate proportion theory, or technique for proving results about proportions, between the anthyphairetic theory and Euclid's theory, relying on exhausting magnitudes by magnitudes commensurable with a given magnitude and on the postulate of a fourth proportional, as is apparently witnessed in Archimedes and some later sources, helps to explain not only how universal-mathematics proofs of the rules of proportion could be discovered without a miracle, but also the otherwise mysterious production of *Elements* V in particular, with its remarkably (but to many readers bewilderingly) successful demonstrations but also with its faults. I would like to conclude with an (even) more speculative suggestion. If anything like Knorr's hypothesis is right, Archimedes had direct access to Eudoxus, not mediated through Euclid. What form did Eudoxus' writing on these things take? A number of writers before Euclid are credited with *Elements*, but Eudoxus is not among them. I have suggested that for Eudoxus, the famous theorems now embedded in *Elements* XII and the new proof of VI,1 and the new proofs of the rules of proportion were all part of the same enterprise, using the same method of exhaustion: it was presumably the investigation of the pyramid and the cone, inherited from Democritus, which led him to this method, but once discovered it could be applied to other subjects as well, yielding something more satisfactory (and more universal) than the anthypairetic proofs, although Eudoxus had no need to reject the anthyphairetic definitions. So in what form did he express these proofs? I would suggest: in a two-book treatise looking very much like Archimedes' *On the Equilibrium of Planes* and *On Floating Bodies*. In those treatises, Book II is a presentation of remarkable new discoveries, drastically harder and also drastically more specialized (about segments of parabolas, resp. paraboloids of revolution) than the general discussions of Book I: Book I serves, not to prove propositions that were not already obvious, but to show that this new domain for mathematics can be turned into an axiomatic science, demonstrating by *reductio ad absurdum*—and thus to establish elementary propositions for Book II to draw on.

---

as was to be proved. V,25 is of no intrinsic interest and is not used in the rest of the *Elements*. Its only significance is for the case where $B = C$; that is, if $A : B :: B : D$, with $A > B > D$, then $(A + D) > 2B$. In other words, $((A + D)/2) > B$; or in other words, the arithmetic mean of A and D is greater than the geometric mean. Ian Mueller in *Philosophy of Mathematics and Deductive Structure in Euclid's Elements* (MIT Press, 1981), p. 134, expresses scepticism about this explanation of why Euclid mentions V,25, since Euclid never mentions the arithmetic mean. On the other hand, we know that the theory of the arithmetic, geometric and harmonic means was an important part of early Greek mathematics, and the three means are implicitly present in *Elements* X, since the medial, binomial and apotome are respectively the geometric, arithmetic and harmonic means of two rational lines incommensurable in length (and it would be very surprising if that were not how they were originally introduced). So, even though V,25 has no significance on *Euclid's* version of the *Elements*, it would have on an earlier version. It is a clever flourish, showing that the abstract, and to some mindsets tedious, theorems of Book V have a practical use in giving a logical foundation for the famous and "concrete" theorem that the arithmetic mean is greater than the geometric mean (assuming, that is, that there *is* a geometric mean between the two given quantities, which cannot be proved by the methods of Book V).

What Eudoxus did for areas and volumes, building up to the volume of the cone, Archimedes aims to do for centers of gravity and equilibrium positions, for segments of parabolas and paraboloids as well as the more familiar figures (and elsewhere he will go beyond Eudoxus on the volumes of the sphere and its segments and the areas of their surfaces, and the volumes of segments of other solids of revolution). The suggestion, then, was that Eudoxus inspired Archimedes to imitate and surpass him, not only in the content of his theorems and his method of exhaustion, but also in the form of his treatise. Book I would establish the rules of proportion, and at the end of the book, the basic results about areas of rectilinear figures in the plane (*Elements* VI,1 and what can be immediately derived from it by use of the rules of proportion): this would give the foundations, at the beginning of Book II (or very end of Book I) to prove *Elements* XII,1–2, and then into the wonder-inducing theorems of *Elements* XII, analogous to the even more wonder-inducing theorems of *On Equilibrium of Planes* II and *On Floating Bodies* II. Archimedes was a far greater mathematician than Euclid (although, contra Knorr, I do not think Euclid was a fool either), and I am not pretending to "explain" him. But thinking about Eudoxus, for all the limitations of our evidence, offers a perspective from which we can hope to understand more, not only about Euclid's achievements, but also about Archimedes'.

Frege's Introduction of Value-Ranges: A Reading of *Grundgesetze* §§ 29–31

THOMAS RICKETTS[*]

> ABSTRACT: In *Grundgesetze der Arithmetik* §§ 29–31, Frege argues that his stipulations of meaning for simple Begriffsschrift names fix the meaning for all Begriffsschrift names constructed from them, including value-range names. In § 10, having identified two value-ranges with the True and the False, Frege notes that he has determined the meanings of value-range names only "so far as is here possible". This remark strongly suggests that Frege does not take his treatment of the meaningfulness of value-range names in § 31 to demonstrated value-range names to have a fixed meaning. I provide an interpretation of §§ 29–31 that substantiates this suggestion.

The value-range functor is the only primitive Begriffsschrift name that Frege does not introduce by describing straightforwardly in colloquial language the value the named function returns for any argument. In *Grundgesetze* §§ 3 and 9 Frege introduces value-range names by, in effect, stipulating the truth of Begriffsschrift formulas of the following form:

$$('\varepsilon\Phi(\varepsilon) = '\alpha\Psi(\alpha)) = (\forall \mathfrak{a})(\Phi(\mathfrak{a}) = \Psi(\mathfrak{a})),$$

where "$\Phi(\xi)$" and "$\Psi(\xi)$" are names of arbitrary first-level functions.[1] I will call these formulas and the stipulation of their truth the introducing stipulations

---

[*] I presented earlier versions of this paper at the 2010 Midwest Philosophy of Mathematics workshop, at a memorial symposium for Leonard Linsky at the University of Chicago, and at the 2015 meeting of the Society for the Study of the History of Analytic Philosophy at Trinity College, Dublin. My thinking on *Grundgesetze* §§ 29–31 is indebted to the writings of Richard Heck on this topic as well as to the writings of and challenging conversations with Joan Weiner. I have also benefited from conversations with Warren Goldfarb.

[1] I take Frege to assume that every function is nameable so that the stipulation covers first-level function names that might be introduced. I discuss this aspect of Frege's procedure below on p. 242.

for value-range names. In §10, Frege goes on to stipulate the identification of two value-ranges with the two truth-values.

In §§29–31, Frege presents an obscure argument[2] that the specifications of meaning for the primitive Begriffsschrift names fix the meaning of every Begriffsschrift name constructible from them. This anomalous introduction of value-range names coupled with Frege's repeated admission of his reluctance to posit value-ranges has led many commentators on this argument, above all Michael Dummett, to take Frege to argue that the stipulations for value-range names, in fixing the truth-value of a restricted class of Begriffsschrift sentences containing these names, thereby fix the truth-value of every Begriffsschrift sentence containing a value-range name.[3] Frege then applies his context principle to conclude that both the individual value-range names and the value-range functor itself all mean something. In this way, Frege intends to provide a kind of contextualist semantic justification of Basic Law V.

There is another, related motivation for this entrenched interpretive approach. Dummett and many others take Frege in *Foundations of Arithmetic* §§62–67 to be groping after a contextualist justification for the posit of numbers as causally inert, non-spatial objects that answers the epistemic question, "How are numbers as objects to be given us independently of sensation and intuition?" Frege's efforts there are unsuccessful, and he goes on in *Foundations* to introduce extensions of concepts in order to construct numbers from extensions. On this interpretive approach, Frege later realizes, however, that the same question that arose for numbers in §62 arises for extensions. *Grundgesetze* §29–31 present Frege's contextualist answer to this question.[4]

The core of Frege's contextualist treatment of value-range names is an argument allegedly in §31 that if an arbitrary first-level function name "$\Phi(\xi)$" has a meaning, then the corresponding value-range name "$'\varepsilon\Phi(\varepsilon)$" also has a meaning. Frege is supposed to deploy the contextualist strategy in this argument. The difficulty here is that any such argument will surely founder on the impredicativity of Frege's higher-order quantifiers.[5] Frege himself recognized that the impredicativity of these quantifiers was essential to his projected construction of number. I do not think he is likely to have become confused by impredicativity in his argument. Moreover, Frege does not in any pre- or post-paradox

---

[2]Many of these challenges are mentioned by Michael Resnik in "Frege's Proof of Referentiality". My own thinking here benefited from a comprehensive survey of these challenges that Kai Wehmeier presented at a Frege workshop in Leiden in 2001. I discuss only some of those challenges in this paper.

[3]Dummett's most extensive discussion of §§29–31 in relation to Frege's context principle is in *Frege: Philosophy of Mathematics*, chaps. 16–18.

[4]I think this approach to *Foundations* §§62–67 is wrong-headed, because Frege from 1879 onwards is hostile to contextual definitions. I develop an alternative interpretation in "Concepts, Objects, and the Context Principle", §§7–8.

[5]"$\Phi(\xi)$" may itself contain an expression of the form "$(\forall f)\ldots'\varepsilon f(\varepsilon)\ldots$". For a general discussion of the difficulties that arise in framing a contextual justification for Basic Law V, see Charles Parson's classic paper, "Frege's Theory of Number" in Parsons, *Mathematics in Philosophy*, §§iv–v. He discusses what is in effect the difficulty posed by impredicative quantifiers on p. 160.

writings present §§ 29–31 as offering a justification for Basic Law V. Indeed, he almost never mentions these sections.[6] Finally, there is no hint in § 31 of a contextualist argument that individual value-range names are meaningful.

I will present an alternative interpretation of §§ 29–31. On my interpretation, Frege's treatment of value-ranges names in § 31 only repeats points from § 10 and is not intended to demonstrate that the introducing stipulations as supplemented in § 10 suffice to determine meanings for value-range names. Indeed, at the conclusion of § 10 Frege observes that they do not. Before turning to a discussion of §§ 29–31, I need to say something about the difficulties facing Frege in his introduction of value-ranges and to summarize my interpretation of § 10.[7]

I

In *The Foundations of Arithmetic*, at the crucial moment, like a *deus ex machina*, Frege invokes extensions of concepts, remarking in a footnote, "I presuppose that it is known (wissen) what the extension of a concept is." I think that by 1893, Frege more fully appreciates the novelty both of his conception of value-ranges and of the contrast he draws between a function and its value-range. He accordingly recognizes that he cannot presume familiarity with extensions or value-ranges.

Let us survey the conceptual and expositional difficulties Frege faces in elucidating the concept *value-range*. First note that throughout his career, Frege remains highly critical of so-called extensional presentations of the notion of class. He comments in a manuscript, "An aggregation consists of its parts [...]. The extension of a concept has its existence [*Bestand*] in the concept, not in the objects that belong to it; these are not its parts."[8] Second, note that Frege's distinction has little to do with the traditional contrast between a concept and its extension. Here a concept is a representation, and so something psychological, in contrast to the things falling under the concept that comprise the concept's extension.[9] Third, in "Function and Concept", Frege analogizes value-ranges

---

[6] I know of only two places where Frege mentions these sections after 1893. One is *Grundgesetze*, v. 2, § 147. Here Frege mentions § 31 merely as the place in *Grundgesetze* to find a list of the primitive names. The second is in Frege to Russell, 22.6.1902, *Wissenschaftlicher Briefwechsel*, p. 213. I discuss this passage at the end of the essay.

[7] I presented an interpretation of *Grundgesetze* § 10 in "Truth-values and Courses-of-Value in Frege's *Grundgesetze*". The longer paper from which this paper is extracted contains a revised presentation of that interpretation. I discuss issues surrounding the introduction on value-ranges both in these papers and in "Concepts, Objects, and the Context Principle", §§ 7 and 8.

[8] "Über Schoenflies: Die logischen Paradoxien der Mengenlehre", *Nachgelassene Schriften*, p. 199 (183). See also "A Critical Elucidation of Some Points in E. Schröder's *Vorlesungen über die Algebra der Logik*", p. 455.

[9] Hence Frege's distinction in the footnote to *Foundations* § 27 between the objective sense and the subjective sense of the word "idea".

to the curves that provide an intuitive (anschaulich) geometrical representation of the values some real-valued functions return for various arguments.[10] As Frege realizes, this comparison provides only a limited and metaphorical understanding of what value-ranges are. Drawing on Frege's Kantian view of geometry, it depends on the intuitive character of the curve in contrast with the associated function. Furthermore, Frege's generalization of the notion of *object* and the mathematical notion of *function* renders this contrast, taken literally, inapplicable to Frege's functions, which must be total over objects. Fourth, a function might be conceived as a rule or a law correlating each object with an object. We then have the contrast between the particular rule and the correlation the rule determines—the contrast between a function-in-intension and a function-in-extension. Nevertheless, the contrast between rule and correlation does not illuminate the distinction between function and value-range. Given Frege's distinction between sense and meaning, it at best highlights the distinction between the sense and meaning of a function name.[11]

Frege thus finds himself with nothing to say about what value-ranges of functions are except that they are objects satisfying the identity-standards laid down in the right-hand side of Basic Law V: the value-ranges of functions f and g are the same just in case f and g return the same values for the same arguments. I call this the extensionality condition. Frege thus realizes that his conception of value-ranges is in this way purely formal. The value-range function is itself a second-level function from first-level functions to objects that satisfies the third-level extensionality condition. This purely formal conception of value-ranges does not fully determine the meanings of value-range names. I emphasize that I am not making the anodyne observation that Frege is unable to *define* the concept *value-range* in more basic terms. Rather, I am claiming that Frege does not have a sufficient grasp of what value-ranges are to warrant introducing the value-range function name as an indefinably simple name into Begriffsschrift.[12] This point is crucial to my interpretation both of §§ 10 and 29–31.

Frege's difficulty here is illuminated by a bit of counterfactual history.. Suppose Frege had not been so impertinent and had studied Cantor carefully and dispassionately.[13] Suppose further that as a result, Frege had early on recognized the contradiction Basic Law V contains, decided to join Crispin Wright

---

[10]"Function and Concept", pp. 8–9. Burton Dreben pointed out to me the importance of this passage for the elucidation of the notion of a value-range.

[11]In "What is a Function?" Frege uses the notion of a law describing for correlating number to draw his readers attention to there being something to a function besides the numbers designated by mathematical terms like "3(3 − 4)" and "7(7 − 4)". He drops this rhetoric, when he presents his own view of functions at the end of the paper.

[12]I then think that it is this inadequacy in his conception of value-ranges that lies behind the reservations about Basic Law V that Frege voices in the foreword and appendix to *Grundgesetze*.

[13]W.W. Tait documents the extent of Frege's impertinence in "Frege versus Cantor and Dedekind: On the Conception of Number" in W.W. Tait, *Early Analytic Philosophy: Frege, Russell, Wittgenstein* (Open Court Publishing Co: Chicago, 1997), pp. 213–248.

to become a neo-Fregean, and enrolled the Cantor-Hume principle[14] among his logical axioms. Frege would face the same problem in introducing a name for a specific second-level function satisfying the third-level condition of that principle.

Frege's problematic response to this dilemma is to offer an indirect, piecemeal explanation of value-range names. The response is problematic, because Frege polemicizes in many places against both piecemeal definitions in mathematics and the posit of systems of numbers along with their arithmetic without establishing the consistency of these systems—something that Frege thinks requires an existence proof.[15] His own elucidation of value-ranges is uncomfortably close to what he admonishes.[16] In particular, what assures Frege of the consistency of the introducing stipulations?

Frege's presents his most persuasive motivation for Basic Law V in *Grundgesetze* vol. 2, § 147. He there observes that Basic Law V captures the inference-mode of "transforming the generalization of an identity into an identity", and urges that this inference is implicit in both mathematics and logic. It is then Frege's reflections on demonstrative argumentation in mathematics and logic that convince him that there are objects that satisfy the identity-standards laid down in the extensionality condition and so a second-level function whose range they comprise.[17] Frege thinks that a concept is shown to be consistent by establishing that it is nonempty.[18] So, although Frege never says as much, it must be his conviction that there is a second-level function that satisfies the extensionality condition that assures him of the consistency of the introducing stipulations. By reference to this unfortunate conviction, Frege can think of himself by his indirect introduction of value-range names not as inventing something, but as working towards the identification of a single second-level function falling under a third-level concept that he holds to be non-empty.

The inadequacy in Frege's conception of value-ranges displays itself inside Begriffsschrift. In § 2 Frege introduces truth-values as the objects meant by sentences, including assertible Begriffsschrift formulas, directing his readers to "On Sense and Meaning" for a full defense of this strange view. However, once the view has been fully understood, Frege thinks there are no further issues, no need for further elucidation, concerning what truth-values are, saying, "These two objects [the True and the False] are recognized, if only implicitly, by anyone who judges something to be true, and so even by the skeptic."[19] After the introduction of the identity-sign directly in § 7, we can state that a particular value-range is identical with one of the truth-values by an equation

---

[14]This principle states that number of (objects falling under the concept) F = the number of G just in case there is a function that 1-1 associates all of the Fs and the Gs.

[15]For example, see *Foundations*, §§ 92–96.

[16]As Frege recognizes in *Grundgesetze* vol. 2, § 146, p. 147.

[17]For further discussion, see my "Concepts, Objects, and the Context Principle", § 8, especially pp. 213–219.

[18]See *Foundations*, § 95, "On Formal Theories of Arithmetic", p. 103, and Frege's letter to Hilbert, 6 January 1900, *Wissenschaftlicher Briefwechsel*, 75–6.

[19]"On Sense and Meaning", p. 34.

joining a value-range name with an assertible Begriffsschrift formula (truth-value name) containing no occurrences of the value-range functor. The opening of § 10 argues that the introducing stipulations do not determine a truth-value for such mixed equations. Frege goes on to stipulate that the value-range of the horizontal function—the value-range of the concept under which the True alone falls—is to be identified with the True; similarly the value-range of a concept under which the False alone falls is to be the False. Frege makes these stipulations to further clarify what value-ranges, as the title for § 10 shows— "More exact determination of what the value-range of a function is supposed to be". The clarification § 10 insures that filling an argument place in Frege's horizontal functor, negation sign, material conditional sign with a value-range name yields a meaningful name. It also insures that mixed equations mean one of the two truth-values.

To establish the consistency of the identification of the truth-values with different value-ranges, Frege in effect argues:

> (A) If a second-level function satisfies the extensionality condition and $\Lambda(\xi)$ and $M(\xi)$ are distinct first-level functions, then there is a second-level function satisfying the extensionality condition that maps $\Lambda(\xi)$ to the True and $M(\xi)$ to the False.[20]

His idea is this: Basic Law V by itself does not *determine* whether any particular object, identified in terms independently of value-ranges, is the value-range of any particular function. The notion of determination operative here is an informal notion of logical inferability. This notion of logical inferability links (A) to Frege's desired conclusion. Were Frege's identifications of the truth-values with different value-ranges to be refutable from the initial stipulations, that refutation could be generalized into a proof of:

> (B) If a second-level function satisfies the extensionality condition, then either its value for $\Lambda(\xi)$ is not the True or its value for $M(\xi)$ is not the False.

From (A) and (B), one may infer that there is no second-level function satisfying the extensionality condition. As I argued above, Frege believes that there is such a second-level function. As Frege has established (A), he concludes from this result that (B) is not provable. Hence, his stipulative identification of truth-values with two selected value-ranges cannot be refuted using the introducing stipulations.

---

[20] Frege earlier presents a variant of this argument to argue that the introducing stipulations do not decide mixed equations, at least not all of them. His argumentation on both these points is more intricate and tangled than this summary suggests. In particular, as Frege has not yet explicitly introduced the concept of a second-level function, he does not explicitly generalize over them in his argumentation. I think that such a generalization is indicated in Frege's use of tilde-names, where tilde-names are just like value-range names, except that a tilde rather than a Greek smooth breathing mark occurs over the initial Greek vowel.

Of course, the identification of truth-values with value-ranges goes only a very small way toward fully determining the meanings of value-range names. Frege admits as much in the final paragraph of §10, saying: "We have determined value-ranges so far as it is here possible." Frege does anticipate adding further simple names to Begriffsschrift in order to incorporate vocabulary required to formalize well-established theories in the mature special sciences, of which geometry is Frege's paradigm.[21] He goes on to note that in introducing new simple first-level function-names into Begriffsschrift, we will need to specify what values the function returns for value-ranges as arguments. This specification, Frege says, "can be seen equally well as a determination (*Bestimmung*) of value-ranges as of the function [named by the newly introduced name]." For example, suppose we add primitive geometrical expressions to Begriffsschrift, among them the concept name "Point(ξ)". Nothing in the geometer's conception of points specifies whether points are value-ranges. So in adding this new function name, we will have to stipulate whether points are value ranges and, if so, of what functions. In so doing we further elucidate both a conception of points and a conception of value-ranges.[22]

This last point leads to the following question. Is there any prospect that the additional, piecemeal determinations of the meanings of value-range names might add up to a complete determination of their meanings? I think there is. Here, I claim, is a sufficient condition for "'$\varepsilon\Phi(\varepsilon)$" to have a determinate meaning:

the result of filling the argument place of

$$'\varepsilon\Phi(\varepsilon) = \xi$$

with a meaningful name of any object whatsoever yields a true or false equation.[23]

Some concepts—call them "sortal concepts"—are associated with uniform standards of identity for the objects falling under them. I take the concepts of *cardinal number*, *horse*, and *letter in the word "Zahl"* to be examples of

---

[21]See *Begriffsschrift*, p. vi. I elaborate this point in "Frege's 1906 Foray into Metalogic", pp. 171–174.

[22]Ignore for a moment the identification of two value-ranges with the truth-values. Michael Kremer urged that on my view of matters, Frege's specifications of meaning for new simple first-level function signs in a Begriffsschrift would also need to stipulate what values those functions return for truth-values as arguments, and that these stipulations should be seen as both a further elucidation of the truth-values and of the newly introduced function. After all, the formal role of truth-values in Begriffsschrift could be played by any two objects. The difference is that Frege's conception of the truth-values is not purely formal; rather it is embedded in his view of judgment. As a result, I suggest that considerations parallel to those that Frege advances in *Foundations* §61 concerning the ontological standing of numbers apply to truth-values.

[23]I do not give this alleged sufficient condition the attention it requires in this essay. Along with Frege, I assume that every object has a name or is nameable. I will say something about this assumption in the next section of this essay.

such concepts. Some concepts do not: *object Frege likes* and *red object* are examples. Frege himself makes this distinction in *Foundations* § 54.[24] Suppose now that the sortal concepts meant by simple names subsequently introduced into Begriffsschrift are like the concept *point* in that nothing in the colloquial grasp of these concepts specifies whether the objects falling under them are value-ranges. Suppose that we then in introducing the new name, specify that the objects falling under the concept meant are not value-ranges. Finally, suppose that in wake of scientific progress, we recognize that there is a surveyable group of sortal concepts, among them the concept *value-range*, which partition objects: every object falls under exactly one concept in the group. I assume that the elucidation of these concepts guarantees that they are pairwise disjoint, but that the recognition that every object falls under one of the concepts is a piece of substantive knowledge resting in part on non-logical grounds.

Consider now any equation of the form

$$\text{'}\varepsilon\Phi(\varepsilon) = \Delta$$

If "Δ" means a value-range, then the introducing stipulations fix whether the equation is true or false. Note, it does not matter whether "Δ" is a value-range name. Suppose "Δ" is "Frege's favorite object" and further that Frege's favorite object is the True. Then the equation will be true just in case anything falling under $\Phi(\xi)$ also falls under $-\xi$, and vice versa. If Δ is not a value-range, then it falls under some other concept in our partitioning group, one of whose characteristics is that no value-range falls under it. In that case, the equation is false. In this way in the imagined circumstances, the conception of value-ranges provided by our piecemeal elucidations would satisfy my sufficient condition for each value-range name's having a determinate meaning.

This strategy gives a kind of conceptual priority to our conception of value-ranges vis-à-vis our conception of the value-range function. In *Foundations*, using extensions, Frege defines a further second-level function, the Number-of function that satisfies the Cantor-Hume principle, and then identifies numbers with the extensions comprising the range of the Number-of function. Frege's actual construction of number thus gives priority to the Number-of function vis-à-vis the concept *number*. In *Grundgesetze*, these priorities are reversed as regards the elucidation of *value-range*. Frege takes the concept *value-range of a function* to be prior so that its piecemeal elucidation, once complete, enables

---

[24]Frege's sense-meaning distinction in application to predicates creates some terminological awkwardness here. Concepts are the meanings of predicates; analogizing senses of predicates to senses of proper names, we can think of the sense of a predicate as a way of being given a concept. We might then say that the sense of a predicate is a conception of a particular concept. Standards of identity are, properly speaking, an aspect of some conceptions. As a first approximation, we might say that a concept is a sortal concept just in case there a conception of that concept that incorporates a uniform standard of identity for objects falling under the concept. I won't bother about these niceties in my exposition.

us to explain the value-range function as the function that maps each first-level function to its value-range.[25]

II

I now turn to Frege's argument in §§ 29–31 that his specifications of meanings for the nine simple Begriffsschrift names fix a meaning for these names as well as for every Begriffsschrift name properly constructed from them. These sections occur in the second part of the exposition of his formalism, "Definitions", which opens *Grundgesetze*. Notationally speaking, Frege's formal definitions are abbreviations of compound names. As Frege puts it:

> By use of a definition we introduce a new name in that we determine that this new name is to have the same sense and the same meaning as one put together from familiar signs. By this means, the new sign becomes the same in meaning (*gleichbedeutend*) with the explaining name. The definition thus immediately becomes an asserted statement (*Satz*). We are then permitted to cite a definition [in the application of inference-rules].[26]

For Frege, definitions are not colloquial stipulations made outside of the formalism. Rather, defining a new name, like asserting a truth or inferring something from previous assertions, is something done using Begriffsschrift. As Frege specifies his inference rules in notational terms, so in § 33 he states notational constraints on definitions. His central aim here is to insure that, from any defined name via a chain of definitions, we can recover a unique compound name constructed from the original simple names;[27] he takes it to be evident that his rules do this.

Before stating the rules for definition, Frege announces in § 28: "For definitions, I lay down the following highest basic principle: Properly constructed names must always mean something."[28] In "On Sense and Meaning," Frege observes that colloquial language contains proper names that, while expressing a sense, do not have a meaning, and says that the history of mathematics tells of errors that have arisen from the presence of such names. He concludes, "It is not altogether unimportant once and for all to plug the source of these errors,

---

[25] Just as in *Foundations*, Frege takes extensions of concepts as known and so implicitly helps himself to a function mapping each concept to its extension.

[26] *Grundgesetze*, § 27, p. 45. The German word I have translated as "asserted statement" is "Satz". In the previous section, Frege stipulates that a *Begriffsschriftsatz* or *Satz* is the result of prefixing a truth-value name or a Latin-mark of a truth-value with the judgment-stroke.

[27] Unique, as Frege says in rule 1 of § 33, up to inessential choice of Greek and fraktur letters, Frege's bound variables. Most of the details of Frege's discussion concern the definitional introduction of new function names with their empty argument places.

[28] *Grundgesetze*, § 28, p. 45.

at least for science."²⁹ As Frege's definitions introduce notational abbreviations into Begriffsschrift, by showing that every name constructed from the simple Begriffsschrift names has a meaning, Frege shows that his rules of definition do not permit the introduction into Begriffsschrift of names without meaning. In this way, he validates his rules for definition. The placement of the argument indicates that this is its sole purpose. This is why Frege minimizes its importance when he mentions it in the foreword.³⁰

Before considering Frege's argument, we need to consider the odd-looking characterizations of meaningfulness stated in § 29:

> A name of a first-level function with one argument has a *meaning* (*means* something, is *meaningful*), if any proper name formed from it by filling its argument place with a proper name has a meaning on condition that this proper name has a meaning.
>
> A proper name has a *meaning*, if the proper name formed from any meaningful first-level function name with one argument by filling its argument place with the first proper name is meaningful, and if the name of any first-level function with one argument formed by filling the ξ-argument place of a meaningful first-level function name with two arguments with the proper name being tested is itself meaningful and the same holds as well for the ζ-argument place.³¹

Frege goes on to characterize in similar terms conditions for function names of other types to have a meaning. There are two problematic features of these characterizations. The first problem is their circularity: the characterization of meaningfulness for names of any one type rests on the meaningfulness of names of other types. It is then opaque how these characterizations could be applied to establish the meaningfulness of Begriffsschrift names generally. The second problem is the tension between these characterizations and Frege's thesis that first-level incomplete expressions are to mean first-level functions, which, as such, must be total over all objects. The characterization of meaningfulness for first-level function names only requires that the result of filling the argument place in a meaningful first-level function name by any meaningful proper name be a meaningful proper name. This requirement does not guarantee that the first-level function name means a total first-level function—it certainly does

---

²⁹"On Sense and Meaning", p. 41. I take the errors in mathematics Frege that has in mind to be those discussed in *Foundations* in which definite descriptions are used as proper names without first proving that exactly one object satisfies the description. This is what Frege has in mind when he speaks in § 3, p. 4 of statements on which the legitimacy of a definition may rest. Frege's *Grundgesetze* surrogate for definite descriptions introduced in § 11 is designed to guarantee a meaning to each surrogate, but at a price: it is not generally true that the surrogate for "The *F*" means something that is *F*.

³⁰Joan Weiner makes a similar claim in "Section 31 Revisited," p. 172.

³¹*Grundgesetze*, § 29, pp. 45-46.

not, if there are objects that go unnamed in Begriffsschrift. The resolution of the second problem, as we will see, also resolves the first one.

A manuscript from Frege's *Nachlass*, "Begründung meiner strengeren Grundsätze des Definierens" contains the key to understanding these characterizations. Here Frege defends his requirement that functions be total over objects against Peano's practice of partially defining the same sign for different contexts. In explaining the idea of intrinsically unrestricted generality within logical types that underlies this requirement, Frege says:

> A complete sentence preceded by the judgment stroke that contains Latin object-letters says that its content is true whatever meaningful proper name may replace each letter, provided that one and the same letter is replaced by the same proper name wherever it occurs in the sentence. Since proper names are signs that mean an individual determinate object, we can express this as follows: such a sentence says that its content is true whatever objects may be understood by the Latin object-letters.[32]

Frege's Latin-object letters more or less correspond to our free variables. He uses them as an alternative notational device for asserting the universal closure of a Begriffsschrift formula. His usage off them in Begriffsschrift is modeled on the familiar use of variables to express generalized equations in colloquial mathematics. In the first sentence, Frege generalizes over meaningful proper names as in the § 29 characterizations of meaningfulness. In restating his point in the following sentence, Frege speaks of a free variable generalization being true "whatever objects may be understood by the Latin object letters." The unrestricted generalization over objects indicates that in the first sentence Frege is assuming an extension of Begriffsschrift in which every object has a name. On the basis of these passages, I follow Richard Heck[33] in taking the § 29 characterizations of meaningfulness to assume that an extension of Begriffsschrift in which every object and every function has a name.[34] So, to apply Frege's characterization of meaningfulness to show that a first-level function name with one argument position from Frege's 1893 Begriffsschrift has a meaning, we have to show that the result of filling its argument position with the name of any object whatsoever yields a meaningful proper name.

Why, though, does Frege characterize meaningfulness as he does, in linguistic terms, in the formal rather than the material mode? Here is one suggestion. Early on in *Grundgesetze*, Frege notes the difficulties that the 'awkwardness of

---

[32] *Nachgelassene Schriften*, p. 167 (154). Frege uses similar rhetoric in "On Mr. Peano's Begriffsschrift and My Own", p. 374: "Then we stipulate that the occurrence of 'x' [in '$\Phi(x)$'] is to say that the sentence is true whatever meaning we may attach to 'x'."

[33] See *Reading Frege's Grundgesetze* pp. 56–58. I don't agree with everything Heck says about how Frege understands this assumption.

[34] Frege's arguments below could be recast so as only to require that every object, etc. is namable.

language' associated with the function-object distinction creates for his exposition of Begriffsschrift, although he still repeatedly uses language whose point will be grasped only by readers who give him a pinch of salt.[35] These expositional difficulties are heightened in the setting of §§ 29–31, in which Frege's points turn on generalizations involving contrasting differences among logical types. Frege takes proper names and functions names alike to be objects, so that the logical syntactic differences among them can be marked by first-level predicates.[36] Talking then about the meaningfulness of names gives Frege a way here to circumvent expositional difficulties that the function-object distinction creates.

Richard Heck has noted the similarity between Frege's characterizations of meaningfulness and Tarski's notion of satisfaction.[37] Indeed, when Frege states, "[...] a sentence says that its content is true whatever objects may be understood by the Latin object-letters [in the sentence]," he seems to be grasping after something like our notion of assignments of objects to free variables. He does not, however, pursue this idea, and for a good reason. Reflecting on and generalizing the contrasting use in colloquial mathematics between letters like "x" and "y" and constants like numerals, Frege grounds his conception of a proper name in the contrast between the use of proper names to say something about a particular thing and the use of a letter in the position of that name to "confer generality of content" on such a sentence."[38] Neither use is to be explained in terms of the other. It would then unnecessarily muddy Frege's correlative logical syntactic and ontological categories to use the vocabulary of designation or meaning in connection with his Latin letters.[39] This is why Frege prefers to talk in terms of every object (and every function as well) having a name, taking the assumption that every object has a name to be as anodyne as our assumption that any sequence of objects from a universe of discourse may be assigned to a sequence of distinct variables.

Frege's argument in §§ 30–31 is an inductive argument. In § 30, Frege presents two ways in which a name may be formed from Begriffsschrift names and argues that, applied to meaningful names, both of these ways produce meaningful names. Frege's argumentation as regards the second way is sketchy, but raises no special problems concerning value-ranges.

In § 31, Frege argues that his meaning-specifications give each of the simple Begriffsschrift names a meaning. There are two stages to Frege's argument.

---

[35] In a footnote to § 4, Frege refers his readers to "On Concept and Object" for a further discussion of these difficulties.

[36] Frege explicitly says that his predicates are, ontologically speaking, objects in his 29 June 1902 letter to Russell, *Wissenschaftlicher Briefwechsel*, p. 218. This view together with the assumption that every function has a name, along with much else, founders on Cantor's theorem.

[37] *Reading Frege's Grundgesetze*, pp. 57–59.

[38] See *Begriffsschrift*, § 1. See also "Einleitung in die Logik," *Nachgelassene Schriften*, pp. 204 and 206f. (188 and 190) and "On the Foundations of Geometry" (1906), pt. i, p. 307f. For further discussion, see "Concepts, Objects, and the Context Principle", § 1.

[39] Weiner makes the same point in "Section 31 Revisited", p. 171.

The first stage argues that Frege's meaning-specifications secure a meaning for what I shall call the *core names*—the simple names of truth-valued functions, i.e., the horizontal functor, negation, material conditional, identity, and the quantifiers. The second and controversial stage maintains that Frege's stipulations concerning value-range names secure a meaning for any value-range name formed using a meaningful first-level function name, and so for the value-range functor itself.[40]

After listing the primitive names, Frege describes his strategy:

> We proceed from the thesis that names of truth-values mean something, either the True or the False. We then gradually widen the circle of names recognized to be meaningful by proving that further names to be accepted in addition to those already so accepted yield a meaningful name, when the one is completed by the other, provided that the one has an argument place suitable for the other.[41]

He then says that to show the names of the horizontal function, negation function, and material conditional function meaningful, it suffices to show that the names that result from completing these function names by truth-value names are meaningful, adding parenthetically, "here we do not yet know [*kennen*] other objects." Clearly these names satisfy this condition. Still, there is a problem: Frege's argument does not fit the earlier meaningfulness characterizations for first-level function names, for they require that the completion of the argument places in these names by *any* meaningful proper names without restriction produce a meaningful proper name. To understand Frege's procedure here, we need to consider more closely his parenthetical remark.

Although Frege in these sections assumes that Begriffsschrift has been expanded by names for every object and function, this assumption by itself brings with it no commitments as to what objects and what functions there are. I noted that Frege holds that anyone who judges implicitly recognizes the True and the False. It is on this basis that Frege takes "names of truth-values to mean [...] either the True or the False." That is, Frege takes it that the proper names in the expanded Begriffsschrift include names of both truth-values. He does not allude to the prospect that there may be (names of) objects other than truth-values. This silence does not show that Frege assumes at this stage of the argument that truth-values exhaust the objects, but only that his argument at this stage makes no commitments beyond the two truth-values as to what objects there are.

At this stage then, we are restricted to considering the meaningfulness of first-level function names putatively for truth-valued functions. Moreover to

---

[40] Having established that value-range functor is meaningful, Frege argues that his surrogate for definite descriptions, whose meaning-specification mentions value-ranges, is meaningful. This final step of the argument raises no further problems, and I will not say anything further about it.

[41] *Grundgesetze*, §32, p. 48.

know the two truth-values is to have an adequate conception of each of them and so an adequate conception of truth-values. Hence, the distinction between the two truth-values as well as the distinction between truth-values and non-truth-values is available for use at this stage of the argument, as it was in the original meaning-specifications. This is the crucial point. The meaning-specification of "$-\xi$" is framed to handle proper names of non-truth-values uniformly—completing this sign with one of them, if there are any, yields a name of the False. So, at this stage of the argument, Frege needs only to confirm that the meaning specification handles proper names of the two recognized objects in order to show that the specifications cover any arbitrary meaningful proper name. What holds for the horizontal sign holds as well for the negation and material conditional signs. The meaning specification for the identity-sign is disquotational. Frege takes it that with the general notion of an object comes a grasp of identity as that relation in which every object stands only to itself. So the notion of identity is available to use in both meaning-specifications for names and in arguments validating those specifications.[42] Trivially then, filling the two argument positions in "$\xi = \zeta$" with meaningful proper names yields a name of either the True or the False, depending on whether the objects meant by the proper names are identical.

Frege next turns his attention to the two quantifiers he deploys in *Grundgesetze*. As the quantifiers are to mean truth-valued functions, Frege's confirmation of their meaning specifications still recognizes among objects only the two truth-values. The universal quantifier over objects is to mean a second-level function that maps any first-level function to the True, if that first-level function maps every object to the True; it maps the function to the False otherwise. Suppose "$\Phi(\xi)$" is an arbitrary, meaningful first-level function name. Either filling its argument place with any meaningful proper name whatsoever yields the True, or not. In the first case, using "$\Phi(\xi)$" to fill the argument place of the universal quantifier over objects means the True, and in the second case, the False. A parallel argument confirms the meaningfulness of the quantifier over first-level functions with a single argument-place.

I find it significant that Frege here generalizes without restriction over both function names and proper names so that his arguments are clearly applications of the earlier characterization of meaningfulness for second-level function names. This feature of his exposition bolsters my claim that the arguments for the meaningfulness of the core first-level simple function names just considered apply the parallel characterizations for first-level function names. Note also how, once we appreciate Frege assumption of an expanded Begriffsschrift, his arguments that each of the core simple names is meaningful are straightforward and appropriately trivial applications of his meaningfulness characterizations.

We now reach the controversial stage of the argument: Frege's discussion of value-range names. Like the quantifier over objects, the value-range functor is a second-level function name with an empty place for first-level function names

---

[42]For further discussion, see "Concepts, Objects, and the Context Principle", §§ 1–3.

of one argument. Unlike any of the core simple names, the value-range functor does not mean a truth-valued function so that in introducing the value-range functor,

> [...] we do not merely introduce a new function name, but also at the same time for every first-level function with one argument we introduce a new proper name (value-range name), and not only for already familiar functions but also in advance for any that might be introduced.[43]

I emphasized that the stipulations introducing value-range names are indirect: they do not use the notion *value-range*. Similarly, Frege's validation of those stipulations may not and does not use the notion of *value-range*: neither the stipulations nor their validation rests on the recognition of value-ranges, as the earlier ones rested on the recognition of truth-values. This is the price levied by Frege's inadequate conception of value-ranges. His strategy is first to show that each *regular* individual value-name, one formed using a meaningful first-level function name, is meaningful. From this result, it follows from the relevant clause of the §29 meaningfulness characterization that the value-range functor is meaningful. Frege says that to investigate the meaningfulness of value-range names, we must test whether the result of using one of these in an argument place in each of the core simple first-level function names yields a meaningful name. He now repeats points from §10 to show that value-range names pass this test. On this basis, he asserts: "Regular value-range names may [*dürfen*] be admitted into our circle of meaningful names."[44]

Frege's argument is perplexing. Unlike the preceding validations, here Frege does not apply his characterizations of meaningfulness. Remember that in order to apply these to establish the meaningfulness of a proper name, it must be shown that the result of filling an argument place in *any* meaningful first-level function name with the putative name yields a meaningful name, not just the core first-level function names. Frege does not attempt to do this. Furthermore, as regards the identity sign, he does not show that filling one of the argument places in "$\xi = \zeta$" by a value-range name and the other by an arbitrary meaningful proper name is meaningful.[45] Frege does not purport to show these things here; he could not do so without using the notion of a value-range and so assuming their existence.

I suppose that it is the logical gulf between what Frege claims in §10 and what his meaningfulness characterization requires of proper names that prompts commentators to suppose that Frege in §31 must have in mind some undisclosed, unsuggested argument that bridges this gulf. This interpretive tack, however, conflicts with Frege's admission at the end of §10 that his

---

[43] *Grundgesetze*, §31, p. 49.
[44] *Grundgesetze*, §31, p. 50.
[45] As in §10, Frege only shows that filling these argument places by any combination of value-range and truth-value names produces a meaningful truth-value name.

stipulations up to that point have not given value-range names determinate meanings. A different approach fits Frege's text better.

I propose that Frege does not purport to show by application of his meaningfulness-characterizations that the stipulations of §§ 3, 9, and 10 give value-range names determinate meanings. Rather, he thinks that the stipulations and discussion of these sections *permit* treating value-range names as if Frege's stipulations give them determinate meanings, even though (*sotto voce*) they do not. After all, § 10 resolves any indeterminacy concerning truth-values and value-ranges, the only objects mentioned in pure logic, and it validates the consistency of the additional stipulation needed to do so. In the material mode, this conclusion becomes: the conception of *value-range* that Frege's stipulations about value-range names communicate is sufficient, despite remaining haziness, to justify the proleptic recognition of value-ranges as objects. The recognition is proleptic in the expectation that the elucidation that comes from further stipulations as the simple names of Begriffsschrift are expanded to embrace more of science will yield a more nearly adequate conception of value-ranges and perhaps eventually a fully adequate one. Furthermore, once Frege admits the value-range functor into the circle of meaningful names, and so recognizes value-ranges as objects, he is free to use the notion *value-range* in subsequent meaning-specifications and in elucidations of those specifications. He does so in § 11 with the introduction of his final simple name and the trivial validation of its meaningfulness.

I have indicated how my interpretation accommodates what Frege's says §§ 29–31 and how it is supported by the relation between §§ 10 and 31. The interpretation also fits with the ancillary importance the foreword assigns to §§ 28–32 and the fact that Frege does not mention these sections in subsequent substantive discussions of value-ranges. I think that a subtle feature of Frege's rhetoric in § 31 further supports my interpretation. Frege asserts outright that his stipulations give the core first-level function-names meanings. In applying the relevant meaning characterization to the universal quantifier over objects, Frege says: "In general, from the fact that the inserted function name '$\Phi(\xi)$' means something, it follows that '$\forall(\mathfrak{a})\Phi(\mathfrak{a})$' means something."[46] Frege makes no such general claims about the meaningfulness of value-range names outside of the contexts mentioned in § 10. So, he concludes only that they may be admitted into the circle of meaningful names.[47]

---

[46] *Grundgesetze*, § 31, p. 49.

[47] I should note that Frege concludes his discussion of the universal quantifier over objects with: "Hence, the function name '$\forall(\mathfrak{a})\varphi(\mathfrak{a})$' is to be admitted [*ist aufzunehmen*] into the circle of meaningful names." Readers may reasonably take "is *to be* admitted" and "*may* be admitted" to be stylistic alternatives. There is a noteworthy difference in context. In the case of the universal quantifier he first establishes that if "$\Phi(\xi)$" has a meaning then either "$\Phi(\Delta)$" is the True, or it is not, no matter what "Δ". He then advances the conclusion just quoted. As regards value-range names, he presents his conclusion on the basis of what he has said in § 10. My interpretation does not rest primarily on a delicate point of usage but on the way in which the discussion in § 31 of the universal quantifier overtly applies the earlier meaningfulness characterization, while the discussion of the value-range functor does not.

On my interpretation, Frege's special mention of the argumentation in § 31 in his reply to Russell's celebrated letter announcing the paradox—as far as I know this is his only substantive post-1893 mention of it— falls into place. Frege says:

> Accordingly, it seems that the transformation of the generalization of an identity into an identity of value-ranges (§ 9 of my *Grundgesetze*) is not always allowed, that my law V is false (§ 20, p. 36), and my explanations in § 31 do not suffice to secure a meaning to all combinations of my signs.[48]

I maintained that Frege's introduction of value-range names in §§ 3, 9, and 10 presupposes the consistency of the introducing stipulations, and argued that Frege's confidence here is a matter of his conviction that a tacitly employed, unquestioned inference mode presupposes that there are objects whose identity-standards are those required by the extensionality condition. In formalizing Russell's paradox in the appendix to *Grundgesetze*, vol. 2, Frege presents the paradox as a derivation from the quantificational part of his system—the part that employs only the vocabulary whose meaningfulness is validated in the first-stage of the argument in § 31—that there is no second-level function that satisfies the extensionality condition. Russell's paradox thus refutes the conviction on which Frege grounds the introduction of value-range names into Begriffsschrift.

The consideration of a point Richard Heck raises illuminates my understanding Frege's conception of value-ranges in *Grundgesetze*. Heck calls attention to the Begriffsschrift formula that says "Every object is the value-range of some function":

$$(\forall \mathfrak{a})(\exists f)[\mathfrak{a} = \text{'}\varepsilon(f(\varepsilon))].\text{[49]}$$

Although the stipulations of § 10 fix the truth-value of Begriffsschrift equations, they do not on their face fix the truth-value of this equation. Heck maintains that, as Frege claims in § 32 to have just shown that all Begriffsschrift names properly constructed from the simple names mean something, he must take this formula to mean a truth-value. He argues that Frege can view the meaning specifications for the primitive names as determining a meaning for this generalization only if he implicitly restricts his object-quantifiers to value-ranges (including now the two truth-values), in which case the sentence is true.[50]

I hold this suggestion to be thoroughly misguided. There is no hint in Frege's explanations of quantifiers in *Begriffsschrift*, "Function and Concept", or *Grundgesetze* that his quantifiers sometimes have a restricted domain. His rhetoric in these places and elsewhere indicates the opposite. Furthermore, were we to admit quantifiers with different domains, there would be a logically

---

[48] Frege to Russell, 22.6.1902, *Wissenschaftlicher Briefwechsel*, p. 213.

[49] *Reading Frege's Grundgesetze*, pp. 93–96.

[50] Other commentators adopt this view of matters, e.g., Dummett, *The Interpretation of Frege's Philosophy*, p. 416 and Marco Ruffino, "Freges Ontologie der Logik", p. 68.

significant feature of the sense of the sentence that is not notationally marked in Begriffsschrift. A main purpose for devising a Begriffsschrift is to provide a notation in which such features are so marked. Frege argues in *Grundgesetze* vol. 2, §65[51] that this desideratum requires quantifiers to be intrinsically unrestricted within logical types so that any restrictions on the validity of a generalization be made explicit in now familiar ways by the use of predicates.[52]

I offer a different view of the formula Heck brings to our attention. Although the stipulations through §10 do not determine a truth-value for this formula, Frege's treatment of value-range names commits him to treating it as if they did, but a meaning which is unknown at present. Frege in effect treats subsequent supplementary determinations of the meaning of value-range names as though they were consequences of the determinate meanings that those names already have.

References

References to Frege's three books are by section number. References to *Grundgesetze* are to volume one unless otherwise specified. References to Frege's papers are by page number in the original publication. References to items from Frege's posthumously published writings (*Nachgelassene Schriften*) are to the page number in the German edition, Frege (1983), followed by the page number to the English translation Frege (1979). References to Frege's correspondence are by correspondent and date followed by a page reference to the German edition (*Wissenschaftlicher Briefwechsel*), Frege (1976). Translations of quoted passages from Frege's writing are my own made in consulting the most prominent English translations.

Dummett, M. (1981). *The Interpretation of Frege's Philosophy*. Cambridge: Harvard University Press.

Dummett, M. (1991). *Frege: Philosophy of Mathematics*. Cambridge: Harvard University Press.

Frege, G. (1879). *Begriffsschrift*. Halle: Nebert. English trans.: Frege (1972).

Frege, G. (1884). *Die Grundlagen der Arithmetik*. Breslau: W. Koebner. English trans.: Frege (1953).

---

[51] Frege also makes this point in his critique of Peano in "Mr. Peano's Begriffsschrift and My Own," pp. 373–374 and "Begründung meiner strengeren Grundsätze des Definierens", *Nachgelassene Schriften*, p. 165–66 (153–154).

[52] In *Reading Frege's Grundgesetze*, p. 95, Heck suggests that "with only a dollop of anachronism" Frege can be interpreted in §65 to be warning against the confusion of the model-theoretic notion of a universe of discourse with the object language semantic notion of a domain of quantification. To my eye, Heck's dollop of anachronism is a full three scoops.

Frege, G. (1885). Über Formale Theorien der Arithmetik. *Jenaische Zeitschrift für Naturwissenschaft 19*, 94–104. English trans.: On Formal Theories of Arithmetic. In Frege (1984), pp. 112–121.

Frege, G. (1891). *Funktion und Begriff.* Jena: H. Pohle. English trans.: Function and Concept. In Frege (1984), pp. 137–156.

Frege, G. (1892a). Über Begriff und Gegenstand. *Vierteljahrsschrift für wissenschaftliche Philosophie 16*, 192–205. English trans.: On Concept and Object. In Frege (1984), pp. 182–194.

Frege, G. (1892b). Über Sinn und Bedeutung. *Zeitschrift für Philosophie und Philosophische Kritik 100*, 25–50. English trans.: On Sense and Meaning. In Frege (1984), pp. 157–177.

Frege, G. (1893). *Die Grundgesetze der Arithmetik, Bd. 1.* Jena: H. Pohle. English trans.: Frege (2013).

Frege, G. (1895). Kritische Beleuchtung einiger Punkte in E. Schröders *Vorlesungen über die Algebra der Logik*. *Archiv für systematische Philosophie 1*, 433–456. English trans.: A Critical Elucidation of Some Points in E. Schröder's *Vorlesungen über die Algebra der Logik*. In Frege (1984), pp. 210–228.

Frege, G. (1897). Über die Begriffsschrift des Herrn Peano und meine eigene. *Berichte über die Verhandlungen der Königlich-Sächsischen Gesellschaft der Wissenschaften 48*, 361–378. English trans.: On Mr. Peano's Conceptual Notation and My Own. In Frege (1984), pp. 234–248.

Frege, G. (1903). *Die Grundgesetze der Arithmetik, Bd. 2.* Jena: H. Pohle. English trans.: Frege (2013).

Frege, G. (1904). Was ist eine Funktion? In *Festschrift Ludwig Boltzmann*, 656–666. Leipzig: J. Barth. English trans.: What is a Function? In Frege (1984), pp. 285–292.

Frege, G. (1906). Über die Grundlagen der Geometrie. *Jahresbericht der Deutschen Mathematiker-Vereinigung 15*, 293–309, 377–403, 423–430. English trans.: On the Foundations of Geometry. In Frege (1984), pp. 273–284, 293–340.

Frege, G. (1953). *The Foundations of Arithmetic.* J. L. Austin (Ed.). Chicago: Northwestern University Press.

Frege, G. (1967). *Kleine Schriften.* I. Angelelli (Ed.). Hildesheim: Olms.

Frege, G. (1972). *Conceptual Notation.* T. W. Bynum (Ed.). Oxford: Oxford University Press.

Frege, G. (1976). *Wissenschaftlicher Briefwechsel*. G. Gabriel et al. (Eds.). Hamburg: Felix Meiner.

Frege, G. (1979). *Posthumous Writings*. H. Hermes et al. (Eds.). Chicago: University of Chicago Press.

Frege, G. (1983). *Nachgelassene Schriften*. H. Hermes et al. (Eds.). Hamburg: Felix Meiner.

Frege, G. (1984). *Collected Papers on Mathematics, Logic, and Philosophy*. B. McGuinness (Ed.). Oxford: Basil Blackwell.

Frege, G. (2013). *The Basic Laws of Arithmetic, Vols. I-II*. P. Ebert & M. Rossberg (Eds.). Oxford: Oxford University Press.

Heck, R. (2012). *Reading Frege's Grundgesetze*. Oxford: Oxford University Press.

Parsons, C. (1983). Frege's Theory of Number. In *Mathematics in Philosophy*, pp. 150–175. Ithaca: Cornell University Press.

Resnik, M. (1986). Frege's Proof of Referentiality. In L. H. . J. Hintikka (Ed.), *Synthesizing Frege. Essays on the Philosophical and Foundational Work of Gottlob Frege*, pp. 177–196. D. Reidel.

Ricketts, T. (1997). Truth-Values and Courses-of-Values in Frege's *Grundgesetze*. In W. W. Tait (Ed.), *Early Analytic Philosophy: Frege, Russell, Wittgenstein*, pp. 187–211. Chicago: Open Court.

Ricketts, T. (2010). Concepts, Objects, and the Context Principle. In M. P. . T. Ricketts (Ed.), *Cambridge Companion to Frege*, pp. 149–219. Cambridge: Cambridge University Press.

Ruffino, M. (2000). Freges Ontologie der Logik. Die Natur und Notwendigkeit logischer Gegenstände. In G. G. . U. Dathe (Ed.), *Gottlob Frege: Werk und Wirkung*, pp. 57–70. Paderborn: Mentis.

Tait, W. W. (1997). Frege versus Cantor and Dedekind: On the Concept of Number. In W. W. Tait (Ed.), *Early Analytic Philosophy: Frege, Russell, Wittgenstein*, pp. 213–248. Chicago: Open Court.

Weiner, J. (2002). Section 31 Revisited. In E. Reck (Ed.), *From Frege to Wittgenstein*, pp. 149–182. Oxford: Oxford University Press.

Dedekind's Structuralism: Creating Concepts and Deriving Theorems

WILFRIED SIEG AND REBECCA MORRIS*

ABSTRACT: Dedekind's structuralism is a crucial source for the *structuralism of mathematical practice*—with its focus on abstract concepts like groups and fields. It plays an equally central role for the *structuralism of philosophical analysis*—with its focus on particular mathematical objects like natural and real numbers. Tensions between these structuralisms are palpable in Dedekind's work, but are resolved in his essay *Was sind und was sollen die Zahlen?* In a radical shift, Dedekind extends his mathematical approach to "the" natural numbers. He creates the abstract concept of a *simply infinite system*, proves the existence of a "model", insists on the stepwise derivation of theorems, and defines structure-preserving mappings between different systems that fall under the abstract concept. Crucial parts of these considerations were added, however, only to the penultimate manuscript, for example, the very concept of a simply infinite system. The methodological consequences of this radical shift are elucidated by an analysis of Dedekind's metamathematics. Our analysis provides a deeper understanding of the essay and, in addition, illuminates its impact on the evolution of the axiomatic method and of "semantics" before Tarski. This understanding allows us to make connections to contemporary issues in the philosophy of mathematics and science.

---

*This paper had been in the works for a very long time; it was started in the summer of 2010 and completed in April of 2015. It is most appropriately dedicated to Bill Tait whose perspective on Dedekind influenced us deeply. For WS, Bill's proof theoretic work has been inspiring and his philosophical reflections illuminating; I also treasure the enduring friendship we have formed.

Introduction

Dedekind's structuralism has strongly influenced both mathematical practice and philosophical analysis that is concerned with the foundations of mathematics. On the mathematical side, Dedekind was keenly aware of the importance of creating new concepts in mathematics. So he writes in the Preface to his famous essay *Was sind und was sollen die Zahlen?*, "...the greatest and most fruitful advances in mathematics and other sciences have invariably been made by the *creation and introduction of new concepts*, rendered necessary by the frequent recurrence of complex phenomena which could be mastered by the old notions only with difficulty" (*WZ*, 792, emphasis added).[1] Dedekind's work on algebraic number theory is characterized by its focus on abstract concepts together with structure-preserving mappings; it impacted modern algebra directly through David Hilbert and Emmy Noether. This development found its expression in van der Waerden's influential monograph *Moderne Algebra*, the first volume of which appeared in 1930 and the second in 1931. More than thirty years later van der Waerden wrote in the Preface to the 1964 edition of Dedekind's *Algebraische Zahlentheorie*:

> Evariste Galois and Richard Dedekind are the ones who gave to modern algebra its structure. The supporting skeleton of this structure derives from them.[2]

Moreover, the general methodological aspects that constitute the "supporting skeleton of this structure" have shaped not just modern algebra, but modern mathematics as exemplified in Bourbaki's work.

On the philosophical side, contemporary discussions on structuralism often consider Dedekind's work, and he is taken to represent a variety of different, sometimes conflicting, positions. Dedekind's comments about the free creation of new mathematical objects, as found in his *SZ*, *WZ*, and letters to Weber, feature prominently in such discussions. Some may seek to downplay these remarks, but a number of philosophers, including William Tait (Tait 1996), Erich Reck (Reck 2003) and Audrey Yap (Yap 2009), have argued that Dedekind's comments about creation should not be dismissed. Instead, they maintain that creating (systems of) particular mathematical objects is a crucial feature of Dedekind's structuralism. Notice, then, that Dedekind's focus on the creation of concepts is of central importance for mathematical practice, while his emphasis on the creation of objects has been significant within philosophy. This

---

[1] Dedekind's *Was sind und was sollen die Zahlen?* will be referred to as *WZ* throughout this paper, and his earlier essay *Stetigkeit und irrationale Zahlen* will be referred to as *SZ*.

[2] Here is the German text: "Evariste Galois und Richard Dedekind sind es, die der modernen Algebra ihre Struktur gegeben haben. Das tragende Skelett dieser Struktur stammt von ihnen." Starting with this remark of van der Waerden's, Mehrtens gives in his (1979) an informative overview of Dedekind's practice of creating mathematical concepts. Van der Waerden's and our perspective on Dedekind's role in the development of modern structural algebra is not shared by Leo Corry in his (2004a).

split in emphasis between concepts, on the one hand, and objects, on the other, is reflected in Dedekind's foundational work. In the penultimate manuscript of *WZ*, Dedekind explicitly created a system of new mathematical objects as *the* abstract natural numbers. This step is no longer taken in the published version. In a *radical shift*, Dedekind introduces instead the concept of a simply infinite system and utilizes metamathematical considerations to ensure that "the definition of the *concept of numbers* given in (73) is completely justified" (*WZ*, 823, emphasis added).

This shift is documented in the Appendix below: it is in the transition from the penultimate version of *WZ* to the publication that the concept of a simply infinite system is introduced for the first time and that the concern with the logical existence of such a system is raised, also for the first time. Taking seriously this radical shift, we interpret Dedekind's new approach as an *extension of his abstract mathematical ways* to "the" natural numbers. We describe his position post-radical-shift in section A, and, in addition, sharpen his views concerning the creation of abstract concepts. Then, in section B, we elaborate the role of structure-preserving mappings between systems falling under such concepts, and explain the character and methodological significance of his metamathematical work. The latter work is primarily, but not exclusively, concerned with categoricity. Relying on a quasi-formal notion of derivation that takes as its starting points the *characteristic conditions* (Merkmale) of abstract concepts, Dedekind proves in *WZ* that all the statements derived from the conditions for the concept of a simply infinite system hold in all particular simply infinite systems. Here we have an incipient model theory (without a Tarskian truth definition) joined with proof theory (without a precise notion of formal derivation).

Equipped with this deeper understanding of Dedekind's axiomatic standpoint, we can compare *WZ* with the earlier essay *SZ*: the treatment of real numbers is already structured in such a way that it can be easily and naturally expanded to parallel the treatment of natural numbers in *WZ*; see section C.1 below. Our interpretation allows us to reemphasize that Hilbert's *axiomatic method*—as formulated and practiced in (Hilbert 1899) and (Hilbert 1900)—is fully in Dedekind's mold. Not surprisingly, as Zermelo was deeply influenced by Hilbert, we can highlight striking parallels between Dedekind's work in *WZ* and Zermelo's formulation of set theory in 1908 as well as its metamathematical investigation in 1930. In the *Concluding Remarks* we point out deep connections with contemporary issues in philosophy of science that are rooted in the co-evolution of mathematics and science in the 19th century, with corresponding changes towards a more scientific philosophy. The underlying intellectual perspective is characteristic of the Göttingen mathematicians: *vide* Gauss, Dirichlet, and Riemann, with all of whom Dedekind had close relations and in whose tradition Hilbert saw himself.[3]

---

[3]This general background is described in detail and with great sensitivity in the first two chapters of Leo Corry's book (2004b).

A. Simply Infinite System: *Concept of Numbers*

In his letter to Keferstein, Dedekind made the following remarkable comment on the genesis of his essay: "... it is a synthesis constructed after protracted labor, based upon a prior analysis of the sequence of natural numbers just as it presents itself, in experience, so to speak, for our consideration." (Dedekind 1890, 99). The steps of analysis address the questions that his essay *WZ* is supposed to answer:

> What are the mutually independent fundamental properties of the sequence N, that is, those properties that are not derivable from one another but from which all others follow? And how should we divest these properties of their specifically arithmetic character so that they are subsumed under more general concepts and under activities of the understanding, *without* which no thinking is possible at all, but *with* which a foundation is provided for the reliability and completeness of proofs, as well as for the formation of consistent definitions of concepts?[4]

We will trace Dedekind's analytic steps and thus obtain a clearer view of how the fundamental properties of N are "subsumed under more general concepts and under activities of the understanding, without which no thinking is possible". After all, these fundamental properties will be the basis for the systematic, stepwise development of number theory or, what Dedekind calls, its *synthetischer Aufbau*.

A.1. Logical Framework

The questions mentioned in the above quotation clearly express Dedekind's goal of isolating properties that serve to define a higher-order concept of *natural number*. The crucial properties are combined into the concept of a *simply infinite system*. In order to build up to this concept, Dedekind begins the essay by introducing the more general notions of *thing* and *system*. Of the former he writes, "In what follows I understand by *thing* [*Ding*] every object of our thought" (*WZ*, 796), and of the latter he explains, "It very frequently happens that different things, a, b, c, ... for some reason can be considered from a common point of view, can be associated in the mind, and we say that they form a system S ..." (*WZ*, 797). After defining a *part* of a system (a

---

[4](Dedekind 1890, 99–100). The German text is as follows: "Welches sind die von einander unabhängigen Grundeigenschaften dieser Reihe N, d. h. diejenigen Eigenschaften, welche sich nicht aus einander ableiten lassen, aus denen aber alle anderen folgen? Und wie muß man diese Eigenschaften ihres spezifisch arithmetischen Charakters entkleiden, der Art, daß sie sich allgemeineren Begriffen und solchen Tätigkeiten des Verstandes unterordnen, *ohne* welche überhaupt kein Denken möglich ist, *mit* welchen aber auch die Grundlage gegeben ist für die Sicherheit und Vollständigkeit der Beweise, wie für die Bildung widerspruchsfreier Begriffserklärungen?"(Sinaceur 1974, 272).

subset, in modern terms) and operations on systems (such as, in modern terms, taking unions and intersections), Dedekind establishes some basic results. For example, he proves the transitivity of the subset relation (#7) and that if $A, B, C, \ldots \subseteq S$ then $\bigcup (A, B, C, \ldots) \subseteq S$ (#10).

Then he introduces a new and extremely important class of mathematical "entities": *mappings*. They are described in Erklärung #21 as follows: "By a mapping [[*Abbildung*]] φ of a system S we understand a law according to which, to every determinate element s of S, there belongs a determinate thing which is called the image [[*Bild*]] of s and which is denoted by $\varphi(s)$ ..." (*WZ*, 799). Dedekind uses the notation $\varphi(S)$ to denote the system of all images $\varphi(s)$ and later writes, "If φ is a ...mapping of a system S, and $\varphi(S)$ part of a system Z, then φ is said to be a mapping of S into Z ..." (*WZ*, 802). Of special interest for Dedekind is the case when the system Z is S, and then we have a "... mapping of the system S into *itself* ..." (*WZ*, 802).[5]

Before continuing, we should note that Dedekind distinguishes between function (Funktion) and mapping (Abbildung).[6] This is indicated in Remark #135, where Dedekind refers to applying " ... the definition (set forth in theorem (126)) of a mapping ψ of the number-sequence N (or of the *function* $\psi(n)$ determined by it) to the case where the system (there denoted by Ω) in which the image $\psi(N)$ is to be contained is the number series N itself".[7] Dedekind does not explicate the difference between functions and mappings. However, clarifying this distinction will bring out the special and novel character of mappings.

It is clear that *mapping* is a dramatic generalization of *function*, but the precise, distinctive character of function is not as transparent. (The explica-

---

[5]Here is the full German text concerning mappings: "Unter einer Abbildung φ eines Systems S wird ein Gesetz verstanden, nach welchem zu jedem bestimmten Element s von S ein bestimmtes Ding gehört, welches das Bild von s heißt und mit $\varphi(s)$ bezeichnet wird; wir sagen auch, daß $\varphi(s)$ dem Element s entspricht, daß $\varphi(s)$ durch die Abbildung φ aus s entsteht oder erzeugt wird, daß s durch die Abbildung φ in $\varphi(s)$ übergeht. Ist nun T irgendein Teil von S, so ist in der Abbildung φ zugleich eine bestimmte Abbildung von T enthalten, welche der Einfachheit wegen wohl mit demselben Zeichen φ bezeichnet werden darf und darin besteht, daß jedem Elemente t des Systems T dasselbe Bild $\varphi(t)$ entspricht, welches t als Element von S besitzt; zugleich soll das System, welches aus allen Bildern $\varphi(t)$ besteht, das Bild von T heißen und mit $\varphi(T)$ bezeichnet werden, wodurch auch die Bedeutung von $\varphi(S)$ erklärt ist. Als ein Beispiel einer Abbildung eines Systems ist schon die Belegung seiner Elemente mit bestimmten Zeichen oder Namen anzusehen."

[6]Ansten Klev offers an alternative interpretation of Dedekind's distinction between function and mapping. See (Klev 2014).

[7]The fuller German text is: "135. Erklärung. Es liegt nahe, die im Satze 126 dargestellte Definition einer *Abbildung* ψ der Zahlenreihe N oder der durch dieselbe bestimmten *Funktion* $\psi(n)$ auf den Fall anzuwenden, wo das dort mit Ω bezeichnete System, in welchem das Bild $\psi(N)$ enthalten sein soll, die Zahlenreihe N selbst ist ...". Jeremy Avigad brought this quotation to our attention and raised the question "What is the difference between mappings and functions for Dedekind?". See (Sieg and Schlimm 2014), where the development of the notion of mapping in Dedekind's work is analyzed, in particular, its structure-preserving variety.

tion we offer here is most closely associated with Dirichlet and Riemann.[8]) Historically we can say, that Dedekind uses the function concept extensively already in his *Habilitationsrede* (1854), whereas he introduces the notion of a mapping only in the manuscript (1872/78). Though *function* is used in the title of (1854), Dedekind discusses at first only *operations*, when treating the seven algebraic functions of elementary arithmetic (addition, subtraction, multiplication, division, exponentiation, taking roots and logarithms); that is also done in (*SZ*, §1, §6) and in Schröder's (1873). The term function is used in the later part of (1854, 434–438), when Dedekind discusses, for example, the trigonometric and elliptic functions. Obviously, that distinction between operations and functions is no longer made in *WZ* as a mapping from N to N "determines" now a function from N to N. So it seems that, in those later considerations, functions are mappings that satisfy two conditions: their domains and co-domains are identical; their domains are particular systems of *numbers*. Dedekind's introduction, in #36, of the notion of "a mapping of S in itself" thus satisfies the first of these conditions, but not the last as it does not restrict the domain to particular systems of numbers.

This understanding of Dedekind's distinction between mappings and functions is certainly in accord with his mathematical work (e.g., with Weber in their (1882)). It also reflects the general mathematical understanding of the times, sustained in the second half of the 19th and into the early years of the 20th century. Indeed, the part of mathematics that was called in German "Funktionentheorie" is not a theory of abstract functions (mappings), but rather complex analysis, i.e., the theory of *differentiable* functions of a complex variable. In the 1922 textbook *Funktionentheorie* by Hurwitz and Courant, the term *function* is introduced specifically for calculation procedures leading from a subset Σ of complex numbers to complex numbers as follows: "Now if to each value z may take, i.e. to each number in Σ, is assigned a complex number-value $w = u + iv$ according to a determinate law, we call w a *function* of z."[9]

---

[8] This evolution has received significant attention, but still deserves a deeper treatment. See (Avigad and Morris 2014) and (Sieg and Schlimm 2014), but also the rich secondary literature mentioned in those papers.

[9] (Hurwitz and Courant 1922, 17–18). The original German is as follows: "Wenn nun jedem Wert, den z annehmen darf, also jeder Zahl von Σ, nach einem bestimmten Gesetz ein komplexer Zahlenwert $w = u + iv$ zugeordnet ist, so nennen wir w eine *Funktion* von z." Notice that "w ist eine Funktion of z" might best be translated as "w is functionally dependent on z". In his textbook *Funktionentheorie I* (1930), Knopp introduces the concept of "the most general" function in a very similar way, with z and w both ranging over complex values. He adds the notational device $w = f(z)$ and remarks that f stands for "the somehow given calculation procedure [Rechenvorschrift]" (Knopp 1930, paragraph 5); the latter allows the determination of w for all z in its domain of variability. We should note, however, that despite Knopp's use of the term "calculation procedure", he does not require the function to be given by an explicit expression. In particular, for domain of variability $\mathfrak{M}$, he claims, "All that is required is that the value w of the function be made to correspond, on the basis of the definition, to each z of $\mathfrak{M}$ in a completely unambiguous manner." (Knopp 1930, paragraph 5)

The general notion of mapping was absolutely central for Dedekind's foundational thinking; that is obvious from the announcement of the third edition of (Dirichlet 1879), but also from a footnote in that very work. Indeed, Dedekind refers back to these considerations in a note to section 161 of the 1894 fourth edition of Dirichlet's lectures:

> It is stated already in the third edition of the present work (1879, footnote on p. 470) that the entire science of numbers is also based on this intellectual ability to compare a thing a with a thing a', or to let a correspond to a', without which no thinking at all is possible. The development of this thought has meanwhile been published in my essay *Was sind und was sollen die Zahlen?* (Braunschweig, 1888).[10]

Remark #135, which we quoted earlier, is the only place in *WZ* where Dedekind uses the term *function*, when he considers recursively defined mappings from N to N and asserts that they determine functions. When discussing mappings he always refers explicitly to domain and co-domain (or, if not the co-domain, at least the image of the domain) which, moreover, are allowed to be different.[11] The metamathematical work in *WZ* crucially depends on this general concept of mapping: without it he would be unable to form the notion of *similarity* of different systems, which, as we will see, plays an essential role in the essay. A mapping is called *similar*, or in modern terms *injective*, when it maps distinct elements of its domain to distinct elements of its co-domain. This allows Dedekind to introduce the notion of *similarity between systems* in #32: two systems R and S are similar when there is a similar mapping φ from S to R such that $\varphi(S) = R$, i.e., two systems are similar when there is a bijection from one to the other.

Within this set-up, Dedekind introduces two important notions in #37 and #44, namely, that of a *chain* and that of a *chain of a system*. If S is a system, φ a mapping from S to S and K a part of S, then K is a chain if $\varphi(K) \subseteq K$. If S is a system, φ a mapping from S to S and A a part of S, then the commonality (in modern terminology, the intersection) of all of those chains which have A as a part is called the chain of the system A and denoted by A_0. A_0 enjoys a number of properties which, as Dedekind remarks in #48, are

---

[10] Here is the German text: "Schon in der dritten Auflage diese Werkes (1879, Anmerkung auf S. 470) ist ausgesprochen, dass auf dieser Fähigkeit des Geistes, ein Ding a mit einem Ding a' zu vergleichen, oder a auf a' zu beziehen, oder dem a ein a' entsprechen zu lassen, ohne welche überhaupt kein Denken möglich ist, auch die gesammte Wissenschaft der Zahlen beruht. Die Durchführung dieses Gedankens ist seitdem veröffentlicht in meiner Schrift *Was sind und was sollen die Zahlen?* (Braunschweig 1888)."(Dirichlet and Dedekind, 1894, footnote on page 456).

[11] Already in #21, when introducing the notion of a mapping φ of a system S, Dedekind describes the image, $\phi(S)$, of S under the mapping. In #25 he is careful to pay attention to the images when composing two mappings. Most importantly, similar observations can be made for §3, where Dedekind defines the similarity of two systems, and for §9, where recursively defined mappings from N to an arbitrary system Ω are introduced.

sufficient to characterize it completely as *the* "smallest" chain containing A. The characteristic properties that allow the direct proof of this claim are the following facts, formulated as #45, #46, and #47, namely that: (i) $A \subseteq A_0$, (ii) $\varphi(A_0) \subseteq A_0$, and (iii) If $A \subseteq K$ and K is a chain, then $A_0 \subseteq K$. In #64, Dedekind calls a system *infinite* just in case it is similar to a proper part of itself. Then, completing the presentation of the logical framework, Dedekind attempts to establish the existence of an infinite system by a purely logical proof.[12] That sets the stage for the final analytic steps and the subsequent synthetic development of number theory; the former are described in section A.2, whereas the latter is detailed in A.3.[13]

A.2. Analysis

Dedekind's manuscript (1872/78) has the subtitle *An attempt at analyzing the number concept from the naïve standpoint* (*Versuch einer Analyse des Zahlbegriffs vom naiven Standpuncte aus*). By emphasizing that the attempt is made from the naïve standpoint, Dedekind indicates that the data of ordinary mathematical experience are to be analyzed without philosophical preconceptions. That perspective is also taken many years later in his letter to Keferstein, when Dedekind asserts that the synthetic development is "based on a prior analysis of the sequence of natural numbers as it presents itself, in experience, so to speak, for our consideration". The analysis itself proceeds in steps Dedekind details for Keferstein, as facts (1) through (6). So the analysis from the naïve standpoint allows Dedekind to define the central concept of his investigations, that of a simply infinite system. We quote this definition, in German *Erklärung*, in full:

> 71. Definition [Erklärung][14]. A system N is said to be *simply infinite* when there exists a similar mapping φ of N into itself such that

---

[12]Though his proof is recognized as highly problematic, given that it appeals to " ... the totality S of all things, which can be objects of my thought ... " (*WZ*, 806), it is nevertheless useful to emphasize why Dedekind tries to obtain such a proof: it is to guarantee the internal coherence of the concept of an *infinite system* and, as we will see later, of a *simply infinite system*.

[13]The separation of "analysis" from "synthesis" is a classical methodological approach that goes back to Aristotle. It is described with great sensitivity and acumen in the Introduction to the third book of Lotze's *Logik*, in particular, sections 297–299. This opposition is also formulated in §117 of Kant's *Logik*; there one finds informatively, "Die analytische Methode heißt auch sonst die Methode des *Erfindens*." Beaney's entry "Analysis" for the Stanford Encyclopedia of Philosophy gives an excellent and broad survey, (Beaney 2012).

[14]Dedekind uses the word *Erklärung* frequently, in particular, in *WZ*. It is prominently used to introduce the notion of a *simply infinite system*. (Hilbert also makes regular use of the term in his *Grundlagen der Geometrie* when giving his axiom system for geometry or, rather, when introducing the notion of *Euclidean space*.) It is an open question for us, whether Dedekind's sense of *Erklärung* and *Definition* reflects to a certain extent the distinctions Kant made in his *Kritik der reinen Vernunft* (*Erstes Hauptstück* of the *Transzendentale Methodenlehre*, A709–A739, especially A730). For Kant, the *Erklärung* of a concept splits into *Exposition*, *Explikation*, *Deklaration*, and *Definition*.

N appears as the chain (44) of an element not contained in $\varphi(N)$. We call this element, which we shall denote in what follows by the symbol 1, *the base-element* of N, and say the simply infinite system N is *ordered* [[*geordnet*]] by this mapping φ. If we retain the earlier convenient symbols for images and chains (§4) then the essence of a simply infinite system N consists in the existence of a mapping φ of N and an element 1 which satisfy the following conditions α, β, γ, δ:

α. $N' \prec N$.[15]

β. $N = 1_0$.[16]

γ. The element 1 is not contained in N'.

δ. The mapping φ is similar. (*WZ*, 808)

We emphasize most strongly that this is *not* the definition of a particular structure in the sense of modern model theory: the base element and successor operation are quantified out, existentially. I.e., a system of objects N is called *simply infinite* just in case there is an object 1 in N and a function φ from N to N that satisfy the characteristic conditions (α)–(δ).[17] Dedekind thus provides a higher-order, *structural definition* under which particular simply infinite systems fall.

The systems that fall under the concept are viewed as having their own "canonical language" in terms of which the characteristic conditions are expressed. This idea of a "canonical language" is important to Dedekind's project. The role it plays will be discussed in detail in section B.2, but for now we simply remark that Dedekind was careful to distinguish between language and that which it is used to speak about. Indeed, *WZ* begins: "In order to be able conveniently to speak of things, we designate them by symbols, e.g., by letters, and we venture to speak briefly of the thing a or of a simply, when we mean the thing denoted by a and not at all the letter a itself" (*WZ*, 796–797).[18]

Note that Hilbert's early conception of axiomatics coincides with Dedekind's methodological approach.[19] The axiomatization of geometry in (Hilbert 1899) and of analysis, the theory of real numbers, in (Hilbert 1900b) *are* structural

---

[15] N' denotes $\varphi(N)$ under the mapping φ, and '$\prec$' is (close to) Dedekind's symbol for the subsystem relationship, which is today symbolized by '$\subseteq$'.
[16] 1_0 is short for $\{1\}_0$, i.e., the chain of the system $\{1\}$.
[17] "Characteristic conditions" (charakteristische Bedingungen) is Dedekind's terminology for the conditions (α)–(δ) at the end of #72.
[18] Aki Kanamori pointed out the relevance of this remark.
[19] This perspective is now largely shared by José Ferreirós; see (Ferreirós 2007, Epilogue, 458–463). Ansten Klev has subjected the position to a detailed critique, presented in (Klev 2011). Klev maintains that Dedekind and Hilbert, around 1900, "held divergent views on the structure of a deductive science". The two main points of the divergence are concisely formulated in his abstract: "Firstly, ... Dedekind sees the beginning of a science in concepts, whereas Hilbert sees such beginnings in axioms." However, this conflicts with Hilbert's own

definitions in Dedekind's sense, introduced also under the heading *Erklärung*. The only difference from Dedekind's structural definition lies in the fact that Hilbert calls the characteristic conditions *axioms*. Indeed, Frege in his letter of 27 December 1899 criticized Hilbert's use of the word "axioms" in the *Erklärung* that introduces the geometric principles. In his immediate reply of 29 December 1899 Hilbert writes:

> If you want to call my axioms rather characteristic conditions of the concepts that are given and hence are existent in the 'definitions' [Erklärungen], I would not object to that at all, except perhaps that that conflicts with the custom of mathematicians and physicists; of course I must also be free in giving characteristic conditions.[20]

After some additional remarks, he comes back to this issue, which he considers as the "main issue" (*Hauptsache*), and writes: "... the renaming 'characteristic conditions' instead of 'axioms' etc. is surely a formality and moreover a matter of taste—but in any case it is easily achieved."[21]

In a more principled way, Hilbert remarks later on: "Well, it is surely obvious that every theory is only a framework of concepts or a schema of concepts together with their necessary relations to each other, and the basic elements can be thought in arbitrary ways."[22] This reflects Dedekind's methodological per-

---

words, for example, in the correspondence with Frege that we analyze in our paper below. "Secondly, ... for Dedekind, the primitive terms of a science are substantive terms whose sense is to be conveyed by elucidation, ... " As we will see below, especially in section B.2, this is incorrect for Dedekind.

At the root of the difference between Klev's and our understanding is the fact that he compares Dedekind's logical framework for arithmetic with Hilbert's "axiomatic" presentation of Euclidean geometry, not with Hilbert's assumed set theoretic framework for geometry. When examining the "structure of a deductive science" in Dedekind's and Hilbert's work, the *Erklärung* #71 of the concept of a simply infinite system should be set alongside the *Erklärung* of the concept of a Euclidean space that opens Hilbert's *Grundlagen der Geometrie*.

However, Klev's interpretation and our own agree on the following point: Dedekind is concerned with the development of "higher level" concepts. Indeed, as we have already noted, the concept of "simply infinite system" is such a higher order one: individual systems fall under it.

[20] The German text is: "Wollen Sie meine Axiome lieber Merkmale der in den "Erklärungen" gesetzten und dadurch vorhandenen Begriffe nennen, so würde ich dagegen garnichts einzuwenden haben, ausser etwa, dass das der Gewohnheit der Mathematiker und Physiker widerspricht—freilich muss ich auch mit dem Setzen der Merkmale frei schalten können." Quoted in (Frege 1980, 12).

[21] The German text is: "Die Umnennung "Merkmale" statt "Axiome" etc. ist doch nur eine Äusserlichkeit und überdies Geschmackssache—ist aber jedenfalls leicht zu bewerkstelligen." Quoted in (Frege 1980, 12).

[22] Here is the German text: "Ja, es ist doch selbstverständlich eine jede Theorie nur ein Fachwerk von Begriffen oder ein Schema von Begriffen nebst ihren notwendigen Beziehungen zu einander und die Grundelemente können in beliebiger Weise gedacht werden." Quoted in (Frege 1980, 13). In perhaps the clearest possible way this idea is expressed in (Hilbert 1902, 566). He defines: "*Ein System von Dingen*, welche wir *Punkte und Geraden* nennen und denen wir gewisse gegenseitige Beziehungen auferlegen, wollen wir eine *Geometrie* nennen. *Mit Hilfe der complexen Zahlen* ... können wir nun leicht eine *Geometrie construieren*, in welcher sämmtliche Axiome *I-IV* gelten, *V* aber nicht."

spective perfectly and is also very similar to the modern, algebraic approach to axiomatics which has, of course, its roots in the work of Dedekind and Hilbert. As an example, consider the introduction of the abstract notion of a group. We call any set G equipped with a binary operation ⊛ satisfying certain conditions (axioms) a group; but when considering particular groups, we will use the names associated with the mathematical objects involved, instead of 'G' and '⊛'. Before coming back to Dedekind's essay, we mention that Zermelo's 1908 axiomatization of set theory follows the Dedekind-Hilbert approach.[23]

Immediately after having given the definition (Erklärung) of a simply infinite system, Dedekind demonstrates Theorem #72: *every* infinite system contains a simply infinite system as a part. Thus, if the proof of Theorem #66, which claimed that there is an infinite system, had been successful, it would guarantee, together with Theorem #72, the existence of a simply infinite system. What significance does this have? Dedekind answers this question in his letter to Keferstein as follows:

> After the essential character of the simply infinite system, *whose abstract type is the number sequence N*, had been recognized in my analysis (articles 71 and 73), the question arose: does such a system exist at all in the realm of our thoughts? Without a logical proof of existence it would always remain doubtful whether the notion of such a system might not perhaps contain internal contradictions. Hence the need for such proofs (articles 66 and 72 of my essay)[24].

Thus, the aim of Theorems #66 and #72 is to establish the consistency, respectively, of the concept of an *infinite* and a *simply infinite* system. That is to be achieved by "a logical proof of existence", i.e., by specifying within "logic" a particular example of a system falling under the concept of a simply infinite system. We come back a little later to the question, what means are taken for granted as being rooted in "logic"; here we just point out that, with this much completed, Dedekind can define natural numbers in *WZ* as follows:

> 73. Definition [Erklärung]. If in the consideration of a simply infinite system N set in order by a mapping φ we entirely neglect the special character of the elements, simply retaining their distinguishability and taking into account only the relations to one another in which they are placed by the ordering mapping φ, then these elements are called *natural numbers* or *ordinal numbers* or simply *numbers*, and the base-element 1 is called the *base-number*

---

[23]Cf. (Zermelo 1908, 201). This is discussed in greater detail in C.3.
[24](Dedekind 1890, 101). Here is the German text: "Nachdem in meiner Analyse der wesentliche Charakter des einfach unendlichen Systems, dessen abstrakter Typus die Zahlenreihe N ist, erkannt war (71, 73), fragte es sich: *existiert* überhaupt ein solches System in unserer Gedankenwelt? Ohne den logischen Existenzbeweis würde es immer zweifelhaft bleiben, ob nicht der Begriff eines solchen Systems vielleicht innere Widersprüche enthält. Daher die Notwendigkeit solcher Beweise (66, 72 meiner Schrift)." (Sinaceur 1974, 275)

of the *number series* N. With reference to this liberation of the elements from every other content (abstraction) we are justified in calling the numbers a free creation of the human mind.[25]

There is an important point to note here that corresponds to our claim that, in the final version of *WZ*, Dedekind was primarily concerned with the concept of natural number, rather than identifying natural numbers as particular logical objects. Indeed, Dedekind refers to the *elements of the original simply infinite system* as the natural numbers; for observe that the natural numbers are "these elements", referring back to the elements of the simply infinite system he begins with. Thus, for Dedekind, natural numbers can be taken to be the elements of *any* simply infinite system when viewed from a more abstract perspective, neglecting the "special character of the elements, simply retaining their distinguishability and taking into account only the relations to one another in which they are placed by the order-setting mapping φ". That fits in well with his letter to Keferstein, where Dedekind refers to the natural number sequence as the "abstract type" of the notion of a simply infinite system.[26] That is to say, the particular natural number sequence is paradigmatic of all simply infinite systems: it exemplifies exactly and completely the structural properties that each has, as Dedekind comes to prove in #131–134 and which we will consider in detail below. It also fits well with Dedekind's remark in his letter to Weber, written in January of 1888, in which he speaks of "my" ordinal numbers "as the abstract elements of the simply infinite system". (The concepts of abstraction that are used here are discussed in section C.2.)

For now, we draw attention to a number of points that will be important for our later considerations. First of all note that the creation of the concepts "simply infinite system" and "natural numbers" does not end Dedekind's analysis. What is still needed is implicit in what Dedekind describes as the subject-matter of arithmetic in #73: "[T]he relations or laws which are derived

---

[25] The German text is as follows: "73. Erklärung. Wenn man bei der Betrachtung eines einfach unendlichen, durch eine Abbildung φ geordneten Systems N von der besonderen Beschaffenheit der Elemente gänzlich absieht, lediglich ihre Unterscheidbarkeit festhält und nur die Beziehungen auffaßt, in die sie durch die ordnende Abbildung φ zueinander gesetzt sind, so heißen diese Elemente *natürliche Zahlen* oder *Ordinalzahlen* oder auch schlechthin *Zahlen*, und das Grundelement 1 heißt die *Grundzahl der Zahlenreihe N*. In Rücksicht auf diese Befreiung der Elemente von jedem anderen Inhalt (Abstraktion) kann man die Zahlen mit Recht eine freie Schöpfung des menschlichen Geistes nennen."

[26] This notion of "abstract type" and, more generally, abstraction is discussed more fully in section C.2. Here we just note that *type* (in German, Typ or Typus) has a number of different meanings. Two seem to be particularly relevant for our investigations. These are, following the Shorter OED: (1) "A class of people or things distinguished by common essential characteristics", and (2) "A person or thing showing the characteristic qualities of a class; a representative specimen; specifically, a person or thing exemplifying the ideal characteristics of a class, the perfect specimen of something". Type in sense (2) is used in biology and refers to a selected individual that serves as the foundation for the scientific description of a taxon, for example, of an animal species. It is in this second sense that "(abstract) type" is to be understood here as well. That is in harmony with the use of "Typus" in philosophy, for example, in Kant's *Kritik der praktischen Vernunft*, A119–A127, and Lotze's *Logik*, #137.

entirely from the conditions α, β, γ, δ in (71) and therefore are always the same in all ordered simply infinite systems, whatever names may happen to be given to the individual elements (compare 134), form the first object of the *science of numbers* or *arithmetic*." Note in particular his claim that these laws are *the same* for each simply infinite system, and his reference to Remark #134. The significant foundational issues will be explained in B.2, where we discuss this very remark.

As is clear from the letter to Keferstein, two general and yet highly "practical" questions still need an answer before we can conclude the analysis: (i) the justification of the method of proof by induction, and (ii) the consistent definition of recursive operations for all natural numbers. Attending to (i), Dedekind proves in #80, "... that the form of argument known as complete induction (or the inference from n to $n+1$) is really conclusive ..." (*WZ*, 791). This result is then used to prove Dedekind's recursion theorem: if Ω is a system, θ a mapping of that system in itself (not necessarily similar), and ω a determinate element of Ω, then there exists a unique mapping ψ from N to Ω such that (a) $\psi(1) = \omega$ and (b) $\psi(n') = \theta\psi(n)$, where n' is Dedekind's notation for $\varphi(n)$, the successor of n. This allows the definition of the standard arithmetic operations like addition, multiplication, and exponentiation when $\Omega = N$ and thereby addresses (ii).

This possibility of explicitly defining recursive functions for the number series N is crucial for the development of arithmetic, but the *general* theorem allows Dedekind to prove an important metamathematical result: all simply infinite systems are similar (indeed, isomorphic). In Part B, we will see that this result plays a crucial role in Dedekind's *complete justification* of his concept of numbers. With the consistent definition of recursive functions (from N to N) given by #126 the analysis ends, as Dedekind very clearly emphasizes in his letter to Keferstein. It is followed by the "synthetic development" (*synthetischer Aufbau*) of arithmetic. As to it, Dedekind admits to Keferstein:

> And yet, it [the synthesis] has taken me a great deal of effort! ... The reader of my essay does not have an easy task either; in order to work completely through everything, not only sound common sense is required, but also a strong determination.[27]

We now come to consider the question: How is the *synthesis* to be accomplished?

A.3. Synthesis

The *Preface* to the first edition of *WZ* provides an answer to the question we posed at the end of section A.2. The very first sentence articulates what

---

[27] Here is the German text: "Er [der synthetische Aufbau] hat mir doch noch Mühe gemacht! ... Auch der Leser meiner Schrift hat es wahrlich nicht leicht; außer dem gesunden Menschenverstande gehört auch noch ein sehr starker Wille dazu, um Alles vollständig durchzuarbeiten." (Sinaceur 1974, 276)

Dedekind considered as the motto for his essay, namely, "In science nothing capable of proof ought to be believed without proof." Lamenting the current state of affairs, he continues: "Though this demand seems reasonable, I cannot regard it as having been met even in the most recent methods of laying the foundations of the simplest science; viz., that part of logic which deals with the theory of numbers" (*WZ*, 790). Dedekind's work is intended to remedy this. He continues in the next paragraph:

> But I know that, in the shadowy forms which I bring before him, many a reader will scarcely recognize his numbers which all his life long have accompanied him as faithful and familiar friends; he will be frightened by the long series of simple inferences corresponding to our step-by-step understanding [[Treppenverstand]], by the matter-of-fact dissection of the chains of reasoning on which the laws of numbers depend ... (*WZ*, 791).

To gain additional insights into this key element of the synthetic development, we explore the central issue: what does Dedekind consider to be a *rigorous proof*? Or, more concretely, how are the "facts" of ordinary arithmetic inferred from the conditions that characterize simply infinite systems? Let us first state very clearly that the facts are not obtained by semantic means. When Dedekind claims in #73 that what we study in arithmetic is what is derived entirely from the conditions α, β, γ, δ, this is to be understood in a quasi-formal manner (noting, as Frege very pointedly did, that Dedekind does not specify rules according to which such derivations are made). This understanding is already suggested by the above passage from the preface in which Dedekind speaks of the "long series of simple inferences corresponding to our step-by-step understanding". Further evidence can be found in numerous places in *WZ* and other of Dedekind's works.[28]

As a first and general observation, we note that Dedekind uses words such as *Ableitung* and *ableiten* and their cognates in *WZ*. These words have the connotation of *being derived* in the formal sense, i.e., as being given by a syntactic proof, rather than as a semantic consequence, for which words such as *Folgerung* and *folgen* are appropriate. More specific evidence can be found by considering Dedekind's descriptions of his work and his opinions about mathematics. Already in *SZ*, Dedekind emphasizes his interest in identifying certain

---

[28]There is a general philosophical position underlying the insistence on rigorous proof. It is expressed in Dedekind's (1854, 428–429) when he compares a man with "unbounded understanding" with us who are subject "to all the imperfections of [our] mental powers". He claims then that "there would essentially be no more science" for such a man, i.e., "a man for whom the final conclusions, which we obtain through a long chain of inferences, would be immediately evident truths." Dedekind adds, "and this would be so even if he stood in exactly the same relation to the objects of science as we do." Cf. Lotze's remark in *Logik*, section IX of the Introduction to the first book; there Lotze contrasts a mind "that stands in the center of the world and of everything real" with the human mind "that does not stand in this center of things", but rather has its "modest place somewhere in the last ramifications of reality".

Dedekind's Structuralism: Creating Concepts and Deriving Theorems 265

properties, "... that can serve as the basis for *proper*[29] *deductions*" (*SZ*, 771). Moreover, in his letter to Lipschitz, Dedekind explains that, if a mathematical theory has been constructed correctly, then replacing the "terms of art" by other, newly invented terms should not result in the collapse of the theory, see (Sieg and Schlimm 2005, 155). In other words, the proofs of arithmetical theorems should not appeal to intuitions connected with the "terms of art" which fail to constitute part of their definitions, since such "proofs" will crumble after the replacement. Finally, in his 1890 letter to Keferstein, Dedekind rejects one of Keferstein's suggested improvements to his work by asserting: "The danger would immediately arise that from such an intuition ... we perhaps take as obvious *propositions that must rather be derived quite abstractly from the logical definition of N.*"[30]

All of these remarks fit well with a quasi-formal conception of proof. Indeed, to see that this idea has been thoroughly realized, one needs only to follow the development of "elementary set theory" in §1–5 or of arithmetic proper in §11–13 of *WZ*.[31] A quasi-formal conception of proof also underlies Dedekind's argument showing that the laws of arithmetic are *the same* in all simply infinite systems; see subsection B.2. While we wish to emphasize that Dedekind's notion of proof is quasi-formal, we do not wish to imply that he was unconcerned with semantic issues. Indeed, he explores them in a variety of places, including his considerations about consistency of concepts (which, as we have seen, for Dedekind involves exhibiting an appropriate model), categoricity, and independence of characteristic conditions.

Before formulating the central questions for Part B, we want to re-emphasize three points. (1) In *WZ*, Dedekind is concerned with creating an abstract *concept* of number and *not* with identifying (systems of) objects as *the* numbers. This concern with concepts as opposed to objects is reflected in various places in Dedekind's writings. In earlier manuscripts of *WZ* Dedekind does create abstract mathematical objects. The move from creating objects to creating concepts is the fundamental part of the *radical shift* in Dedekind's position, which we explore in later sections, in particular, section B.3. (2) Dedekind does not have the modern distinction between syntax and semantics, but his notion of inference with respect to arithmetic theorems is nonetheless much more *syntactic* than semantic. It is only *quasi*-formal, however, as the inferential principles that can be appealed to are not explicitly enunciated. This pertains also to Hilbert, whose proofs proceed, around the turn from the 19th to the

---

[29]Instead of "proper", Ewald uses "valid". The German adjective is *wirklich* and the original German phrase is: "*als Basis für wirkliche Deduktionen*". (Dedekind 1932, 322).

[30]The fuller German text is: "Dem würde ich mich mit grösster Bestimmtheit widersetzen, weil sofort die Gefahr entsteht, aus einer solchen Anschauung vielleicht unbewusst auch Sätze, als selbstverständlich zu entnehmen, die vielmehr ganz abstrakt aus der logischen Definition von *N* abgeleitet werden müssen."(Sinaceur 1974, 277)

[31]Whereas in the case of arithmetic the starting points of proofs are the characteristic conditions of the concept of a simply infinite system, in the case of "set theory" Dedekind uses the definition of operations, like union and intersection for systems or composition and inversion for (similar) mappings, as beginnings of proofs.

20th century, by "finitely many logical steps"; the latter are, as in Dedekind's case, not specified. (3) Much of Dedekind's work in *WZ* is metamathematical. Indeed, it is only in §§11–13 that Dedekind comes to consider the arithmetic operations of addition, multiplication and exponentiation and rigorously proves their basic properties. The crucial and quite novel metamathematical work was missing from the earlier manuscripts of *WZ*, but it is precisely this work that allows Dedekind, as he puts it, to "completely justify" the concept of numbers. We explore next Dedekind's line of argumentation.

B. The Concept of Numbers: *Completely Justified*

In the previous part we have drawn attention to the fact that Dedekind is primarily concerned with the creation of the *concept* of numbers. This understanding is further supported by his letter to Keferstein in which Dedekind responds to Keferstein's objection that he has not given an adequate definition of the number 1. He writes:

> I define the 〚ordinal〛 number 1 as the basic element 〚Grundelement〛 of the number sequence without any ambiguity in articles 71 and 73, and, just as unambiguously, I arrive at the 〚cardinal〛 number 1 in the theorem of article 164 as a consequence of the general definition in article 161. *Nothing further may be added to this at all if the matter is not to be muddled.* (Dedekind 1890, 102, emphasis added)

For Dedekind the number 1 is fully specified as the base element of the number sequence N. How should it be possible to identify it to any further degree, if—as we described already in A.2—the elements of *any* simply infinite system can take on the role of the natural numbers? In subsection B.1 we will present a *striking fact* established by Dedekind as Theorem #133: any system Ω that is similar to a simply infinite one is itself simply infinite. The order setting mapping of Dedekind's simply infinite system N (with 1 and φ), for example, can be used (together with the bijection ψ from N to Ω) to fix the base element ω in Ω and an order setting mapping θ on Ω.

In this part we examine Dedekind's metamathematical considerations that *justify* his approach to the natural numbers, as the elements of *any* simply infinite system. Thus, Dedekind must ensure that arithmetic theorems do not depend upon which particular simply infinite system is being considered. He does this by proving a central *representation theorem*, which we describe in B.1; as a consequence one obtains that the concept of a simply infinite system is *categorical*.[32] That is to say, we can map bijectively the elements of one

---

[32] The emergence of the notion of categoricity is described in (Awodey and Reck 2002a;b). We are applying the term "categorical" here to an abstract notion or structural definition. Awodey and Reck provide a modern perspective on (or rather, framing of) Dedekind's and Hilbert's work concerned with the foundations of number theory and geometry, respectively. Their approach yields insights, but does not highlight the novelty and precision of Dedekind's work and its influence on Hilbert.

Dedekind's Structuralism: Creating Concepts and Deriving Theorems   267

simply infinite system to the elements of any other, whilst respecting the order setting mappings. We will examine in B.2 the consequences of categoricity, including crucially the reason for Dedekind's claim made in #73 that the laws of arithmetic are the same in all simply infinite systems. B.3, finally, is concerned with a very important and quite radical shift away from abstract objects to abstract concepts. (That shift is visible in the penultimate manuscript of *WZ*.)

B.1. A Representation Theorem

Before analyzing Dedekind's metamathematical work in detail, we pause to make two additional observations. The first concerns an example of what falls under the concept of *mapping*. Dedekind considers *naming* the elements of a particular system to constitute a mapping of that system; he writes: "As an example of a mapping of a system we may regard the mere assignment of determinate symbols or names to its elements" (*WZ*, 800). More explicitly, given any system S the following is a mapping of the system: $\chi(s) = \mathbf{s}$, where $\mathbf{s}$ is a symbol that is used to denote the element s of S. The second observation concerns "the number system N" that is being used in Dedekind's investigations above and below, in particular in the representation theorem. *Any* simply infinite system would do, but in Dedekind's case one may think of the particular system he obtained in #72 from the infinite system whose existence he had "proved" in #66. The reader can replace it by a personally favored modern one, say the von Neumann or Zermelo ordinals, where the distinguished element is the empty set and the successor operation $\varphi(x)$ is given by $x \cup \{x\}$ or $\{x\}$, respectively.

As already indicated, Dedekind's recursion theorem (Theorem #126, stated in the penultimate paragraph of subsection A.2) has important consequences for our purposes. Indeed, it allows us to establish directly the following central result:

> 132. Theorem. All simply infinite systems are similar to the number series N and consequently by (33)[33] also to one another.

Given any simply infinite system Ω (with θ and ω), Dedekind appeals for the proof of #132 to the recursion theorem and obtains a similar mapping ψ from N to Ω satisfying the equations $\psi(1) = \omega$ and $\psi(\varphi(n)) = \theta(\psi(n))$. He then demonstrates that ψ is bijective and, as it maps 1 to ω and commutes with the mappings φ and θ, it is an isomorphism, indeed, a *canonical* one. This establishes the crucial first part of the theorem, which is a *representation theorem*: each simply infinite system is isomorphic to the number series N. To demonstrate that any two simply infinite systems Ω (with θ and ω) and Ω' (with θ' and ω') are isomorphic, Dedekind observes that we have isomorphisms ψ and ψ' from

---

[33]Theorem #33 states that if R and S are similar systems, then any system similar to R will be similar to S.

N to Ω and Ω', respectively. But then the composition of $\overline{\psi}$, the inverse of ψ, with ψ' is an isomorphism between Ω and Ω'. This is depicted in Figure 1.

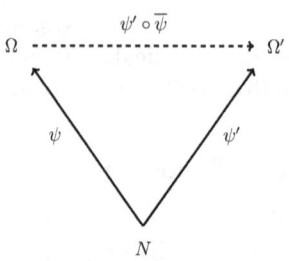

Figure 1

The second part of Theorem #132 consequently formulates the categoricity of the concept of a simply infinite system. It is followed by a striking observation, Theorem #133:

> 133. Theorem. Every system that is similar to a simply infinite system, and therefore by (132), (33) to the number series N, is itself simply infinite.

For any given system Ω whose elements can be bijectively mapped to a simply infinite system Ω', we can obtain a mapping that sets the elements of Ω in order as a simply infinite system. The next two diagrams reflect the idea that is involved in the proof.

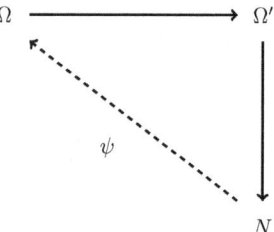

Figure 2

Figure 2 depicts Ω as being similar to Ω'. On account of #132, Ω' is similar to N and thus, as the similarity relation is transitive and symmetric, the system N is similar to the system Ω through the mapping ψ. Figure 3 shows how this fact allows us to construct a mapping that sets the elements of Ω in order.
More specifically, with ψ as the similar mapping from N to Ω, we can put $\psi(1) = \omega$ and if $\overline{\psi} : \Omega \to N$ denotes, as above, the inverse of ψ, then the mapping $\theta(\nu) = \psi(\varphi(\overline{\psi}(\nu)))$ sets Ω in order, where φ is the successor function on N.

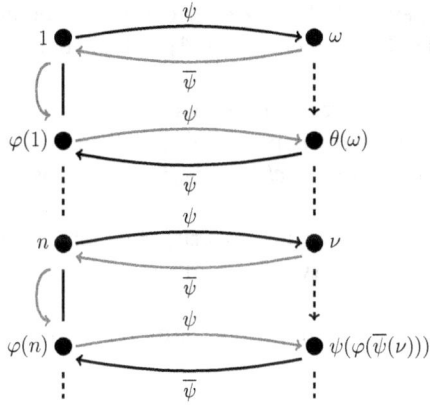

Figure 3

This is then *the striking fact*: the elements of *any* system that can be mapped bijectively to (any simply infinite one and thus to) the natural numbers can be taken to *be* natural numbers. We only have to fix appropriately a base element and an order setting mapping. For example, the system E of even natural numbers is in bijective correspondence with the natural number series. Taking 2 to be the base element and the addition of 2 as the "successor operation", E is a simply infinite system and E's elements can be taken to be the natural numbers. More generally, Dedekind points out at the beginning of Remark #134 that the simply infinite systems form a "class" (*Klasse*) as defined in #32. Considering any simply infinite system T, the systems similar to T are according to Theorems #132 and #133 exactly all the simply infinite systems. Viewing T as the representative of the class, Dedekind remarks "the class is not changed by taking as representative any other system belonging to it" (*WZ*, 802). So it is irrelevant from a "structuralist perspective" which simply infinite system is chosen as a representative.

Before turning in subsection B.2 to Dedekind's unusual and distinctive justification in the next subsection, we recall Dedekind's remarks in *SZ*, §2, and in a letter to Lipschitz that adumbrate the considerations here. In §1 of *SZ*, Dedekind discusses the order relation between the rational numbers whose system R falls under the concept of a field (*Zahlkörper*); he formulates the following laws (*Gesetze*):

(I) If $a > b$ and $b > c$, then $a > c$. We say that b lies between a and c.

(II) If $a \neq c$, then there are infinitely many different numbers between a and c.

(III) If a is a definite number then all numbers of R fall into two classes A_1 and A_2. A_1 contains all numbers a_1 such that $a_1 < a$ and A_2 contains all numbers a_2 such that $a_2 > a$. a can either be added to A_1 or it can be added to A_2. Every element of A_1 is less than every element of A_2.

These laws, Dedekind asserts, remind us of the mutual positional relations (*Lagenbeziehungen*) between the points of the straight line L. In regard to this difference in position (*Lagenverschiedenheit*) analogous laws hold:

(I′) If p is to the right of q and q is to the right of r, then p is to the right of r. We say that q lies between p and r.

(II′) If $p \neq r$, then there are infinitely many different points that lie between p and r.

(III′) If p is a definite point on L then all points that lie in L fall into two classes P_1 and P_2. P_1 contains all points p_1 which lie to the left of p and P_2 contains all numbers p_2 which lie to the right of p. p can either be added to P_1 or to P_2. Every element of P_1 lies to the left of every element of P_2.

The analogy of the laws can be turned into a "real connection" (*wirklicher Zusammenhang*) via a correspondence between the rational numbers a and the associated points p_a on the line that preserves the relationship, namely, if $a < b$ then p_b is to the right of p_a. (We note that this *informal* correspondence was not "mathematically" expressible at the time, as Dedekind introduces general mappings only later.) In sections 4 and 5, Dedekind defines an order relation $\prec$ between cuts and shows that it satisfies the analogous laws. In addition, the system of cuts is continuous or complete. For our considerations here, we just observe that for rational numbers a and b, $a < b$ implies $c_a \prec c_b$, where c_q is the cut associated with the rational number q.

Dedekind emphasizes in section 2 that the fundamental laws for L "are in complete accord" (*entsprechen vollständig*) with those for the rational numbers. The letter to Lipschitz on which we reported in A.3 can be read as allowing the extension of the "complete accord" from the basic laws to the theorems that can be inferred, stepwise, from them. In *SZ*, when discussing in section 4 the corresponding order relations for the rational numbers, the points of the geometric line and the real numbers, Dedekind used the phrase *transfer from one area to another* (*Übertragung aus einem Gebiet in ein anderes*). This phrase, as we will see, is being used by Dedekind in *WZ* to describe the same phenomenon between different simply infinite systems. So we have uncovered two pillars on which the accord rests, namely, the structure-preserving correspondence between systems of mathematical objects satisfying basic laws (or falling under a structural definition) and the stepwise proof of theorems from these basic laws (or from the characteristic conditions of the structural definition). We have been discussing the first pillar and have seen shadows of the second; a more thorough treatment of the second is the topic of the next section.

B.2. Transfer By Proofs

Dedekind's remarks in #134 are often taken to be a sketch of the proof that any two structures satisfying the Dedekind-Peano axioms satisfy the same sentences, i.e., are elementarily equivalent. While Dedekind's remark can be understood this way, in this section we will develop an alternative interpretation that is faithful to his metamathematical work. More specifically, instead of interpreting Dedekind's remarks as being about the modern, semantic notion of elementary equivalence, we suggest that he was working with a *syntactically grounded* notion of "proof theoretic equivalence". On this interpretation, two structures are proof theoretically equivalent just in case they fall under the same structural definition and "satisfy" the same set of sentences, namely those that can be deduced step-by-step from the characteristic conditions of the definition. We will now present our analysis of #134. Before proceeding, we want to emphasize that Dedekind was grappling with subtle methodological issues, but without the now familiar modern tools of mathematical logic.

Dedekind, as pointed out in section B.1, begins Remark #134 with the observation that the simply infinite systems form a class. He continues:

> At the same time, with reference to (71), (73) it is clear that every theorem regarding numbers, i.e., regarding the elements n of the simply infinite system N ordered by the mapping φ (and indeed every theorem in which we leave entirely out of consideration the special character of the elements n and discuss only such notions as arise from the arrangement φ) possesses perfectly general validity for every other simply infinite system Ω ordered by a mapping θ and its elements ν ... (*WZ*, 823).

How is it that *every* theorem regarding numbers, i.e., the elements of N with mapping φ and distinguished element 1, "possesses perfectly general validity" for any other simply infinite system Ω with mapping θ and distinguished element ω? To answer this question, we follow Dedekind and look in detail at #71 and #73. According to *Erklärung* #71, quoted in full in section A.2, the essence of a simply infinite system consists in the existence of a mapping φ of N and an element 1 satisfying the "characteristic conditions" $\alpha, \beta, \gamma, \delta$. The laws that are exclusively derived from these conditions, Dedekind asserts in #73, constitute the *science of numbers* or *arithmetic*. Laws thus obtained are "therefore in all simply infinite systems always the same, no matter what the accidentally given names of the individual elements might be".[34]

The remarks in #73 aim to explain why the laws obtained from the characteristic conditions for N have general validity for *all* simply infinite systems:

---

[34]The full German text is: "Die Beziehungen oder Gesetze, welche ganz allein aus den Bedingungen $\alpha, \beta, \gamma, \delta$ abgeleitet werden und deshalb in allen einfach unendlichen Systemen immer dieselben sind, wie auch die den einzelnen Elementen zufällig gegebenen Namen lauten mögen (vgl. 134), bilden den nächsten Gegenstand der *Wissenschaft von den Zahlen* oder *Arithmetik*."

one appeals in proofs only to the characteristic conditions and they, together with the statements obtained by stepwise deduction, clearly hold in all such systems. For the understanding of this and the further claim that the laws are "always the same, no matter what the accidentally given names of the individual elements might be", we have to move back to #134 and directly continue the above long quotation replacing the ellipsis at its end by the following "conjunct":

> and that the transfer from N to Ω (e.g., also the translation of an arithmetic theorem from one language into another) is effected by the mapping ψ considered in (132) and (133), which transforms every element n of N into an element ν of Ω, i.e., into $\psi(n)$.[35]

The transformation (*Verwandlung*) of the elements of N into elements of Ω underlies the "transfer from N to Ω" and is achieved by the mapping ψ associating $\nu = \psi(n)$ with each n. "This element ν", Dedekind continues, "can be called the nth element of Ω and accordingly the number n is itself the nth number of the number series N." The crucial reason for the transfer, we emphasized already earlier, lies for Dedekind in the fact that ψ is structure-preserving or, to connect this fact most directly to laws, ψ transforms φ into θ and θ into φ; thus, the mappings φ and θ have the "same significance" for the laws in the domains N and Ω. Dedekind expresses that most clearly as follows:

> The same significance which the mapping φ possesses for the laws in the domain N, [in so far as every element n is followed by a determinate element $\varphi(n) = n'$], is found, after the transformation effected by ψ, to belong to the mapping θ for the same laws in the domain Ω, [in so far as the element $\nu = \psi(n)$ arising from the transformation of n is followed by the element $\theta(\nu) = \psi(n')$ arising from the transformation of n']. We are therefore justified in saying that φ is transformed by ψ into θ, which is symbolically expressed by $\theta = \psi\varphi\overline{\psi}, \varphi = \overline{\psi}\theta\psi$.[36] (*WZ*, 823).

---

[35] Ewald's translation used "passage" instead of "transfer" for *Übertragung*, and "changes" instead of "transforms" for "verwandelt." The latter holds also for the next long quotation. The full German text is: "und daß die Übertragung von N auf Ω (z.B. auch die Übersetzung eines arithmetischen Satzes aus einer Sprache in eine andere) durch die in 132, 133 betrachtete Abbildung ψ geschieht, welche jedes Element n von N in ein Element ν von Ω, nämlich in $\psi(n)$ verwandelt."

[36] Recall that $\overline{\psi}$ is Dedekind's notation for the inverse of ψ. We put the parallel phrases starting with "insofar as" in parentheses in order to clarify the structure of this long sentence. The German text is: "Dieses Element ν kann man das $n$te Element von Ω nennen und hiernach ist die Zahl n selbst die $n$te Zahl der Zahlenreihe N. Dieselbe Bedeutung, welche die Abbildung φ für die Gesetze im Gebiete N besitzt, insofern jedem Elemente n ein bestimmtes Element $\varphi(n) = n'$ folgt, kommt nach der durch ψ bewirkten Verwandlung der Abbildung θ zu für dieselben Gesetze im Gebiete Ω, insofern dem durch Verwandlung von n entstandenen Elemente $\nu = \psi(n)$ das durch Verwandlung von n' entstandene Element $\theta(\nu) = \psi(n')$ folgt; man kann daher mit Recht sagen, daß φ durch ψ in θ verwandelt wird, was sich symbolisch durch $\theta = \psi\varphi\overline{\psi}, \varphi = \overline{\psi}\theta\psi$ ausdrückt."

Before Dedekind moves on to §11 and further applications of the recursion theorem #126, he writes, "These remarks, as I believe, completely justify the definition [*Erklärung*] of the concept of numbers given in #73."

To be perfectly clear about the *complete justification* that has been achieved, we have to further clarify the role the canonical isomorphism ψ plays for Dedekind, not only in connecting the elements of N and Ω, but also in translating laws "from one language into another". This is where Dedekind's distinction between an object and its name, and, more particularly, his use of canonical languages, plays a striking role. So let us consider a "language" $\mathfrak{L}$ in which the concept of a simply infinite system can be formulated: X is a simply infinite system if and only if there is a mapping f from X to X and an element d in X, satisfying the characteristic conditions. Consider now an arbitrary such system and instantiate the existential quantifiers (e.g., via there-is elimination of natural deduction), in order to obtain a version of the conditions parameterized with f and d. Let M stand for the conjunction of the characteristic conditions. Replacing the variables f and d by particular symbols $\varphi, \mathbf{1}$, and $\boldsymbol{\theta}, \boldsymbol{\omega}$, denoting $\varphi, 1$ and θ, ω respectively, one obtains "canonical languages" $\mathfrak{L}^N$ and $\mathfrak{L}^\Omega$ for the particular simply infinite systems N (with φ and 1) and Ω (with θ and ω). If A is a statement in $\mathfrak{L}$, then A^N and A^Ω are the statements in these canonical languages obtained from A by the obvious replacements; indeed, there is also a direct translation τ from $\mathfrak{L}^N$ into $\mathfrak{L}^\Omega$ associating A^Ω with A^N. (In a certain way, Pettigrew's Aristotelian perspective in his (2008) joins these approaches by considering the particular symbols φ and $\mathbf{1}$ as parameters.)

Basic for the translation τ is the mapping from (closed) terms in $\mathfrak{L}^N$ to those in $\mathfrak{L}^\Omega$ that satisfies the recursion conditions $\tau(\mathbf{1}) = \boldsymbol{\omega}$ and $\tau(\boldsymbol{\varphi}(\boldsymbol{n})) = \boldsymbol{\theta}(\tau(\boldsymbol{n}))$ as τ is induced by ψ. We can describe τ as a composition of mappings. Let χ_N be the mapping which associates to each element of the number sequence N its name; it maps 1 to $\mathbf{1}$ and $\varphi(n)$ to $\boldsymbol{\varphi}(\boldsymbol{n})$. Similarly let χ_Ω be the mapping which maps the elements of Ω to their names. We take the inverse $\overline{\chi_N}$ of χ_N which has as its domain the names of the natural numbers and observe that $\tau = \chi_\Omega \circ \psi \circ \overline{\chi_N}$. This is illustrated in Figure 4.

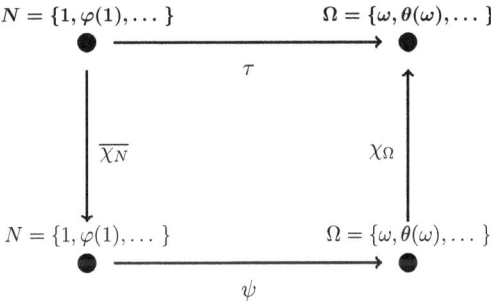

Figure 4

This observation tells us that the way in which we name the elements of one simply infinite system and the way in which we name the elements of another simply infinite system are "equivalent up to a translational isomorphism". Given our particular simply infinite system we can use its associated language to talk about the corresponding elements of any simply infinite system, and we can call both n and $\psi(n)$ the nth element of their respective systems and denote them both by $\boldsymbol{n}$. With this set-up we can understand the "transfer from N to Ω" as depicted in Figure 5, where $M \vdash A$ (and the variants) indicate the stepwise deducibility of A from M.

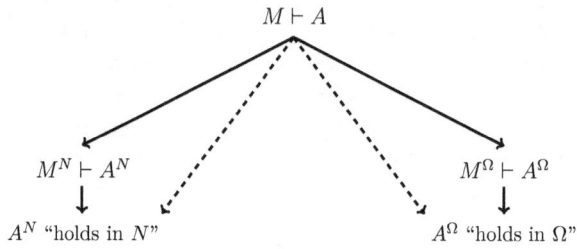

Figure 5

The claim "A^N 'holds in N' " is now simply expressed by the statement in which the names of the form $\varphi(\dots(\varphi(1))\dots)$ have been replaced by $\varphi(\dots(\varphi(1))\dots)$ and similarly for the claim involving Ω. This seems then to be the context, pedantically reconstructed in more modern terms, that allows the following understanding of the part of #134 we sought to clarify: ν and n are the nth elements of Ω and N, respectively, and ψ preserves the order set by θ and φ for the elements in these systems. The structure-preserving feature (and nothing more) is expressed by the sentence that ends with the observation $\theta = \psi\varphi\overline{\psi}$, $\varphi = \overline{\psi}\theta\psi$. So, the theorems in the parameterized theory M are exactly the statements that "hold" in all simply infinite systems. Through these considerations we can come to understand in what sense the laws obtained by stepwise logical deduction are "always the same" in all simply infinite systems.[37] This analysis can be extended to provide a Dedekindian treatment of complete ordered fields. There is, however, a complication because there is no "canonical naming scheme" as for the natural numbers. After all, given any (countable) language, there are always real numbers not named in it. But in a restricted context, e.g., in the proof of a particular theorem, only finitely many real numbers are named and the proof theoretic equivalence can be established as before. Indeed, the issue was addressed by Dedekind in his letter to Weber of January 1888; Dedekind remarked, "Whether the sign language suffices to denote all individuals that have been created does not matter; it [the sign

---

[37]This understanding of the fundamental "semantic" relationship—a statement holds in a system—is closely related to Gödel's perspective in his thesis concerning the completeness of first order logic. See note 53.

language] is always sufficient to denote the individuals occurring in a (limited) investigation."[38]

We should finally remark that Dedekind's approach in #134 links in closely with the following passage from the Preface to the first edition of *WZ*:

> If we scrutinize closely what is done in counting a set or number of things, we are led to consider the ability of the mind to relate things to things, to let a thing correspond to a thing, or to represent a thing by a thing, an ability without which no thinking is possible. Upon this unique and therefore absolutely indispensable foundation ... the whole science of numbers must, in my opinion, be established.[39]

These considerations emerged already much earlier in the 1870s and are reemphasized in the note to section 161 of (Dirichlet and Dedekind 1894), which we quoted in A.1. Thus it is quite clear that Dedekind has mappings in mind when he refers to "the ability of the mind to relate things to things ... an ability without which no thinking is possible". And as we have seen above, it is the use of various different kinds of mappings that make possible his approach to the natural numbers. Dedekind's comments in Remark #134 indeed justify his attitude towards the natural numbers. For he established (i) that the concept of simply infinite system is *categorical* and (ii) that there is an *isomorphic translation* between any two of the canonical *languages* associated with two simply infinite systems. These results ensure that Dedekind's definition of natural number is not dependent upon the elements of the simply infinite system he starts with. Moreover, they allow Dedekind to explicate the sense in which the theorems proved for one simply infinite system are "the same" as those proved for another: they are proved in a "logically abstract" way from the *Merkmale* M or their instantiation M^N, but leaving "entirely out of consideration the special character of the elements n" of N.

B.3. A Radical Shift

We have argued that Dedekind is in *WZ* primarily concerned with *creating* concepts. However, there are important passages in *SZ* and *WZ* that seem to indicate without any ambiguity that (systems of) objects are created. The considerations in those passages can be supported by remarks Dedekind made

---

[38] Here is the German text: "Ob ferner die Zeichensprache ausreicht, um alle neu zu schaffenden Individuen einzeln zu bezeichnen, fällt nicht ins Gewicht; sie reicht immer dazu aus, um die in irgend einer (begrenzten) Untersuchung auftretenden Individuen zu bezeichnen." (Dedekind 1932, 490).

[39] The German text is as follows: "Verfolgt man genau, was wir bei dem Zählen der Menge oder Anzahl von Dingen tun, so wird man auf die Betrachtung der Fähigkeit des Geistes geführt, Dinge auf Dinge zu beziehen, einem Dinge ein Ding entsprechen zu lassen, oder ein Ding durch ein Ding abzubilden, ohne welche Fähigkeit überhaupt kein Denken möglich ist. Auf dieser einzigen, auch sonst ganz unentbehrlichen Grundlage muß nach meiner Ansicht ... die gesamte Wissenschaft der Zahlen errichtet werden." (Dedekind 1932, 336)

in letters to Lipschitz (1876) and to Weber (1878). Consider first a remark in
SZ directly following Dedekind's proof that there are infinitely many cuts not
engendered by rational numbers:

> Thus, whenever we have a cut (A_1, A_2) that is not engendered by
> a rational number, we *create* a new number, an *irrational* number
> α, which we regard as completely defined by this cut (A_1, A_2); we
> shall say that the number α corresponds to this cut, or that it
> engenders this cut. From now on, therefore, to every definite cut
> there corresponds exactly one definite rational or irrational number
> ...[40]

That view is re-emphasized in the letter of 19 November 1878 to Weber, in
which Dedekind points to the fact that the phenomenon of the cut can be
used for the introduction of new *numbers*: "so many cuts, so many numbers"
(*soviel Schnitte, soviel Zahlen*). He claims that the standard operations of
addition, subtraction etc. can be defined for these new numbers. Then he asks
rhetorically:

> You also want that students learn how to deal with $\sqrt{2}, \sqrt{3}$ etc.;
> do you really want that they always view them just as symbols for
> calculating approximations? Or don't you also think it would be
> better, if they viewed them as symbols for new numbers that are in
> complete parity with the old ones?[41]

The introduction of "new numbers that are in complete parity with the old
ones" is at the core of *SZ*, and the necessity of such an extension is discussed
in two long and deep letters to Lipschitz written on 10 June and 27 July
1876. In them, Dedekind defends his position against Lipschitz's claim that
the Euclidean axioms alone, without bringing in the principle of continuity,
can serve as the foundation for a complete theory of the real numbers viewed
as ratios of magnitudes. At the end of the first letter he asserts:

> ... my theory of irrational numbers creates the perfect pattern of a
> *continuous* domain which is [on account of its continuity] capable

---

[40]The German text is found in (Dedekind 1932, 486): "Jedesmal nun, wenn ein Schnitt (A_1, A_2) vorliegt, welcher durch keine rationale Zahl hervorgebracht wird, so *erschaffen* wir eine neue, eine *irrationale* Zahl α, welche wir als durch diesen Schnitt (A_1, A_2) vollständig definiert ansehen; wir werden sagen, daß die Zahl α diesem Schnitt entspricht, oder daß sie diesen Schnitt hervorbringt. Es entspricht also von jetzt ab jedem bestimmten Schnitt eine und nur eine bestimmte rationale oder irrationale Zahl ..."

[41]The German text is found in (Dedekind 1932, 486): "Du willst doch auch, daß die Schüler mit $\sqrt{2}, \sqrt{3}$ u.s.w. umgehen lernen; willst Du nun, daß sie darin immer nur Symbole für Näherungsrechnungen sehen? Oder hälst Du es nicht auch für besser, daß sie darin Symbole für neue, mit den alten ganz gleichberechtigte *Zahlen* sehen?"

of characterizing every ratio of magnitudes by a determinate individual number that is contained in it [the continuous domain].[42]

In a long footnote to (Dedekind 1877), Dedekind describes how such extensions are to be achieved in a methodologically satisfactory way.[43] However, for a paradigmatic, rigorous execution he refers back to *SZ* and sees at its center "the introduction or the *creation* of new arithmetical elements". He emphasizes that the creation has to be done uniformly in one step and that it has objective boundaries: "The irrational numbers thus defined together with the rational numbers form a continuous domain R without gaps; ... it is impossible to put additional new numbers into the domain R." (Dedekind 1932, 269)

After the publication of *WZ*, in apparent conflict with our understanding of the essay, Dedekind continues to defend this "creative" perspective. At least he seemingly does so in a letter to Weber from 24 January 1888. He responds to Weber's proposal of treating cardinal numbers as classes[44] and writes:

> But if one were to take your route—and I would recommend that it be explored to the very end—then I would advise that by [cardinal] number [Anzahl] one rather not understand the *class* itself (the system of all similar finite systems) but something *new* (corresponding to this class) which the mind *creates*.[45]

How is this remark together with the immediately following observations on the analogous creation of irrational numbers to be understood? The observations seem to paraphrase the problems and their solution from the 1870s. Are they to be viewed as expressing Dedekind's position at this time in 1888? Our answer to the last question is "No". The observations are rather to be understood in the context of the exchange with Weber: *if* one were to follow Weber's proposal, *then* one should do what he, Dedekind, had done for the real numbers in *SZ* and create something new that corresponds to the class. The new mathematical object is presumably, as he put it in *SZ*, *completely defined* by its corresponding class. Appealing to our *Schöpfungskraft* (creative power), Dedekind explains,

---

[42]The German text is found in (Dedekind 1932, 474): " ... durch meine Theorie der irrationalen Zahlen [wird] das vollkommene Muster eines *stetigen* Gebiets erschaffen, welches eben deshalb fähig ist, jedes Grössen-Verhältnis durch ein bestimmtes in ihm enthaltenes Zahl-Individuum zu charakterisieren."

[43]Dedekind connects the general considerations with Kummer's introduction of ideal numbers. See the Note in (Dedekind 1932, 481) based on a report by Bernstein after visiting Dedekind on 6 March 1911; Dedekind related to Bernstein also some informative aspects of his first encounter with Kummer. Note that Hilbert speaks, in the context of extending the field of algebraic numbers to a complete ordered field, of an extension by "ideal" or "irrational" elements; see (Hilbert 1899, 475).

[44]The definition of "class" is given also in *WZ* #32 and subsection B.1 above.

[45](Dedekind 1932, 489). The German text is: "Will man aber Deinen Weg einschlagen— und ich würde empfehlen, ihn einmal ganz durchzuführen—so möchte ich doch rathen, unter der Zahl (Anzahl, Cardinalzahl) lieber nicht die *Classe* (das System aller ähnlichen endlichen Systeme) selbst zu verstehen, sondern etwas *Neues* (dieser Classe Entsprechendes) was der Geist *erschafft* ..."

this path can be taken in a perfectly coherent way. However, when working on the 1887 manuscript, Dedekind had recognized that the creation of an abstract system of abstract new objects is unnecessary, if one focuses on the laws that govern the *relations* between numbers. The manuscript itself ends abruptly with Remark #107 about the "creation of the pure natural numbers":

> *Creation of the pure natural numbers.* It follows from the above that the laws regarding the relations between numbers are completely independent of the choice of that infinite system N, which we have called the number series, as well as independent of the mapping from N, by which N is *ordered* as a simple series.[46]

Here Dedekind formulates the *basic insight* that had prompted a radical revision of section 5 of the manuscript with the introduction of the notion of a *simply infinite system*. That definition is the foundation for the sequence of thoughts we analyzed and that provide the "complete justification" of the concept of numbers in *WZ*. In the letter to Weber, Dedekind's own *new* perspective is indicated only by one short phrase, namely, when "my ordinal numbers" are characterized as "the abstract elements of the ordered simply infinite system".[47]

While revising the 1887 penultimate manuscript, Dedekind not only introduces the notion of a simply infinite system, but eliminates at the same spot an important passage. He had introduced "abstraction" in almost exactly the same way as in *Erklärung* #73 which we quoted in A.2, but included more expansive remarks which are absent from the published version (see the Appendix for more details):

> By this abstraction, the originally given elements n of N are turned into new elements $\mathfrak{n}$, namely into numbers (and N itself is consequently also turned into a new abstract system $\mathfrak{N}$). Thus, one is justified in saying that the numbers owe their existence to an act of free creation of the mind. For our mode of expression, however, *it is more convenient to speak of the numbers as of the original elements of the system N and to disregard the transition from N to $\mathfrak{N}$*, which itself is a similar mapping. Thereby, as one can convince oneself using the theorems regarding definition by recursion ... nothing essential is changed, nor is anything obtained surreptitiously in illegitimate ways. (Emphasis added)[48]

---

[46](Dedekind 1887, 19). The German text is as follows: "*Schöpfung der reinen natürlichen Zahlen.* Aus dem Vorhergehenden ergiebt sich, daß die Gesetze über die Beziehungen zwischen den Zahlen gänzlich unabhängig sind von der Wahl desjenigen einfach unendlichen Systems N sind, welches wir die Zahlenreihe genannt haben, sowie auch unabhängig von der Abbildung von N, durch welche N als einfache Reihe *geordnet* ist."

[47]The German phrase is: "die abstracten Elemente des geordneten einfach unendlichen Systems."

[48]The German text is as follows: "Da durch diese Abstraction die ursprünglich vorliegenden Elemente n von N (und folglich auch N selbst in ein neues abstraktes System $\mathfrak{N}$) in

So it seems quite clear that the reflections on proof theoretic equivalence led Dedekind to abandon the creation of the abstract domain 𝔑. What positive role could 𝔑 really play in mathematical practice? After all, due to our intellectual limitations emphasized in (1854, 428–9; cf. Note 28) and the nature of our step-by-step understanding (*Treppenverstand*) described in (*WZ*, 791), we have to give proofs. The fact that the objects in 𝔑 are to have no other properties and relations than those specified by the concept under which they fall is quite directly captured by restrictions on the sequence of thoughts that constitute a *derivation*. Derivations proceed by logic and start either from the characteristic conditions M of the concept or from their instantiations M^N. Thus, in either case, derivations yield a truly "sober dissection of the sequence of thoughts on which the laws of numbers are based".[49] This shift to a schematic, formal approach even for natural numbers is a dramatic change in perspective. It provides, in the end, the complete justification of the *concept* of numbers.

C. Creating by Abstraction

The radical shift in Dedekind's perspective we discussed in B.3 is literally visible on one page of the penultimate manuscript of *WZ*; see our Appendix. In the first half of section C.1 we argue that Dedekind's shift from abstract objects to abstract concepts is the endpoint of a quite natural evolution. It allows us to reinterpret *SZ* from the perspective of that endpoint. We discuss in the second half of C.1 different contemporary perspectives that lead us to the central issues of C.2, the creation of concepts by abstraction. Finally we describe in C.3 how Dedekind's "axiomatic standpoint", as Emmy Noether called his perspective on mathematics, was transformed into structural axiomatics in the hands of Hilbert and Zermelo.

C.1. From Objects to Concepts

Dedekind's motivation for introducing the irrational numbers *corresponding to* or being *completely determined by* cuts is quite straightforward: for the laws

---

neue Elemente n, nämlich in Zahlen umgewandelt sind, so kann man mit Recht sagen, daß die Zahlen ihr Dasein einem freien Schöpfungsacte des Geistes verdanken. Für die Ausdrucksweise ist es aber bequemer, von den Zahlen wie von den ursprünglichen Elementen des Systems N zu sprechen, und den Übergang von N zu 𝔑, welcher selbst eine deutliche Abbildung ist, außer Acht zu lassen, wodurch, wie man sich mit Hilfe der Sätze über Definition durch Recursion ... überzeugt, nichts Wesentliches geändert, auch Nichts auf unerlaubte Weise erschlichen wird." (Dedekind 1887, §5)

[49](Dedekind 1932, 336–7) and (Ewald 1996, 791). The German phrase is: "nüchterne Zergliederung der Gedankenreihen, auf denen die Gesetze der Zahlen beruhen". Dedekind emphasizes the importance of derivations in the very next sentence: "Ich erblicke dagegen in der Möglichkeit, solche Wahrheiten auf andere, einfachere zurückzuführen, mag die Reihe der Schlüsse noch so lang und scheinbar künstlich sein, einen überzeugenden Beweis dafür, daß ihr Besitz oder der Glaube an sie niemals durch innere Anschauung gegeben, sondern immer nur durch eine mehr oder weniger vollständige Wiederholung der einzelnen Schlüsse erworben ist."

real numbers satisfy it is irrelevant that, for example, 1 is an element of the left part of the cut determining $\sqrt{2}$. The *basic insight for natural numbers* described in B.3, and associated with #107 in the penultimate manuscript, was formulated by Dedekind as follows: "[T]he laws regarding the relations between numbers are completely independent of the choice" of the system N, the base element 1, and the mapping φ that orders N as a simple series. That independence, Dedekind realized, does not necessitate the creation of abstract objects, but can be achieved by introducing an appropriate abstract concept and focusing on proofs that use as starting points only characteristic conditions of this abstract concept and proceed otherwise by "logical steps".

This realization is not far-fetched and did not come out of the blue. There is, first of all, Dedekind's general experience in algebraic number theory and his work with abstract, structural concepts. This general experience is complemented, secondly, by an important aspect in the evolution of Dedekind's thought on "ordinary" numbers. We are alluding to Dedekind's move away from the genetic conception of natural numbers and their generative expansion to integers as well as rational numbers. The genetic conception can be seen in his *Habilitationsrede* of 1854, but also in *SZ*. Right after the publication of *SZ*, he began to examine those presuppositions in a critical way. Sieg and Schlimm describe in their (2005, section 3, *Extending Domains*) the increasingly structural approach Dedekind explores for the expansions even before having completed *WZ*. Later on he uses *WZ* as the basis to show, in a perfectly modern way, how to introduce the standard arithmetic operations on (equivalence classes of) pairs of natural numbers and to obtain a system of mathematical objects that falls under the abstract notion of a field; see (Sieg and Schlimm 2005, section 4, *Creating Models*).

This structural approach was extended in *WZ* to the natural numbers themselves; here let us briefly show how it can be used to *reinterpret SZ*. Recalling our discussion at the end of subsection B.1 we observe, in a first step, that *SZ* introduces the structural definition of a "fully continuous domain", which has the system of all cuts of rational numbers (with its induced ordering) as a model.[50] This observation can be broadened, through a second step, to the concept "complete ordered field" by following *SZ* in defining the arithmetic operations on cuts. The final third step, rounding off the metamathematical treatment of complete ordered fields in analogy to that of simply infinite systems, requires Dedekind's general concept of mappings. The latter makes it possible to prove a *representation theorem* ("any complete ordered field is isomorphic to the system of all rational cuts") and then to establish that any two complete ordered fields are isomorphic; i.e., the notion is categorical. The point of these reinterpreting and extrapolating remarks is that the architecture of *SZ* is already refined to such a degree that the analogy to *WZ* is not forced, but rather direct and natural.

---

[50]Expanding the "order axioms" by the principle of continuity yields the concept of "a fully continuous domain". The main proposition of *SZ* is Theorem *IV* in section 5 stating that the system of all cuts of rational numbers is continuous.

The data from *WZ*, i.e, metamathematical results and methodological observations, that we have been appealing to have been used by other commentators in different ways to support particular "structuralist positions". Stewart Shapiro, for example, sees Dedekind as presenting an incipient version of his *ante rem* structuralism (Shapiro 1997), while Charles Parsons considers interpretations of Dedekind as an *eliminative* and *non-eliminative* structuralist; see (Parsons 1990) for the terminology. Parsons highlights, see p. 307, that *Erklärung* #73 and Theorem #132 invite an eliminative structuralist reading, though he does not endorse such an interpretation. More specifically, if we allow $\Gamma(N, 0, S)$ to stand for the characteristic conditions of a simply infinite system, where $N, 0, S$ are variables for the system, base element and successor mapping, respectively, and if we allow $A(N, 0, S)$ to stand for an arithmetic statement, we can consider $A(N, 0, S)$ to be elliptical for:

$$(\forall N)(\forall 0)(\forall S)\,(\Gamma(N, 0, S) \to A(N, 0, S))$$

Note that all occurrences of $N, 0, S$ are "quantified out" and thus removed in a sense that would satisfy an eliminative structuralist.[51]

Erklärung #73 and Theorem #132 are also used to support Tait's non-eliminative interpretation, the crucial aspect of which is "Dedekind abstraction".[52] This abstraction is the process by which, given a particular structure, one introduces a new structure of the same type along with an isomorphism from the original system to the newly introduced one, see (Parsons 1990, 308). On Tait's interpretation, Dedekind starts with a particular simply infinite system, and, as the concept of a simply infinite system is categorical, he can introduce a new system and corresponding isomorphism. The latter system is then taken to contain *the* natural numbers. However, it is precisely this abstracting move that is removed in the transition from the 1887 manuscript to the published version of *WZ*. The justification for removing it is already indicated in the manuscript. After all, Dedekind claims there that by disregarding the abstraction, " ... as one can convince oneself using the theorems regarding definition by recursion ... nothing essential is changed, nor is anything obtained surreptitiously in illegitimate ways." This highlights Dedekind's "indifference

---

[51] It is important to note that this description of the logical situation is in direct conflict with both Dedekind's and Hilbert's way of articulating their structural definitions of, e.g., simply infinite system and Euclidean space. They give an explicit structural definition applying to systems. This is the reason why Hilbert and Bernays call this way of formulating an axiomatic theory *Existentiale Axiomatik*; they view, as Dedekind did, giving a proof that there is a system falling under the structural definition as a crucial task. That is beautifully and extensively discussed in (Bernays 1930, 20–21).

[52] Linnebo and Pettigrew discuss Dedekind abstraction in their recent paper "Two Types of Abstraction for Structuralism" (Linnebo and Pettigrew 2014). They argue that this notion of abstraction is unable to satisfy both of two structuralist goals, namely, to ensure that: (i) there is a unique natural number structure, and (ii) the elements of the natural number structure have no "foreign properties". Their arguments are convincing. However, the second of these goals can be achieved by different means for any simply infinite system as we argued in section B above; the first goal is given up by Dedekind.

to identification" with respect to natural numbers, using Burgess' apt phrase (Burgess 2011, 8).

Reck and Yap suggest a further, novel interpretation. More particularly, they interpret Dedekind as possessing his own, unique brand of structuralism which they call *logical structuralism*. This is deeply concerned with the creation of mathematical *objects*, in particular natural numbers and their system. Let us sketch the considerations presented in (Reck 2003) and (Yap 2009). Their anchor is Dedekind's remark in the very first section of WZ, "A thing is completely determined by all that can be affirmed or thought concerning it" (WZ, 797); this is apparently Leibniz's principle of the identity of indiscernibles. According to Reck's interpretation, any legitimate attempt to create mathematical objects must consequently ensure that they are completely determined. Reck suggests that, to do just this, Dedekind developed a procedure in which his metamathematics plays a significant role. The procedure, as applied to the natural numbers, aims to establish that they are "completely determined" in the sense that (i) it specifies precisely what statements can be formed about them, and (ii) it ensures, given such a statement ϕ, that exactly one of ϕ or $\neg \phi$ follows *semantically* from Dedekind's "characterizing conditions".[53] Furthermore, each step appeals to purely logical means, hence the name of *logical* structuralism. The procedure consists of the following steps, see (Reck 2003, 395):

(i) Designing a language with which to speak about the natural numbers. This language includes a name for the base element, and a function symbol for the successor function.

(ii) Articulating "the basic definitions and principles" from which all arithmetical truths should follow.

(iii) Establishing that the notion of a simply infinite system is categorical.

Together these steps ensure that the (system of) natural numbers is completely determined: "By (iii), for any sentence ϕ in the language of arithmetic, as specified in (i), either ϕ or $\neg \phi$ follows (semantically) from the basic definitions and principles, as specified in (ii); *tertium non datur*" (Reck 2003, 395). This procedure "creates", according to Reck, the system of natural numbers in the

---

[53]In WZ there is no formal language, nor is there a concept of semantics. Thus, a notion of *semantic consequence*, as is presupposed here, is not available: it emerges only slowly. The development culminates in Gödel's doctoral dissertation (Gödel 1929), but not in the contemporary Tarskian way via a truth-definition. The fundamental "semantic" relation is rather a structure "falling under" or "satisfying" a second-order concept; see (Sieg and Schlimm 2014, note 17). This difference from the modern concept is discussed in (Schiemer and Reck 2013) with respect to Gödel, but also Hilbert and Ackermann whose terminology Gödel followed. However, Schiemer and Reck do not indicate the path from Dedekind's WZ through Hilbert's *Grundlagen der Geometrie* to the Hilbert Lectures of 1917/18 and, thence, to Hilbert and Ackermann. Dedekind's grappling with "model theoretic" problems is being analyzed in the present section.

following sense: "It is *identified* as a new system of mathematical objects, one that is neither located in the physical, spatio-temporal world, nor coincides with any of the previously constructed set-theoretic simple infinities ... what has been done is to *determine uniquely* a certain "conceptual possibility", namely a particular simple infinity ... It is that simple infinity whose objects only have arithmetical properties, not any of the additional, "foreign" properties objects in other simple infinities have" (Reck 2003, 400).

Our interpretation is evidently quite different from Reck's. As we have discussed above, Dedekind is concerned with theorems obtained by stepwise derivations, not as semantic consequences. Most importantly, Dedekind is ultimately creating concepts rather than objects, and "the" natural numbers are in *WZ* explicitly taken as the elements of a previous logical simply infinite system. It is to an understanding of "creation of concepts" that we turn next. This is an intricate, complex subject. We focus on Dedekind and what seem to us the most direct influences on him originating from Kant and Lotze.

C.2. Formation of Concepts (Begriffsbildung)

The emphasis on concepts in Dedekind's reflection on mathematics goes back to his *Habilitationsrede* of 1854, where he claimed:

> The introduction of such a concept ... is, as it were, a hypothesis which one puts to the inner nature of the science; only in the further development does the science answer it; the greater or lesser *efficacy* of such a concept determines its worth or unworth.[54]

In the Preface to *WZ* he refers back to (1854) and asserts that the most fruitful advances in mathematics as well as in other sciences are made by the "creation and introduction of new concepts". One traditional method of creating new concepts is *abstraction*, and Dedekind appeals to abstraction, when explaining in #73 what natural numbers are, which he had done also in the 1887 manuscript.[55]

But there, as we saw, he introduces new abstract objects that then constitute the abstract system 𝔑. Exactly that step to the system 𝔑 is taken back in the transition from the 1887 draft to *WZ*. And yet, Dedekind refers in

---

[54](Dedekind 1932, 429) and (Ewald 1996, 756). The German text is as follows: "Die Einführung eines solchen Begriffs ... ist gewissermaßen eine Hypothese, welche man an die innere Natur der Wissenschaft stellt; erst im weitern Verlauf antwortet sie auf dieselbe; die größere oder geringere *Wirksamkeit* eines solchen Begriffs bestimmt seinen Wert oder Unwert." Juliet Floyd argues in her (2013, 1025–1027) that the initial discussion of the *Habilitationschrift* "echoes Kant's account of reflective judgement in his introduction to *The Critique of Judgment*".

[55]Other mathematicians, prominently Cantor, also used "abstraction"; see (Deiser 2010, 60–62) and (Hallett 1984, 119 ff.). Let us mention Ernst Cassirer who in his *Substanzbegriff und Funktionsbegriff* refers to "freie Produktion bestimmter Relationszusammenhänge" (pp. 15–16); he discusses abstraction in general terms on p. 6 and with direct reference to *WZ* on p. 50ff.

his 1888 letter to Weber to "abstract elements" of *the* ordered simply infinite system, and in his 1890 letter to Keferstein he calls the "Zahlenreihe N" the "abstract type" of a simply infinite system. Here Dedekind understands "abstract" in the traditional context of conceptual abstraction—no new abstract objects are introduced, but familiar objects fall under more abstract concepts. This "abstraction *from*" is described in Kant's *Logik* (*Elementarlehre* §6, A 145–148): dropping some characteristic conditions or *Merkmale* from a given concept leads to a more abstract one. That's how we read #73 of *WZ*, quoted in full towards the end of section A.2. Dedekind considers there a particular number sequence N; he "retains" only the distinguishability of its elements and "takes into account" only their relations to each other that are due to the mapping φ. The elements of N are now called numbers, and the abstraction allows one to call *them*, when viewed from this abstracting standpoint, a free creation of the human mind. This is the sense in which we also understand Dedekind's brief remark in the letter to Weber when he describes his ordinal numbers as "the abstract elements of the ordered simply infinite system".[56]

This reading can be much more directly supported, if Dedekind's particular way of defining simply infinite systems is explicitly taken into account: *any* system N of objects can be considered to be simply infinite, if there is some element 1 and some mapping φ on N that satisfy the notion's characteristic conditions. This definition reflects a form of abstraction that is radically different from traditional abstraction as discussed above following Kant's *Logik*. Lotze introduced it in the 1843 edition of his *Logik* and viewed it as a most significant contribution. It had the clear goal of reflecting the actual practice of ordinary and scientific thinking. Lotze describes it with pertinent examples in the second edition of his *Logik*, §23, *Die Lehre vom Begriffe*. This abstraction usually does not drop *Merkmale*, but rather replaces some by more general ones and yields in this way "more abstract" concepts.[57] That is indeed done in *WZ*, as we saw in A.2, when the structural definition of a simply infinite system is introduced: any particular system together with its base element and order setting mapping can fall under the more abstract concept. This standpoint clarifies further Dedekind's assertion that the ordered system N is the *abstract type* of simply infinite systems, cf. Note 26; but it also allows us to point to a real difference between the two abstractions, when we join the above considerations with those for deducibility in subsection B.2. In the first case, we accept as principles the interpreted characteristic conditions M^N; in the

---

[56]This might seem to be a round-about interpretation. We feel, obviously, it is coherent and allows an informed understanding of Dedekind's evolving perspective on mathematics. However, we face here the additional task to explain, how Dedekind could retain the formulation of the manuscript in #73 despite the fact that the context had been changed by the introduction of the concept of a simply infinite system. The issue is also illuminated by the circumstances of the preparation of the final manuscript as described in the Appendix.

[57]Lotze's abstraction contains the Kantian one as a special case. We have thus isolated two different kinds of conceptual abstraction. Both forms are different from the abstraction used by Frege and Neo-logicists. The latter is, in Kantian terminology, "abstraction *to*". That is made marvelously clear in Chapter 1, *Philosophical Introduction*, of (Fine 2002).

second case, we work directly with M. Though this is a difference, it is one without consequence as far as the "laws of arithmetic" are concerned; that was established in B.2.

It seems clear then that Lotzean abstraction is at work and leads to the concept of a simply infinite system. However, for it to be applicable, there must be a particular such system from which to abstract. Dedekind has thus to prove the existence of a suitable system; that proof ensures at the same time that the notion has no internal contradictions. For the proof, Dedekind needs objects, a distinguished element, and an order setting mapping that together constitute a simply infinite system. The objects, including his Ego as the base element, and the order setting mapping are obtained in logic. Indeed, he calls his proof of #66 in *WZ*, when writing to Keferstein, a "logical existence proof".[58] The need for such a proof is recognized, it seems, only in the revisions to the penultimate 1887 draft of *WZ*; the emendations are described at great length in (Sieg and Schlimm 2005, section 6.1). That only reemphasizes the dramatically changed perspective underlying the structural definition of a simply infinite system.

It is truly fascinating, given our understanding of *WZ*, that Dedekind extended his structuralist approach from algebraic number theory to the fundamental objects of traditional mathematics, i.e., to number systems beginning with the natural numbers and ending with the real and complex ones. The character of individual numbers is no longer at issue, but rather that of systems whose elements stand in particular relationships. That is made clear in *WZ* last but not least by the systematic metamathematical investigation of simply infinite systems. Dedekind describes in the preface to the first edition of *WZ* (pp. v–vi) how, on this basis, a program of "the stepwise extension of the number *concept*" can be carried out, covering negative, fractional, irrational and complex numbers. The extended notions are obtained always by "a reduction to the earlier *concepts*" without mixing in "foreign conceptions" like that of "measurable magnitudes". Indeed, Dedekind claims that the latter magnitudes can be understood in "complete clarity" only through the science of numbers. That is for Dedekind, it seems, the ultimate point of the arithmetization of analysis.

The formation of abstract concepts and the insistence on stepwise argumentation from their characteristic conditions locate Dedekind's structuralism within the logic of his time, in particular, that of Lotze and Schröder. The principles of unrestricted comprehension and extensionality, both used in *WZ* and viewed as logical principles, form the framework for founding number theory—with one crucial addition, the notion of *mapping*. The latter is also part of logic.[59] Dedekind asserted in 1879, long before the completion of *WZ*, "the entire science of numbers is based on this intellectual ability ... with-

---

[58] Frege found Dedekind's argument, with the emendation that "thoughts exist independently from our thinking", fully convincing; see the note on pp. 147–8 of (Frege 1969).

[59] Recall our remarks in section A.1. It was only Zermelo who introduced, in his 1908 paper, mappings as set theoretic objects—under the very name of *Abbildungen*; see (Sieg and Schlimm 2014, section 2.2).

out which no thinking at all is possible". The intellectual ability he pointed to is that of establishing correspondences, and that is of course reflected in *WZ* mathematically by mappings. Abstract concepts and structure-preserving mappings are also the crucial tools of Bourbakian mathematics concerned with the *principal structures* (structures-mères) of algebra, topology and order. We alluded in the Introduction to this connection with contemporary mathematics; it emerged out of the transformation of the subject that is brilliantly reflected in Dedekind's mathematical and foundational work.[60] In the next section we explore how Dedekind's methodological perspective underlies crucial developments in the early part of the 20th century, namely, the emergence of structural axiomatics as opposed to its later formal variety; cf. (Sieg 2014). A good starting point is Hilbert's Paris address of 1900 challenging mathematicians with twenty-three problems. The second of these problems is most closely related to the methodological issues we have been discussing.

C.3. Structural Axiomatics

In the Introduction to his Paris talk, Hilbert views the "arithmetic comprehension of the *concept* of the continuum" as one of the most important developments in 19th-century mathematics. At the same time, he wants to refute the view "that only the *concepts* of analysis or even those of arithmetic alone are susceptible to a completely rigorous treatment". On the contrary, he thinks, emphasizing *concepts* throughout:

> ... wherever, from epistemology or in geometry or from scientific theories, mathematical concepts emerge, mathematics has the task to investigate the principles on which these concepts are based and to fix them [the concepts] by a simple and complete system of axioms in such a way that the exactness of the new *concepts* and their use *in deductions* is in no way inferior to the old arithmetic concepts.[61]

---

[60]These developments led naturally to category theory. As far as Bourbaki's work is concerned, it began in the early 1930s; see the papers (Bourbaki 1950) and (Dieudonné 1970) that describe the emergence of their work and the influence of Dedekind, Hilbert and Noether. What were the motivating and guiding ideas at that time? Helmut Hasse isolated general methodological features in a talk entitled *Die moderne algebraische Methode* that was given at the annual meeting of the German Association of Mathematicians in September 1929. The methodological features are realized through abstract domains and overarching concepts; they are illustrated by many examples, of course, from algebra. However, Hasse emphasizes that the modern algebraic method permeates all of mathematics: "Everywhere one can apply its principle to find the simplest conceptual foundations for a given theory and to have, in this way, a unifying and systematizing effect" (Hasse, 33). This is supported according to Hasse by a certain philosophical position reflected in the axiomatic standpoint (p. 28).

[61](Hilbert 1900, 49). Here is the German text: " ... wo immer von erkenntnistheoretischer Seite oder in der Geometrie oder aus den Theorien der Naturwissenschaft mathematische Begriffe auftauchen, erwächst der Mathematik die Aufgabe, die diesen Begriffen zu Grunde liegenden Prinzipien zu erforschen und dieselben durch ein einfaches und vollständiges System

Dedekind's Structuralism: Creating Concepts and Deriving Theorems   287

For geometry this goal had been pursued and, in Hilbert's view, successfully reached through his *Festschrift* (Hilbert 1899). That work gives, after the addition of the completeness principle, a structuralist foundation for geometries, in particular for Euclidean geometry, in the same way Dedekind's work does for number systems, in particular for the system of real numbers. The emphasis on concepts and their use in deductions in the above quote is noteworthy. Similarly close to Dedekind's view is Hilbert when explaining his views on geometry to Frege:

> If I think of my points as any arbitrary system of things, for example, the system: love, law, chimney sweeps ... and then only assume all my axioms as relations between these things, then my theorems, e.g. the Pythagorean one, hold of these things as well. In other words, each and every theory can always be applied to infinitely many systems of basic elements. *For one merely has to apply an invertible transformation and stipulate that the axioms for the transformed things be the correspondingly same ones.* (Emphasis added.)[62]

This expresses, in a general and informal way, Dedekind's striking insight formulated as #133 (and analyzed in section B.1).

The centrality of *concepts* for modern mathematicians is not special to Dedekind and Hilbert, but seems to be equally important to "classical" constructive mathematicians such as Kronecker: Kronecker's sole extended foundational essay (1887) has the revealing title "On the number concept" (*Über den Zahlbegriff*). However, let us turn to Hilbert's (1900) and his formulation of the second problem that asks one to find a *direct* proof of the consistency of analysis. Hilbert's mathematical formulation and his optimism for finding a direct proof through a suitable modification of the "familiar inference methods in the theory of irrational numbers" are well known; the extended remark he added to his formulation was to characterize the significance of the consistency problem from a different, more philosophical perspective. Here we point out that the consistency of *concepts* is to be established and that their "mathematical existence" is to be guaranteed:

---

von Axiomen derart festzulegen, daß die Schärfe der neuen Begriffe und ihre Verwendbarkeit zur Deduktion den alten arithmetischen Begriffen in keiner Hinsicht nachsteht."

[62]The full German text is as follows "Wenn ich unter meinen Punkten irgendwelche Systeme von Dingen, z.B. das System: Liebe, Gesetz, Schornsteinfeger ...denke und dann nur meine sämmtlichen Axiome als Beziehungen zwischen diesen Dingen annehme, so gelten meine Sätze, z.B. der Pythagoras auch von diesen Dingen. Mit andern Worten: eine jede Theorie kann stets auf unendliche viele Systeme von Grundelementen angewandt werden. Man braucht ja nur eine umkehrbar eindeutige Transformation anzuwenden und festzusetzen, dass die Axiome für die transformirten Dinge die entsprechend gleichen sein sollen. Thatsächlich wendet man auch diesen Umstand häufig an, z.B. Dualitätsprinzip etc, und ich in meinen Unabhängigkeitsbeweisen." Quoted in (Frege 1980, 14).

> If one succeeds in proving that the characteristic conditions [*Merkmale*] given to the concept can never lead to a contradiction through a finite number of logical inferences, then I say that the mathematical existence of the concept ... has been proved in this way.[63]

This can be taken as definitional: a concept *exists mathematically* if and only if it is consistent in the sense that no contradiction can be inferred from its characteristic conditions in a finite number of logical steps. If however, as Hilbert clearly does, the mathematical existence of a (second level) concept is taken as the basis for inferring the existence of a system of mathematical objects falling under that concept, then a step is taken that is not justified in general (and no longer maintained in the Hilbert school during the 1920s). The issue is still a topic of wide ranging discussion, frequently in the context of the Frege-Hilbert correspondence.[64] The penetrating reflections of Bernays are articulated in the essay "Mathematische Existenz und Widerspuchsfreiheit" (Bernays 1950) and involve, centrally, a notion of *bezogene Existenz*, i.e., the view that mathematical existence claims are related to a *methodological frame*. A connection to such a broader frame is implicit already in Dedekind's and Hilbert's work.

In the background of Hilbert's considerations concerning arithmetic is *pure logic*, and that was also central for Dedekind.[65] Zermelo sharpened the *logical* framework of Dedekind and Hilbert to a *set theoretic* one and formulated its principles as axioms in his (1908) following Hilbert's structural ways. He considers a *domain* (Bereich) **B** of individuals, "which we call simply *objects* and among which are *sets*". The objects in **B** stand in *fundamental relations* of the form $a \in b$, and *sets* are those objects that have an element; the null or empty set is the only exception. The step from this set-up to the formulation of the characteristic conditions for the domain is taken with the sentence, "The fundamental relations of our domain **B**, now, are subject to the following axioms, or postulates." This classic paper then presents "the axioms of set theory" and

---

[63](Hilbert 1900, 55–56). Here is the German text: "Gelingt es jedoch zu beweisen, daß die dem Begriffe erteilten Merkmale bei Anwendung einer endlichen Anzahl von logischen Schlüssen niemals zu einem Widerspruche führen können, so sage ich, daß damit die mathematische Existenz des Begriffes ... bewiesen worden ist."

[64]See, for example, (Shapiro 2009). It seems that Hilbert's views around 1900 are complex and sometimes conflicting. An analysis of his lecture notes should provide valuable insights. However, there are a few significant remarks, e.g. (Hilbert 1900a, 1095), (Hilbert 1900b, 1105), and (Hilbert 1905, 137–138). In (Hilbert 1922, 158–9) one finds a slightly modified formulation that is, nevertheless, of real interest. Hilbert describes the axiomatic method as usual and asserts, "The continuum of real numbers is a system of things that are connected to each other by certain relations, so-called axioms. ... [C]onceptually a real number is just a thing of our system." He completes the discussion by adding, "The standpoint just described is altogether logically completely impeccable, and it only remains thereby undecided, whether a system of the required kind *can be thought [ist denkbar]*, i.e., whether the axioms do not lead to a contradiction." Here "Denkbarkeit" not "Existenz" (in some sense) is to be guaranteed by consistency.

[65]Logicist statements are found in Hilbert's early lectures throughout the 1890s; see (Sieg 2013, 83–4).

develops from them a theory of equivalence that avoids the formal use of cardinal numbers. Dedekind's way of metamathematically investigating simply infinite systems is taken up for *normal domains* (*Normalbereiche*) in (Zermelo 1930) or, to focus on the mathematical work without making a historical connection, Zermelo's analysis can be viewed as being carried out in parallel with Dedekind's. Any domain (of objects) is called *normal* if and only if it satisfies the axioms ZF', namely, the principles that allow constructing segments of the cumulative hierarchy along suitable ordinals (Separation, Pairing, Power Set, Union and Replacement) together with Extensionality and Foundation.[66] Zermelo's distinction between the abstract concept of a normal domain and concrete, particular normal domains reflects that between the abstract concept of a simply infinite system and concrete, particular instantiations.[67]

Zermelo's remarkable metamathematical investigations of normal domains do not, of course, establish a categoricity result akin to Dedekind's for simply infinite systems. But what they do show is that, given two normal domains $\boldsymbol{N}_1$ and $\boldsymbol{N}_2$ (with sets of *urelements* of the same size), either $\boldsymbol{N}_1$ is isomorphic to an "initial segment" of $\boldsymbol{N}_2$ or vice versa. It would lead us too far astray to look at the issues Zermelo raises concerning a new axiom that postulates "the existence of an unbounded sequence of boundary numbers" (*die Existenz einer unbegrenzten Folge von Grenzzahlen*), at Gödel's program of ever stronger axioms of infinity, or at Friedman's work on the effect of such axioms on ordinary mathematics. However, set theory can be viewed, and was viewed originally, from a less exalted perspective as a uniform framework for mathematics. In this spirit Zermelo wrote in (1908):

> Set theory is that branch of mathematics whose task is to investigate mathematically the fundamental notions "number", "order", and "function", taking them in their pristine form, and to develop thereby the logical foundations of all of arithmetic and analysis; thus it constitutes an indispensable component of the science of mathematics. (p. 200)

---

[66] The axiom of infinity is not viewed as part of *general* set theory, whereas the axiom of choice is being considered as a logical principle on the basis of which one can assume that every set is well-ordered.

[67] It is of interest to note that a very closely related distinction is extracted in (Martin 2010) from Gödel's work on set theory. Martin contrasts two senses of the "concept of set": the instances of the concept of set in "my [i.e., in Martin's] sense" are not sets, but rather concepts of set in "the straightforward sense"; the latter pertain to particular objects considered as sets. That distinction is explicated, most clearly on p. 358 and p. 361, and Martin argues that *his* sense of the concept of set "fits better than the straightforward one with much of what Gödel says about set theory". Relatedly, Parsons' "Platonism and Mathematical Intuition in Kurt Gödel's Thought", in particular sections 4 and 5 (pp. 167–185), is an illuminating discussion of the intricacies of Gödel's thinking on the issue. Feferman presents in his (2009) and (2014) a position he calls *conceptual structuralism*; it is prima facie close to Dedekind's structuralism. However, he does not sustain the distinction that Martin makes for set theory and that we have analyzed in Dedekind, but rather shifts between concepts and conceptions, i.e., abstract concepts and their instantiations.

Hilbert, who had stimulated Zermelo's interest in set theory and encouraged him to publish his investigations, made this connection later in a refined way:

> Set theory encompasses all mathematical theories (such as number theory, analysis, geometry) in the following sense: the relations that hold between objects of one of these mathematical disciplines are represented in a completely corresponding way by relations that obtain [between objects] in a sub-domain of Zermelo's set theory. (Hilbert 1920, 356)

In our way of speaking, the abstract concepts can be instantiated with sets in a segment of the cumulative hierarchy. From work in proof theory we know that, e.g., for the practice of analysis, only a very small part of even the first step in building the cumulative hierarchy is needed, when one takes natural numbers as urelements. So, (fragments of) set theory or other theories of "canonically generated" objects, such as the elements of constructive number classes, can serve as *methodological frames* in the sense of Bernays.[68] Bernays emphasized that his reflections on methodological frames apply not only to mathematics but also to the sciences.

Concluding Remarks

Lotze considered mathematics in his (1989, #18) to be rooted in logic, as did Dedekind, who according to our understanding was deeply influenced by Lotze's views on concept formation, in particular, on abstraction. Those views can be taken as a basis for integrating mathematical with ordinary and scientific knowledge. Let us look at one example in Dedekind, namely, the precise investigation of our *representations* (*Vorstellungen*) *of space*. Already in (*SZ*, 11), he remarks that space, "if it has a real existence", need not be continuous; however, nothing can prevent us from completing it in thought. Coming back to this very idea, he presents in (*WZ*, 793) an analytic model of Euclidean geometry that is nowhere continuous. So, for Dedekind the precise investigation of space, as asserted in (*WZ*, 791), is made possible by relating our representations to the "continuous number-realm"; the latter is obtained in arithmetic, which in turn has a "purely logical development" (*rein logischer Aufbau*).[69]

---

[68] See (Sieg 2013, Essay III.2) for an indication of the rich material that supports such a *reductive structuralism* in which methodological frames provide, in particular, the basis for consistency proofs, be they semantic or syntactic. In the spectrum of current structuralisms, it combines "relativist" and "predicate" structuralism as characterized in sections 4 and 8 of (Reck and Price 2000). It does so in a principled, but open-ended way, as the domains of "canonically generated" objects can be obtained by a variety of (constructive) operations. In Dedekind's case, the methodological framework is logic in the broad sense of, e.g., Lotze. The latter emphasizes in #18 of his *Logik* that "die Grundbegriffe und Grundsätze des Mathematik ihren systematischen Ort in der Logik haben".

[69] These matters are reflected in Hilbert's early lectures; see (Sieg 2013, 84–5).

Dedekind's Structuralism: Creating Concepts and Deriving Theorems 291

The idea of considering systems of objects structured in particular ways as corresponding to another realm was taken up by Hilbert already in 1893 for geometry, which he viewed as "the most perfect natural science"; using the notions "*System*" and "*Ding*" so prominent in *WZ*, he formulated the central question for the foundations of geometry as follows:

> What are the necessary and sufficient and mutually independent conditions a system of things has to satisfy, so that to each property of these things a geometric fact corresponds and conversely, thereby making it possible to completely describe and order all geometric facts by means of the above system of things? (Hilbert 1894, 72–3)

Four years later, in the notes for Hilbert's 1898–99 lectures, this remark is almost literally repeated; however, it is now explicitly connected to Hertz's *Prinzipien der Mechanik* and the last part of the above sentence (after "conversely") is replaced by, "thereby having these things provide a complete 'image' of geometric reality".[70]

This broad and open perspective on *theory, model* and *reality*, informed by logical, mathematical and scientific developments, was sustained in Hilbert's work in physics through the 1920s as well as in his comprehensive and ambitious Königsberg lecture of 1930, "Naturerkennen und Logik". It also pervades the reflections of philosophers of science. We are thinking, in particular, of Ernest Nagel and Patrick Suppes.[71] The work of Suppes is perhaps most directly shaped by the structuralist, Bourbakian approach of modern mathematics: the methodological framework is set theory; scientific theories are given as "set theoretic predicates" or, as we would say, by structural definitions; models of a theory are set theoretic structures that satisfy the conditions of the predicate articulating the theory; structure-preserving mappings are crucial for formulating and proving representation theorems. What is left quite open,

---

[70]This is followed by, "The axioms, as Hertz would say, are images or symbols in our mind, such that consequences of the images are again images of the consequences, i.e., what we derive logically from the images is true again in nature." To see how that is connected to parts of the contemporary discussion in philosophy of science, we recommend reading the "Introduction: the 'picture theory of science'" in van Fraassen's *Scientific Representation*. The general direction of Hilbert's considerations is further expounded in lectures he gave in 1902; see (Hallett and Majer 2004, 540–602). The description of the goals on pp. 540–1 is illuminating: for the construction of the "logisches Fachwerk von Begriffen" for geometry one assumes only the "Gesetze der reinen Logik und Zahlenlehre". This raises immediately the question, "Welche Sätze müssen wir diesem Bereich [der Logik und Zahlenlehre] adjungieren, damit wir die Geometrie aufbauen können?" The main question (Hauptfrage) is then formulated in almost exactly the same way as in (1894).

[71]See Nagel's remarks on the "construction of scientific concepts" (Nagel 1961, 14) and the two chapters of the same book on models (Chapters 5 and 6). As to Suppes, we are referring to his *Representation and Invariance of Scientific Structures*. His earlier expositions of this view, articulated in this book systematically and with a great number of examples, has deeply influenced the emergence of scientific structuralism; see, for example, Chapter 12 of (Suppes 1957).

in both Dedekind's and Hilbert's remarks, is the correspondence to "reality"; in Suppes, there is a crucial theory of measurement and measurement structures.

Dedekind's and Hilbert's introduction of abstract concepts had a transformative impact on the foundations and practice of mathematics in the 20th century. Our remarks in the last three paragraphs are only intended as an indication of the far-reaching consequences the radical transformation of mathematics in the 19th century had both on scientific practice and on informed philosophical reflections about the scientific enterprise. Conversely, the transformation of mathematics was, undoubtedly, also motivated by the co-evolution of mathematical and scientific developments. We are thinking especially, and very narrowly, of the work of Göttingen mathematicians—Gauss, Dirichlet, and Riemann—that exemplifies this co-evolution. Dedekind was very deeply associated with each of them, and Hilbert stood quite consciously in their tradition. Here is one important locus, in a significantly broader context, of a truly scientific philosophy: free from foundational dogma and open to both intellectual experience and experimentation.

Appendix. Manuscripts of *WZ*

Dedekind started to work intermittently on the foundations for natural numbers right after having completed *SZ*. The result of his work is assembled in a "First Draft" (Dedekind 1872/78); it is published in (Dugac 1976, 293–309), and its mathematical content is described in (Sieg and Schlimm 2005, 140–144). In the Preface to *WZ*, Dedekind remarks on p. iv that this earlier manuscript contains "all essential basic thoughts of my present essay". He mentions as critical points "the sharp distinction between the finite and the infinite", the concept of cardinal, and the justification for the proof principle of induction as well as for the definition principle of recursion. Though Dedekind does treat proof by induction in this manuscript, there are only very brief, almost cryptic remarks concerning definition by recursion.

The second draft was written in June and July of 1887; the third draft was written during the period from August to October 1887. (The dates are in Dedekind's hand on the cover pages of the documents; it should be mentioned that the third draft is still quite different from the published version.) In the second draft, which is incorporated into the third one, a dramatic shift occurs: the concept of a simply infinite system is introduced and major additional changes are made. This happens on page 5 of the manuscript that is fully reproduced below.

The page reproduced below reflects a perfectly standard pattern of writing manuscripts: a sheet is folded in the middle and one writes only on the left half, leaving the right half blank for later additions or corrections. So, the original text of §5 has the heading "*Die Reihe der natürlichen Zahlen*" and begins essentially with the text of #73 of *WZ*. Then, framed by Dedekind in the middle, the creation by abstraction of the elements $\mathfrak{n}$ of $\mathfrak{N}$ is described; we

Dedekind's manuscript

quoted this passage at the very end of B.3. On the right half, the new §5 has the title "*Die einfach unendlichen Systeme (Reihe der natürlichen Zahlen)*". The section begins with the definition (*Erklärung*) of the concept of a simply infinite system, essentially what is presented in #71 of *WZ*. The framed part of this page concerns the creation of $\mathfrak{N}$ and is left out in the published version.

Major changes from the second to the third draft are also found in §4, entitled "*Das Endliche und Unendliche*", where Dedekind formulates a new theorem with number 40*** that states: "*Es giebt unendliche Systeme.*" He added in parentheses that he had made "remarks on the supplementary sheet" (*Bemerkungen auf dem Beiblatt*). Unfortunately, the supplementary sheet seems not to have been preserved. All of these changes were thus made very late: Dedekind dated his Preface to *WZ*, "Harzburg, 5. Oktober 1887"; the final manuscript was delivered to the publishing house *Vieweg und Sohn* on 17 October 1887. That delivery date is mentioned in a letter Dedekind wrote on 30 October of that year to the publisher; here is the relevant passage:

> For several reasons it matters a great deal to me that the essay *Was sind und was sollen die Zahlen?*, I submitted as a manuscript on the 17th of this month, be delivered at Christmas; thus, I would like to express my strong desire that printing be started very soon. The setting up [of the manuscript], I believe, does not present any difficulties and, as far as I am concerned, the proofreading will be done by me as quickly as possible.[72]

We mention these details, as it is quite clear that Dedekind made significant, indeed dramatic changes and major additions in a very short period. In particular, the shift in perspective that is reflected in #73 and #134 (with the supporting metamathematical work) was made during that time. So it is perhaps not too surprising that some "tensions" remain between the old and the new perspective. Finally, it seems that Dedekind sent copies of the book to Cantor and Weber. His letter to Weber of 24 January 1888 is responding to remarks on *WZ* that Weber had sent him (and mentions that Cantor had also responded). As to the general reception, there is a report by Hilbert who, from 9 March to 8 April 1888, made his *Rundreise* from Königsberg to other German universities. He arrived in Berlin, his first stop, just as *WZ* had been published; Hilbert recounts that in mathematical circles everyone, young and

---

[72]Vieweg-Archive, V1D:17. The German text is as follows: "Da mir aus mehreren Gründen sehr daran gelegen sein muß, daß die am 17. d[ieses] M[onats] im Manuskript von mir eingelieferte Schrift "Was sind und was sollen die Zahlen?" zu Weihnachten ausgegeben werden kann, so erlaube ich mir hiermit meinen dringenden Wunsch auszusprechen, daß in diesen Tagen mit dem Druck begonnen werden möge. Der Satz bietet, wie ich glaube, gar keine Schwierigkeiten dar, und soviel an mir liegt, sollen die Korrekturen stets auf das Schnellste von mir besorgt werden". In his next letter of 13 November 1887 to his publisher, Dedekind complains that he still has not received the galleys and emphasizes that he really would like to be able to give two copies of his book as Christmas presents.

old, talked about Dedekind's essay, but mostly in an opposing or even hostile sense.[73]

Acknowledgments

Over the years, we have received detailed comments and critical, as well as supportive, remarks from a number of colleagues. Thus we are deeply grateful to J. Avigad, P. Blanchette, J. Burgess, P. Corvini, J. Floyd, A. Kanamori, A. Klev, E. Reck, G. Schiemer, W.W. Tait, N. Tennant, M. van Atten and M. Wilson.

References

The translations of German texts are mostly our own, unless they are quoted from an English edition. We thank Dr. Helmut Rohlfing from the Niedersächsische Staats- und Universitäts-Bibliothek in Göttingen for access to the unpublished Dedekind manuscripts. Similar thanks go to Klaus D. Oberdiek who helped us get access to Dedekind's correspondence with his publishing house, Vieweg; that correspondence is preserved in the Vieweg-Archive of the Universitätsbibliothek Braunschweig.

Avigad, J. and R. Morris (2014). The concept of "character" in Dirichlet's theorem on primes in an arithmetic progression. *Archive for History of Exact Sciences 68*(3), 265–326.

Awodey, S. and E. H. Reck (2002a). Completeness and categoricity. part I: Nineteenth-century axiomatics to twentieth-century metalogic. *History and Philosophy of Logic 23*(1), 1–30.

Awodey, S. and E. H. Reck (2002b). Completeness and categoricity, part II: Twentieth-century metalogic to twenty-first-century semantics. *History and Philosophy of Logic 23*(2), 77–94.

Beaney, M. (2012). Analysis. *The Stanford Encyclopedia of Philosophy.* Available online at http://plato.stanford.edu/archives/sum2012/entries/analysis/.

Bernays, P. (1930). Die Philosophie der Mathematik und die Hilbertsche Beweistheorie. *Blätter für Deutsche Philosophie 4*, 326–367. Reprinted in (Bernays 1976, 17–61).

Bernays, P. (1950). Mathematische Existenz und Widerspuchsfreiheit. *Études des Philosophie des Sciences*, 11–25. Reprinted in (Bernays 1976, 92–106).

---

[73]That is reported in (Hilbert 1931, 487), but also in Hilbert's diary from his trip; the latter is part of the Hilbert Nachlass, Cod. MS. 741, 1/5. See also (Dugac 1976, 203).

Bernays, P. (1976). *Abhandlungen zur Philosophie der Mathematik.* Wissenschaftliche Buchgesellschaft.

Bourbaki, N. (1950). The architecture of mathematics. *Mathematical Monthly* 57, 221–232.

Burgess, J. (2011). Structure and rigor. Manuscript of a talk given at Oxford University.

Cassirer, E. (1910). *Substanzbegriff und Funktionsbegriff.* Bruno Cassirer.

Corry, L. (2004a). *David Hilbert and the Axiomatization of Physics (1989–1918).* Kluwer Academic Publishers.

Corry, L. (2004b). *Modern Algebra and the Rise of Mathematical Structures, Second Revised Edition.* Springer Science and Business Media.

Dedekind, R. (1854). Über die Einführung neuer Funktionen in der Mathematik. Habilitationsvortrag. Published in (Dedekind 1932, 428–438) and translated in (Ewald 1996, 755–762).

Dedekind, R. (1863–1901). *Correspondence with Vieweg.* Vieweg-Archive der Universitätsbibliothek Braunschweig, V1D: 17.

Dedekind, R. (1872). *Stetigkeit und irrationale Zahlen.* Friedrich Vieweg und Sohn.

Dedekind, R. (1872–1878b). Was sind und was sollen die Zahlen? (Erster Entwurf). Niedersächsische Staats- und Universitätsbibliothek, Göttingen, Cod. Ms. Dedekind III, 1:I. Printed in (Dugac 1976, 293–309).

Dedekind, R. (1876). Letters to Lipschitz. Published in (Dedekind 1932, 464–482).

Dedekind, R. (1877). Sur la théorie des nombres entiers algébriques. *Bulletin des sciences mathématiques et astronomiques 1 (XI)*, 1–121. Partially reprinted in (Dedekind 1932, 262–296). Translated in (Stillwell 1996).

Dedekind, R. (1878a). Letter to Weber. Published in (Dedekind 1932, 485–486).

Dedekind, R. (1887). Manuscript of *Was sind und was sollen die Zahlen?*, third version. Niedersächsische Staats- und Universitätsbibliothek, Göttingen , Cod. Ms. Dedekind, III, 1:III:. This is a continuation written in August through October of 1887; the beginning was written in June and July of the same year and is catalogued as Cod. Ms. Dedekind, III, 1:II.

Dedekind, R. (1888a). Letter to Weber. Published in (Dedekind 1932, 488–490).

Dedekind, R. (1888b). *Was sind und was sollen die Zahlen?* F. Vieweg. References and page numbers are made to the English translation in (Ewald 1996, 790–833).

Dedekind, R. (1890). Letter to Keferstein. The German text is found in (Sinaceur 1974). The letter was translated into English and published in (van Heijenoort 1967, 98–103).

Dedekind, R. (1932). *Gesammelte mathematische Werke*, Volume 3. F. Vieweg. R. Fricke, E. Noether, Ö. Ore (eds).

Dedekind, R. (1964). *Über die Theorie der ganzen algebraischen Zahlen, with an Introduction by B. van der Waerden*. Vieweg.

Dedekind, R. and H. Weber (1882). Theorie der algebraischen Funktionen einer Veränderlichen. *Journal für reine und angewandte Mathematik 92*, 181–290.

Deiser, O. (2010). Introductory note to Zermelo 1901. In H. D. Ebbinghaus, C. G. Fraser, and A. Kanamori (Eds.), *Ernst Zermelo–Collected Works/Gesammelte Werke*, Volume 21 of *Schriften der Mathematisch-naturwissenschaftlichen Klasse der Heidelberger Akademie der Wissenschaften*, pp. 52–70. Springer.

Dieudonné, J. (1970). The work of Nicholas Bourbaki. *American Mathematical Monthly 70*, 134–145.

Dirichlet, J. P. G. L. and R. Dedekind (1863). *Vorlesungen über Zahlentheorie (1st edition)*. F. Vieweg.

Dirichlet, J. P. G. L. and R. Dedekind (1871). *Vorlesungen über Zahlentheorie (2nd edition)*. F. Vieweg.

Dirichlet, J. P. G. L. and R. Dedekind (1879). *Vorlesungen über Zahlentheorie (3rd edition)*. F. Vieweg.

Dirichlet, J. P. G. L. and R. Dedekind (1894). *Vorlesungen über Zahlentheorie (4th edition)*. F. Vieweg.

Dugac, P. (1976). *Richard Dedekind et les fondements des mathématiques (avec de nombreux textes inédits)*. Librairie Philosophique J. Vrin, Paris.

Ewald, W. B. (Ed.) (1996). *From Kant to Hilbert: A Source Book in the Foundations of Mathematics*, Volume II. Oxford University Press.

Ewald, W. B. and W. Sieg (Eds.) (2013). *David Hilbert's Lectures on the Foundations of Arithmetic and Logic 1917–1933*. Springer.

Feferman, S. (2009). Conceptions of the continuum. *Intellectica 51*, 169–189.

Feferman, S. (2014, to appear). Logic, mathematics and conceptual structuralism. In P. Rush (Ed.), *Metaphysics of Logic*. Cambridge University Press.

Feferman, S., J. W. Dawson, S. C. Kleene, G. H. Moore, R. M. Solovay, and J. van Heijenoort (Eds.) (1986). *Kurt Gödel Collected Works*, Volume I. Oxford University Press.

Feferman, S., C. Parsons, and S. Simpson (Eds.) (2010). *Kurt Gödel: Essays for His Centennial*. Cambridge University Press.

Ferreirós, J. (2007). *Labyrinth of Thought: A History of Set Theory and Its Role in Modern Mathematics Second Revised Edition*. Springer Science and Business Media.

Fine, K. (2002). *The Limits of Abstraction*. Oxford University Press.

Floyd, J. (2013). The varieties of rigorous experience. In M. Beaney (Ed.), *The Oxford Handbook of the History of Analytic Philosophy*. Oxford University Press.

Frege, G. (1980). *Gottlob Freges Briefwechsel*. Felix Meiner Verlag.

Gödel, K. (1929). *Über die Vollständigkeit des Logikkalküls*. Ph. D. thesis, University of Vienna. Translated and reprinted in (Feferman et al. 1986, 60–101).

Hallett, M. (1984). *Cantorian Set Theory and the Limitation of Size*. Oxford University Press.

Hallett, M. and U. Majer (Eds.) (2004). *David Hilbert's Lectures on the Foundations of Geometry 1891–1902*. Springer.

Hasse, H. (1930). Die moderne algebraische Methode. *Jahresbericht der Deutschen Mathematiker-Vereinigung 39*, 22–34.

Hertz, H. (1894). *Die Prinzipien der Mechanik in neuem Zusammenhange dargestellt*. Leipzig.

Hilbert, D. (1894). Die Grundlagen der Geometrie. This was published in (Hallett and Majer 2004, 72–127).

Hilbert, D. (1898–99). Lectures on Euclidean Geometry. This was published in (Hallett and Majer 2004, 221–286).

Hilbert, D. (1899). *Grundlagen der Geometrie*. B.G. Teubner, Leipzig. This was published in (Hallett and Majer 2004, 436–525).

Hilbert, D. (1900a). Mathematische Probleme. Lecture given at the International Congress of Mathematicians in Paris. This is partially translated in (Ewald 1996, 1096–1105).

Hilbert, D. (1900b). Über den Zahlbegriff. *Jahresbericht der Deutschen Mathematiker-Vereinigung 8*. This is translated in (Ewald 1996, 1092–1095).

Hilbert, D. (1905). Über die Grundlagen der Logik und der Arithmetik. *Proceedings from the Third International Congress of Mathematicians*, 174–185. This is translated in (van Heijenoort 1967, 130–138).

Hilbert, D. (1920). Probleme der mathematischen Logik. This was published in (Ewald and Sieg 2013, 342–377).

Hilbert, D. (1922). Neubegründung der Mathematik. *Abhandlungen aus dem mathematischen Seminar der Hamburgischen Universität 1*, 157–177. Translated in (Ewald 1996, 1115–1134).

Hilbert, D. (1931). Die Grundlegung der elementaren Zahlenlehre. *Mathematische Annalen 104*, 485–494. Translated in (Ewald 1996, 1148–1157).

Hurwitz, A. and R. Courant (1922). *Vorlesungen über allgemeine Funktionentheorie und elliptische Funktionen*. J. Springer.

Kant, I. (1800). *Logik*. Friedrich Nicolovius. Edited by G. B. Jäsche. Published in volume 3 of (Weischedel 1958, 421–582).

Kant, I. (1992). *Lectures on Logic*. Cambridge University Press. Translated and edited by J. M. Young.

Kant, I. (1998). *Critique of Pure Reason*. Cambridge University Press. Translated and edited by P. Gruyer and A. W. Wood.

Klev, A. (2011). Dedekind and Hilbert on the foundations of the deductive sciences. *The Review of Symbolic Logic 4*, 645–681.

Klev, A. (2014). The nature and meaning of mappings. Manuscript.

Knopp, K. (1930). *Funktionentheorie I: Grundlagen der allgemeinen Theorie der analytischen Funktionen*. de Gruyter.

Linnebo, Ø. and R. Pettigrew (2014). Two types of abstraction for structuralism. *The Philosophical Quarterly 64*(255), 267–283.

Lotze, R. (1989). *Logik, Erstes Buch. Vom Denken*. Meiner Verlag.

Martin, D. A. (2005). Gödel's conceptual realism. *The Bulletin of Symbolic Logic 11*(2). Reprinted in (Feferman et al. 2010, 356–373).

Mehrtens, H. (1979). Das Skelett der modernen Algebra. Zur Bildung mathematischer Begriffe bei Richard Dedekind. In C. Scriba (Ed.), *Disciplinae Novae*, pp. 25–43. Vandenhoeck and Ruprecht.

Nagel, E. (1961). *The Structure of Science: Problems in the Logic of Scientific Explanation*. Routledge.

Parsons, C. D. (1987). Developing arithmetic in set theory without infinity: Some historical remarks. *History and Philosophy of Logic 8*, 201–213.

Parsons, C. D. (1990). The structuralist view of mathematical objects. *Synthese 84*(3).

Parsons, C. D. (1995). Platonism and mathematical intuition in Kurt Gödel's thought. *The Bulletin of Symbolic Logic 1*, 44–74. Reprinted in (Parsons 2014, 153–187).

Parsons, C. D. (2014). *Philosophy of Mathematics in the Twentieth Century— Selected Essays*. Harvard University Press.

Pettigrew, R. (2008). Platonism and Aristotelianism in mathematics. *Philosophia Mathematica 16*(3), 310–332.

Reck, E. H. (2003). Dedekind's structuralism: An interpretation and partial defense. *Synthese 137*(3).

Reck, E. H. and M. P. Price (2000). Structures and structuralism in contemporary philosophy of mathematics. *Synthese 125*(3), 341–383.

Schiemer, G. and E. H. Reck (2013). Logic in the 1930s: Type theory and model theory. *Bulletin of Symbolic Logic 19*(4), 433–472.

Shapiro, S. (1997). *Philosophy of Mathematics: Structure and Ontology*. Oxford University Press.

Shapiro, S. (2009). Categories, structures, and the Frege-Hilbert controversy: the status of meta-metamathematics. *Philosophia Mathematica 13*, 61–77.

Sieg, W. (2013). *Hilbert's Programs and Beyond*. Oxford University Press.

Sieg, W. (2014). The ways of Hilbert's axiomatics: Structural and formal. *Perspectives on Science 22*(1), 133–157.

Sieg, W. and D. Schlimm (2005). Dedekind's analysis of number: Systems and axioms. *Synthese 147*.

Sieg, W. and D. Schlimm (2014). Dedekind's abstract concepts: models and mappings. *Philosophia Mathematica*. doi: 10.1093/philmat/nku021.

Sinaceur, M. (1974). L'infini et les nombres. Commentaires de R. Dedekind à <<Zahlen>>. La correspondance avec Keferstein. *Revue d'histoire des sciences 27*.

Stillwell, J. (Ed.) (1996). *Theory of Algebraic Integers*. Oxford University Press.

Suppes, P. (1957). *Introduction to Logic*. D. van Nostrand Company.

Suppes, P. (2002). *Representation and Invariance of Scientific Structures*. CSLI Publications.

Tait, W. W. (1996). Frege versus Cantor and Dedekind: on the concept of number. In W. W. Tait (Ed.), *Frege, Russell, Wittgenstein: Essays in Early Analytic Philosophy (in honor of Leonard Linsky)*. Lasalle: Open Court Press.

van der Waerden, B. L. (1930). *Moderne Algebra, Teil I*. Springer-Verlag.

van der Waerden, B. L. (1931). *Moderne Algebra, Teil II*. Springer-Verlag.

van Fraassen, B. (2008). *Scientific Representation: Paradoxes of Perspective*. Oxford University Press.

van Heijenoort, J. (Ed.) (1967). *From Frege to Gödel: a source book in mathematical logic, 1879-1931*. Harvard University Press.

Weischedel, W. (Ed.) (1958). *Immanuel Kant Werke in Sechs Bänden*. Insel Verlag.

Yap, A. (2009). Logical structuralism and Benacerraf's problem. *Synthese 171*(1), 157–173.

Zermelo, E. (1908). Untersuchungen über die Grundlagen der Mengenlehre I. *Mathematische Annalen 65*, 261–281. Translated in (van Heijenoort 1967, 200–215).

Zermelo, E. (1930). Über Grenzzahlen und Mengenbereiche. Neue Untersuchungen über die Grundlagen der Mengenlehre. *Fundamenta Mathematica 16*, 29–47. Translated in (Ewald 1996, 1219–1233).

Bibliography of Works by W. W. Tait

Books

Tait, W. W. (2005): *The Provenance of Pure Reason: Essays in the Philosophy of Mathematics and its History*, Oxford University Press: Oxford

(in progress): *Meaning, Understanding, and Knowing: A View from Outside*, manuscript

(in progress): *Lectures on the Infinite*, manuscript

Edited Books

Tait, W. W., ed. (1997): *Frege, Russell, Wittgenstein. Essays in Early Analytic Philosophy (in Honor of Leonard Linsky)*, Open Court: Lasalle, IL

Edited Journal Issues

Tait, W. W., guest ed. (1990): *The Philosophy of Mathematics, Parts 1–2, Synthese* 84, Numbers 2–3 (special issues)

Articles

Tait, W. W. (1959): A counterexample to a conjecture of Scott and Suppes, *Journal of Symbolic Logic* 24, 15–16

(1961): Nested recursion, *Mathematische Annalen* 143, 236–250

(1963): A second-order theory of functionals of higher-type. Appendix A: Intensional functions. Appendix B: An interpretation of functionals by convertible terms, *Stanford Seminar Report*, 171–206; published as Tait (1967)

(1965a): Functionals defined by transfinite recursion, *Journal of Symbolic Logic* 30, 155–174

(1965b): Infinitely long terms of transfinite type I, in *Formal Systems and Recursive Functions*, J. Crossley & M. Dummett, eds., North-Holland: Amsterdam, pp. 176–185

(1965c): The substitution method, *Journal of Symbolic Logic* 30, 175–192

(1966): A non-constructive proof of Gentzen's Hauptsatz for second-order predicate logic, *Bulletin of the American Mathematical Society* 72, 980–983

(1967): Intensional interpretation of functionals of finite type I, *Journal of Symbolic Logic* 32, 198–212

(1968a): Constructive reasoning, in *Logic, Methodology, and Philosophy of Science III*, B.v, Rootselaar & F. Staal, eds., North-Holland: Amsterdam, pp. 185–199

(1968b): Normal derivability in classical logic, in *The Syntax and Semantics of Infinitary Languages*, J. Barwise, ed., Lecture Notes in Mathematics 72, Springer: Berlin, pp. 204–236

(1970): Applications of the cut-elimination theorem to some subsystems of classical analysis, in *Intuitionism and Proof Theory*. Proceedings of the Summer Conference at Buffalo N.Y. 1968, A. Kino, J. Myhill & R. Vesley, eds, pp. 475–488

(1971): Normal form theorem for bar recursive functionals of finite type, in *Proceedings of the Second Scandinavian Logic Symposium*, North-Holland: Amsterdam, J. Fenstad, ed., pp. 353–367

(1981): Finitism, *Journal of Philosophy* 78, 524–556; reprinted as Ch. 1 of Tait (2001), pp. 21–42

(1983): Against intuitionism. Constructive mathematics is part of classical mathematics, *Journal of Philosophical Logic* 12, 175–195

(1986a): Truth and proof: The Platonism of mathematics, *Synthese* 69, 341–370; reprinted as Ch. 3 of Tait (2005a), pp. 61–88

(1986b): Wittgenstein and the 'skeptical paradoxes', *Journal of Philosophy* 83, 475–488; reprinted as Ch. 9 of Tait (2005a), pp. 198–211

(1986c): Plato's second-best method, *The Review of Metaphysics* 39, 455–482; reprinted as Ch. 7 of Tait (2005a), pp. 155–177

(1990b): The iterative hierarchy of sets, *Iyyun: A Jerusalem Philosophical Quarterly* 39, 65–79

(1992): Reflections on the concept of a priori truth and its corruption by Kant, in *Proof and Knowledge in Mathematics*, M. Detlefsen, ed., Routledge: London, pp. 33–64

(1993): Some recent essays in the history of the philosophy of mathematics: A critical review, *Synthese* 96, 293–331

(1994): The law of excluded middle and the axiom of choice, in *Mathematics and Mind*, A. George, ed., pp. 45–70; reprinted as Ch. 5 of Tait (2005a), pp. 105–132

(1995): Extensional equality in the classical theory of types, *Vienna Circle Institute Yearbook* 3, 219–234

(1996): Frege versus Cantor and Dedekind: On the concept of number, in *Frege: Importance and Legacy*, M. Schirn, ed., Berlin: DeGruyter, pp. 70–113; republished in, but originally intended for, Tait, (1997a), pp. 213–248; reprinted as Ch. 10 of Tait (2005a), pp. 212–251

(1998a): Zermelo's conception of set theory and reflection principles, in *The Philosophy of Mathematics Today*, M. Schirn, ed., Clarendon Press: Oxford, pp. 469–483

(1998b): Variable free formalization of the Curry-Howard theory of types, in *Twenty-Five Years of Constructive Type Theory*, G. Sambin & J. Smith, eds., Oxford University Press: Oxford, pp. 265–274

(2000): Cantor's *Grundlagen* and the paradoxes of set theory, in *Between Logic and Intuition. Essays in Honor of Charles Parsons*, G. Sher & R. Tieszen, eds., Cambridge University Press: Cambridge, pp. 269–290; reprinted as Ch. 11 of Tait (2005a), pp. 252–275

(2001a): Beyond the axioms: The question of the objectivity of mathematics, *Philosophia Mathematica* (III) 9, 21–36; reprinted as Ch. 4 of Tait (2005a), pp. 89–104

(2001b): Gödel's unpublished papers on foundations of mathematics, *Philosophia Mathematica* (III) 9, 87–126; reprinted as Ch. 12 of Tait (2005a), pp. 276–313

(2002a): Noesis: Plato on exact science, in *Reading Natural Philosophy. Essays in the History and Philosophy of Science and Mathematics (in Honor of Howard Stein)*, D. Malament, ed., Open Court: Chicago, pp. 11–30; reprinted as Ch. 8 of Tait (2005a), pp. 178–197

(2002b): Remarks on finitism, in *Reflections on the Foundations of Mathematics. Essays in Honor of Solomon Feferman*, Lecture Notes in Logic 15, 410–419; reprinted as Ch. 2 in Tait (2005), pp. 43–53

(2002c): The myth of the mind, *Topoi* 21 (1–2), 65–74

(2003): The completeness of Heyting first-order logic, *Journal of Symbolic Logic* 68, 751–763

(2005b): Constructing cardinals from below; originally published as Ch. 6 of Tait (2005a), pp. 133–154

(2005c): Gödel's reformulation of Gentzen's first consistency proof for arithmetic: The no-counter-example interpretation, *Bulletin of Symbolic Logic* 11, 225–238

(2005d): Introduction; originally published in Tait (2005a), pp. 3–20

(2006a): Gödel's correspondence on proof theory and constructive mathematics, *Philosophia Mathematica* (III) 14, 76–111

(2006b): Gödel's interpretation of intuitionism, *Philosophia Mathematica* (III) 14, 208–228

(2006c): Proof-theoretic semantics for classical mathematics, *Synthese* 148, 603–622

(2008): The five questions, in *Philosophy of Mathematics. Five Questions*, V. Hendricks & H. Leitgeb, eds., Automatic Press, pp. 249–263

(2010a): Gödel on intuition and on Hilbert's finitism, in *Kurt Gödel. Essays for his Centennial*, S. Feferman, C. Parsons & S. Simpson, eds.,

Cambridge University Press: Cambridge, pp. 88–108

(2010b): The substitution method revisited, in *Proofs, Categories, and Computations. Essays in Honor of Grigori Mints*, S. Feferman & W. Sieg, eds., College Publications, pp. 231–241

(2012a): Primitive recursive arithmetic and its role in the foundations of arithmetic: Historical and philosophical reflections, in *Epistemology versus Ontology. Essays in the Philosophy of Mathematics* (in Honor of Per Martin-Löf), P. Dybjer, S. Lindström, E. Palmgren & G. Sundholm, eds., Springer, pp. 161–180

(2012b): The locus of grammatical-logical norms in Wittgenstein's account of meaning and understanding, in *Wittgenstein. Zu Philosophie und Wissenschaft*, P. Stekeler-Weithofer, ed., Hamburg: Meiner Verlag, pp. 150–169

(2015a): Gentzen's original consistency proof and the Bar Theorem, in *Gentzen's Centenary: The Quest for Consistency*, R. Kahle & M. Rathjen, eds., Springer, pp. 213–228

(2015b): First-order logic without bound variables: Compositional semantics, in *Dag Prawitz on Proofs and Meaning*, H. Wansing, ed., Springer, pp. 359–384

(2016): Kant and finitism, *Journal of Philosophy* 113:5/6 (special issue in honor of Charles Parsons), 261–273

(forthcoming a): Kant on 'number', in *Kant's Philosophy of Mathematics: The Critical Philosophy and its Roots*, Vol. I, O. Rechter & C. Posy, eds., Cambridge University Press

(forthcoming b): In defense of the ideal, in *Exploring the Frontiers of Incompleteness*, P. Koellner & C. Parsons, eds.

(in progress): Cut-elimination for deductions in Π_1^1-CA with the omega-rule, manuscript

Co-Authored Work

Edited Books

Sieg, W., Tait, W. W., et al., eds. (in preparation): *Paul Bernays: Essays in the Philosophy of Mathematics, Volumes I–II* (bilingual edition: German/English), Oxford University Press

Articles

Gandy, R., Kreisel, G. & Tait, W. W. (1960/61): Set existence, *Bulletin of the Polish Academy of Science* 8 (1960), pp. 577–582, and 9 (1961), pp. 881–882

Kreisel, G. & Tait, W. W. (1961): Finite definability of number-theoretic functions and parametric completeness of equational calculi, *Archiv für Mathematische Logik* 7, 28–38

Norman, D. & Tait, W. W. (2018): On the computability of the Fan Functional: In memory of Solomon Feferman, in *Feferman on Foundations: Logic, Mathematics, Philosophy*, W. Sieg et al., eds., Springer, pp. 57–69

Information about Contributors

Awodey, Steve
Departments of Philosophy and Mathematics
Carnegie Mellon University
Email: awodey@cmu.edu

Feferman, Solomon
Department of Philosophy
Stanford University
(Prof. Feferman passed away in 2016)

Friedman, Michael
Department of Philosophy
Stanford University
Email: mlfriedm@stanford.edu

Goldfarb, Warren
Department of Philosophy
Harvard University
Email: goldfarb@fas.harvard.edu

Hellman, Geoffrey
Department of Philosophy
University of Minnesota at Minneapolis
Email: hellm001@umn.edu

Howard, William
Department of Mathematics
University of Illinois at Chicago
Email: wahow@uic.edu

Menn, Stephen
Department of Philosophy
McGill University
Email: stephen.menn@mcgill.ca

Morris, Rebecca
Department of Philosophy
Stanford University
Email: email@rebeccaleamorris.com

Parsons, Charles
Department of Philosophy
Harvard University
Email: parsons2@fas.harvard.edu

Reck, Erich
Department of Philosophy
University of California at Riverside
Email: erich.reck@ucr.edu

Ricketts, Thomas
Department of Philosophy
University of Pittsburgh
Email: ricketts@pitt.edu

Sieg, Wilfried
Department of Philosophy
Carnegie Mellon University
Email: sieg@cmu.edu

Name Index

Abel, Niels 102
Acerbi, Fabio 185, 189n.7, 197n.21, 206n.37, 214n.52, 216n.55
Avigad, Jeremy 16n.7, 17n.8, 18n.11, 18n.12, 22n.17, 25, 256n.8, 295
Ackermann, Wilhelm 16, 17, 18, 20, 282n.53
Aczel, Peter 73
Alexander of Aphrodisias 195, 196
Allegri, Linda 205n.36
Anderson, Alan 1, 15n.6
Archimedes 185, 188, 189, 197n.20, 204-208, 209n.42, 212, 213, 218, 220, 229, 230
Aristotle 9, 185, 187n.3, 188, 190-n.10, 194-197, 201-203, 207-n.40, 211, 215, 220n.62, 258-n.13, 273
Awodey, Steve 5n.10, 7n.14, 10, 60, 65, 68n.7, 131n.32, 266n.32

Bachmann, Heinz 47
Barwise, John 15n.4
Baumgartner, James 78n.3
Beaney, Michael 258n.13
Beck, Lewis White 1
Becker, Oskar 194, 196n.18, 197n.21, 198, 199, 200n.26, 201, 202-n.32, 205, 210, 214n.51, 215, 220n.62
Benacerraf, Paul 99-101

Bernays, Paul 6, 15, 97, 100, 105, 281n.51, 288, 290
Bernstein, Felix 277n.43
Blanchette, Patricia 295
Brouwer, L. E. J. 1, 4-7, 19, 34-36, 156
Boolos, George 77, 82, 83, 97
Borelli, Giovanni 202n.32, 205, 209, 212n.48, 216n.55, 219n.58, 220, 222n.64, 225n.70, 225-n.71
Bourbaki, Nicolas 252, 286, 291
Bowen, Alan C. 187n.3
Bryant, Sophie 201n.28
Buchholtz, Ulrich 53
Buchholz, Wilfried 24
Burali-Forti, Cesare 6, 174
Burgess, John 108, 120n.13, 282, 295
Butz, Carsten 60

Cantor, Georg 6, 8, 17, 25, 78, 81, 82, 84, 86-88, 113, 114, 118, 121, 128, 130, 131-135, 234, 235, 238, 242n.36, 283n.55, 294
Carey, Susan 96
Carnap, Rudolf 10, 95, 100n.30, 141-168, 172
Cassirer, Ernst 119, 121, 135, 283-n.55
Chemla, Karine 193n.14
Christianidis, Jean 194n.15

Church, Alonzo 23n.19, 37, 47
Cohen, Paul 104
Conant, James 183
Corry, Leo 252n.2, 253n.3
Courant, Richard 256
Corvini, Pat 295
Creath, Richard 155n.26
Curry, Haskell 5, 41, 47, 55

Davidson, Donald 97
Davis, Martin 2
De Risi, Vincenzo 185, 205n.36, 205-n.36, 212
Dedekind, Richard 6, 8, 10, 25, 54, 104, 113-125, 127, 128, 130, 131, 133-136, 179, 188, 251-267, 269-290, 292-295
Democritus 189, 190n.10, 229
Demopoulos, William 141, 167n.48
Dieudonné, Jean 286n.60
Dijksterhuis, Eduard 196n.18
Dirichlet, Peter Lejeune 253, 256, 257, 275, 276, 292
Dositheus 189n.7
Drake, Frank 78n.1, 85n.19
Dreben, Burton 234n.10
Dummett, Michael 2, 5, 7, 9, 15n.4, 114, 117, 118, 120, 121, 232, 247n.50
Dyson, Verena 15

Ehrenfeucht, Andrzej 15n.4
Eilenberg, Samuel 68
Einstein, Albert 160, 161n.38, 162, 163n.44, 164, 165, 167n.48
Erdös, Paul 85
Euclid 5, 72, 104n.39, 162, 163, 185-188, 190-193, 195-201, 203, 205-230, 276, 287, 290
Eudemus 187n.3
Eudoxus 185-230
Ewald, William 265n.29, 272n.35

Feferman, Solomon 2, 4, 10, 14n.3, 15, 17n.8, 18n.10, 18n.11, 18n.12, 20n.15, 22, 22n.17, 23, 24, 35, 37, 48, 91, 92, 95, 96, 289n.67
Fermat, Pierre 101
Ferreirós, José 259n.19
Field, Hartry 8
Fine, Kit 83n.15, 131n.32
Fitch, Frederick 1, 15n.6
Floyd, Juliet 283n.54, 295
Føllesdal, Dagfinn 91
Fowler, David 189n.7
Fraenkel, Abraham 69, 78n.2, 154, 155
Frascolla, Pasquale 179
Frauchiger, Michael 93n.5
Frege, Gottlob 6, 8-10, 25, 56, 83, 92-98, 101, 102, 113, 114, 117, 118, 119n.12, 121, 125-128, 130-135, 142-144, 146, 147, 153n.16, 155, 157n.29, 158-161, 165, 167n.48, 171, 172, 174, 177, 180-182, 231-248, 260, 264, 285n.58, 287-n.62, 288
Friedman, Harvey 2, 14, 22, 23, 24, 141,
Friedman, Michael 7n.14, 10, 119-n.11, 155n.26
Fogelin, Robert 177

Galilei, Galileo 205n.36, 220n.62
Galois, Evariste 252
Gambino, Nicola 60
Gandy, Robin 2, 13n.1, 15n.4
Garner, Richard 60, 62n.4
Gauss, Carl Friedrich 253, 292
Geach, Peter 177
Gentzen, Gerhard 3-5, 17, 20, 21, 22n.17, 24, 33, 38, 40, 41, 48
George, Alexander 92n.2
Girard, Jean-Yves 18
Giusti, Enrico 205n.36, 220n.62
Gödel, Kurt 2-6, 8, 15, 18, 19, 22-n.17, 25, 33-35, 37-41, 46,

Name Index

47, 81n.11, 83, 92-97, 98-n.19, 103, 105, 106n.45, 107-109, 149, 153-155, 274n.37, 282n.53, 289
Goldberg, Samuel 1
Goldfarb, Warren 7n.14, 10, 231
Gowers, Timothy 102n.34
Gratzl, Norbert 124n.23
Grothendieck, Alexander 62
Grünbaum, Adolph 1, 15n.6
Guillaumin, Jean-Yves 206n.39

Hale, Robert 125
Halmos, Paul 2
Halstead, George 205n.36
Hailperin, Theodore 1, 15n.6
Harrison, Joseph 15
Hartogs, Friedrich 82
Hasse, Helmut 286n.60
Heath, T. L. 187, 189n.8, 194, 196-n.18, 201n.28, 205n.36, 225-n.71
Heck, Richard 231, 241, 242, 247, 248
Heiberg, Johan 187n.4, 188n.6, 193-n.14, 204n.35, 225n.71
Heis, Jeremy 118n.7
Hellman, Geoffrey 6n.13, 10, 82n.12, 86n.20, 123n.20
Helmholtz, Hermann 161n.38
Henkin, Leon 1, 85n.17, 154n.21
Herbrand, Jacques 22n.17, 24
Hertz, Heinrich 291
Heyting, Arend 1, 36, 146
Hilbert, David 3, 4, 6-9, 15, 16, 19, 83, 113, 114, 121, 123-125, 127, 129, 130, 132, 133, 135, 146-148, 157, 159, 160n.36, 162, 163, 165, 235n.18, 252, 253, 258n.14, 259-261, 265, 266n.32, 277n.43, 279, 281-n.51, 282n.53, 286-288, 290-292, 294, 295n.73
Hintikka, Jaakko 2, 14
Hippia 190n.10

Hippocrates of Chios 187n.3, 190-n.9, 190n.10
Hofmann, Martin 62, 289n.67
Howard, William 2, 4, 5, 10, 13, 15, 18, 40, 41, 47, 55
Hume, David 125, 167n.48, 235, 238
Hurwitz, Adolf 256
Husserl, Edmund 5, 7, 118n.9
Hylton, Peter 183

Isaacson, Daniel 91, 92, 95n.10, 98-101, 102n.36, 102n.37, 103, 105

Jäger, Gerhard 24

Kamareddine, Fairouz 55
Kanamori, Akihiro 82n.13, 259n.18, 295
Kant, Immanuel 5, 6, 9, 94, 96, 119, 141-143, 164, 181, 234, 258-n.13, 258n.14, 262n.25, 283, 284
Kapulkin, Chris 69
Keferstein, Hans 254, 258, 261-263, 265, 266, 284, 285
Keller, Pierre 119n.11
Kelley, John L. 88
Kennedy, Juliette 107n.47
Kleene, Stephen 15, 23n.19, 34, 36, 37
Klev, Ansten 114n.2, 255n.6, 259-n.19, 295
Knopp, Konrad 256n.9
Knorr, Wilbur 185, 186, 187n.3, 189-n.7, 190n.9, 190n.10, 193-n.14, 194, 194n.15, 197n.20, 197n.21, 200n.26, 202n.32, 203n.34, 204-207, 209, 211, 212n.48, 213, 214, 216n.55, 219n.58, 220n.62, 222, 226, 227, 229, 230
Koellner, Peter 79n.3, 85, 88, 92n.1, 105n.41
Kotzsch, Hans-Christoph 10

Kreisel, Georg 2, 3, 13n.1, 14, 15, 17, 18, 20, 23n.18, 33-37, 39, 40, 47, 104
Kremer, Michael 237n.22
Kretzmann, Normal 190n.9
Kripke, Saul 9, 85n.17
Kronecker, Leopold 5, 6, 287
Kummer, Eduard 277n.43

Lasserre, François 189n.7
Law, Andrew 10
Lawvere, William 55, 58
Ledent, Jeremy, 73
Legendre, Adrien-Marie 205n.36
Leibniz, Gottfried Wilhelm 6, 182, 282
Leitgeb, Hannes 131n.32
Lenzen, Victor 160-162
Levy, Azriel 15n.4
Lewis, David 156
Licata, Daniel 66, 72
Liebeck, Martin W. 102n.34
Lifschitz, Vladimir 15, 17n.8
Linnebo, Øystein 81n.10, 125, 126, 127n.25, 281n.52
Linsky, Leonard 2, 7, 231
Lipschitz, Rudolf 265, 269, 270, 276
Litland, Jon 127n.27
Lotze, Rudolf 258n.13, 262n.26, 264n.28, 283-285, 290
Lugg, Andrew 183
Łukasiewicz Jan, 156
Lumsdaine, Peter 60n.3, 62n.4, 69

Mac Lane, Saunders 2, 68
MacIntyre, Angus 15n.4
Maddy, Penelope 83n.15
Mahlo, Paul 78
Malament, David 2, 161n.38
Marcus, Ruth B. 2
Martin, Donald A. 91, 92, 95, 96, 98, 101-108
Martin-Löf Per, 5, 18, 40, 55, 135n.36
Maxwell, James 146, 148

McDowell, John 9
McGuinness, Brian 178
McKinsey, J. C. C. 2, 14
Mehrtens, Herbert 252n.2
Mendell, Henry 185, 189n.7, 207n.40
Menge, Heinrich 188n.6
Menn, Stephen 7n.14, 10
Mill, John S. 143
Mints, Grigori 15, 17n.8
Moravcsik, Julius 189n.7
Morley, Michael 15n.4
Morris, Rebecca 10, 120, 121, 134n.34, 256n.8
Morse, Anthony 88
Mueller, Ian 2, 185n.*, 229n.79
Myhill, John 2, 14

Naibo, Alberto 91
Nagel, Ernest 291
Ness, Sally 10
Noether, Emmy 252, 279, 286n.60
Novák, Zsolt 92n.2

Oberdiek, Klaus 295
Ockham, William 99

Panzo, Marco 91
Pappus 204, 205, 207, 208, 211, 212, 218
Parikh, Rohit 15
Parsons, Charles 2, 7n.14, 10, 25, 92n.1, 116n.3, 141, 173, 281, 289n.67
Peano, Giuseppe 17, 20, 105n.41, 114, 122, 155n.24, 157n.28, 158, 161, 167n.48, 241, 248n.51, 271
Pears, David 178
Pettigrew, Richard 125, 126, 127n.25, 273, 281n.52
Platek, Richard 15n.4
Plato 6, 7, 8, 9, 99, 188
Pohlers, Wolfram 24

Name Index

Poincaré, Henri 165n.46, 172-175, 178, 181, 182
Posey, Carl 141
Postnikov, Mikhail 68
Prawitz, Dag 2, 15n.4, 18, 24n.21
Price, Michael 116n.3, 120n.13, 123n.19, 290n.68
Proclus 189n.7
Ptolemy 204
Putnam, Hilary 79, 80, 81, 82n.12, 86, 99n.25
Pythagoras 186, 187n.5

Quine, W. V. O. 7, 8, 92, 93n.5, 97, 141-144, 146n.5, 148n.6
Quillen, Daniel 60, 69

Rathjen, Michael 26
Ravaglia, Mark 20n.15
Rechter, Ofra 141
Reck, Erich 7n.14, 114n.2, 116n.3, 117n.6, 118n.8, 119n.11, 119n.12, 120n.13, 123n.19, 134n.34, 155n.26, 252, 266n.32, 282, 283, 290n.68, 295
Riemann, Bernhard 253, 256, 292
Reinhardt, William 6
Richard, Jules 174
Ricketts, Thomas 7n.14, 10, 177, 180, 183
Roberts, Samuel 79n.4, 81n.11, 82n.12, 86
Rogers, Brian 176, 177
Rohlfing, Helmut 295
Rome, Adolphe 204n.35
Rosen, Gideon 108n.48
Rosser, John Barkley 81n.11
Royden, Halsey Lawrence 2, 14
Ruffino, Marco 247n.50
Rush, Penelope 96n.17
Russell, Bertrand 6, 56, 81, 94, 95n.8, 95n.12, 95n.13, 97, 113, 114, 117, 118, 119n.12, 121-123, 125, 127, 132, 133, 135, 142, 143, 160, 171-177, 180-182, 233n.6, 234n.13, 242n.36, 247

Saccheri, Giovanni 205n.36, 212n.48
Saito, Ken 185, 189n.7, 197n.21
Scott, Dana 2, 13n.1, 14, 55, 122n.16, 124n.21, 132n.33
Schiemer, Georg 124n.23, 282n.53, 295
Schilpp, Paul 142n.1, 142n.2
Schlick, Moritz 165n.46
Schlimm, Dirk 10, 25, 255n.7, 256n.8, 265, 280, 282n.53, 285n.59, 292
Schröder, Ernst 174, 233n.8, 285
Schütte, Kurt 3, 4, 18, 20, 24, 37-41
Shapiro, Stewart 83, 122n.16, 281, 288n.64
Shepherdson, John 105
Shoenfield, Joseph 15n.4
Shulman, Michael 53, 66, 68n.7, 72
Sieg, Wilfried 7n.14, 10, 22n.17, 24, 120, 121, 134n.34, 255n.7, 256n.8, 265, 280, 282n.53, 285n.59, 288n.65, 290n.68, 292
Silver, Jack 85
Simonyi, Andras 92n.2
Simplicius 187n.3
Simpson, Stephen 104, 105
Simson, Robert 227n.75
Sinaceur, Hourya B. 114n.1
Skolem, Thoralf 33
Slaman, Theodore 2
Smith, Charles 201n.28
Sojakova, Kristina 66
Spector, Clifford 3, 4, 5, 15, 18, 34, 35, 39, 40
Spur, Jane 10
Stein, Howard 2, 185, 196n.18, 205n.36, 210n.45
Strahm, Thomas 20n.15
Streicher, Thomas 62
Suppes, Patrick 2, 13n.1, 14, 15, 189n.7, 291

Tait, W. W. 1-10, 13-26, 33, 38-47,
    53, 55, 77, 78n.3, 85n.18,
    86-89, 91, 113, 114, 118,
    119, 121, 130n.30, 131n.32,
    134n.35, 135, 136, 141, 182,
    185, 234n.13, 251, 252, 281,
    295
Takashi, Moto 24n.21
Takeuti, Gaisi 3, 24
Tannery, Paul 187n.5
Tarski, Alfred 2, 14, 148, 149, 150,
    152, 155, 156, 242, 251, 253
Tennant, Neil 295
Theodosius 204n.35, 208
Theon 204n.35, 225n.71
Thorup, Anders 200n.26, 202n.32
Torricelli, Evangelista 205n.36, 212-
    n.48, 220n.62, 222n.64
Troelstra, Anne S. 18, 40, 46
Turing, Alan 36-38

van Atten, Mark 295
van den Berg, Benno 60n.3, 62n.4
van der Waerden, B. L. 189n.8, 252
van Fraassen, Bas 291n.70
van Heijenoort, Jean 15n.4
Väänänen, Jouko 105
Veblen, Oswald 21
Vesley, Richard 34, 36
Vitrac, Bernard 193n.14, 195n.17,
    197n.21, 204n.35, 206n.37,
    206n.39
Voevodsky, Vladimir 54, 60n.3, 62-
    n.5, 65, 66, 69
von Neumann, John 82, 87, 121,
    122, 125, 267

Wang, Hao 95, 98n.19, 107n.47, 108,
    109
Warren, Michael 60, 60n.3
Weber, Heinrich 252, 256, 274, 277,
    278, 284, 294
Wehmeier, Kai 176, 177, 183, 232-
    n.2
Weiner, Joan 231, 240n.30, 242n.39

Whitehead, Alfred North 160
Wilson, Mark 295
Wimsatt, William 2
Wittgenstein, Ludwig 5, 7-10, 142,
    143, 171-182, 234n.13
Wright, Crispin 8, 125, 234

Yablo, Stephen 92n.3
Yap, Audrey 119n.11, 252, 282
Yokoyama, Keita 104, 105

Zach, Richard 16n.7, 17n.8, 20n.15
Zermelo, Ernst 6, 69, 77, 78, 79,
    79n.5, 82, 88, 103, 105, 154,
    253, 261, 267, 279, 285-
    n.59, 288-290
Zeuthen, H. G. 196n.18
Zucker, Jeffrey 48

www.ingramcontent.com/pod-product-compliance
Lightning Source LLC
Chambersburg PA
CBHW071654160426
43195CB00012B/1460